NUCLEAR ENERGY

NUCLEAR ENERGY

Editor

Frej Nagel

scitus academics

Nuclear Energy
Edited by **Frej Nagel**

ISBN: 9781681178769
Library of Congress Control Number: 2017948903

SCITUS Academics LLC,
616, Corporate Way, Suite 2, 4766,
Valley Cottage, NY 10989

www.scitusacademics.com

Notice

Printed in United States of America.

Preface

Nuclear energy is attracting revived interest as a potential alternate for electric power generation in the event of increased concerns about global warming. Nuclear energy has one of the lowest impacts on the environment of any energy source since it does not discharge air pollution, isolates its waste from the environment and need a relatively small amount of land. Nuclear energy stations do not produce air pollutants or greenhouse gases when they generate electricity. Today's nuclear reactors are safe and highly efficient energy systems that provide electricity and a multitude of co-generation energy products ranging from potable water to heat for industrial applications. However, terrible earthquake and tsunami events in Japan resulted in the nuclear accident that obligatory us to rethink our approach to nuclear safety, design requirements and facilitated growing interests in advanced nuclear energy systems, next generation nuclear reactors, which are essentially capable to withstand natural disasters and avoid catastrophic consequences without any environmental impact. Nuclear power has been called a clean source of energy because the power plants do not release carbon dioxide. While this is true, it is unreliable. Nuclear power plants may not emit carbon dioxide during operation, but high amounts of carbon dioxide are emitted in activities related to building and running the plants. Nuclear power plants use uranium as fuel. The process of mining uranium releases high amounts of carbon dioxide into the environment. Carbon dioxide is also released into the environment when new nuclear power plants are built. Finally, the transport of radioactive waste also causes carbon dioxide emissions. Nuclear power plants constantly emit low levels of radiation into the environment. There is a differing of opinion among scientists over the effects caused by constant low levels of radiation. Various scientific studies have shown an increased rate of cancer among people who live near nuclear power plants.

Radioactive waste is a huge concern. Waste from nuclear power plants can remain active for hundreds of thousands of years. Currently, much of the radioactive waste from nuclear power plants has been stored at the power plant.

The aim of this volume, 'Nuclear Energy' is to bring nuclear power to our readers as one of the promising energy sources that has a inimitable prospective to meet energy demands with minimized environmental impact, near-zero carbon trace, and competitive economics through vigorous prospective applications. This book intends to provide useful information for scientists and those in technical fields in several areas in nuclear power plants including: nuclear systems protection, design and modeling of critical parameters in nuclear power plants, nuclear waste management and safety and reliability assessment.

Table of Contents

CHAPTER 1

Nuclear Energy for Sustainable Economic Development

F. D'Auria, A. Bousbia Salah

Dipartimento di Ingegneria Meccanica, Nucleare e della Produzione, Pisa, Italy

ABSTRACT

The discovery and the application of nuclear energy constitute the most important technological achievement of the past century. However, the development and the exploitation of this technology have been remarkably smaller than foreseeable. An overview of the significant features of the nuclear technology including the comparison with competitive energy sources is made. The "embedded" safety engineering and the pollution are discussed and the main features are mentioned. Indeed, nuclear technology can be applied for the sustainable society development by producing substantial amount of clean water from the ocean. The idea is to build up nuclear power plant sites that produce desalinated water and pump it several tens of kilometers away to form a lake into a desert region. This could help to establish the conditions for an agriculture-based civilization.

Keywords: Nuclear Energy, Sea Water Desalination, Economic Development

1. INTRODUCTION

The nuclear energy constitutes the most relevant discovery of the past century. This can be compared with the discovery of the fire and is also sometimes referred as "nuclear fire" to distinguish it from the "chemical fire" available to the humanity several millennia before. The chemical fire can be used to burn the house of the neighbor, or to prepare food, or to construct cars, or to produce electricity. The first utilization does not justify the ban for the other uses. The "non-limited" and much more powerful "nuclear fire" (compared with the chemical fire) is not adequately exploited by this civilization because of its initial use as a mass-destructive weapon. The same civilization, for its overwhelming energy needs, appears satisfied for burning in a few tens of years amounts of fossil fuels accumulated by the nature during the course of millions of years. The production of pollutants, that accompanies the chemical burning, might be seen as a minor problem compared with the disappearance of such a natural resource.

The number of fatalities and the environmental impact, unavoidably associated with any energy source, must be at the basis for designing a consistent share of energy sources. The actual cost of energy should have a role inside a broader context where the available natural resources and the level of technological development of individual countries are properly considered.

Other than the number of fatalities and the environmental impact, two major considerations should drive the industrialized countries toward a more profitable consumption of nuclear energy:

- The use of nuclear energy preserves the consumption of fossil materials produced on the earth during hundreds million years, avoiding their full burning in less than a century. Power from fissile elements ensures the energy survival for a period of the order of 10^4 years.
- Fossil fuel burning should be left to the less developed countries. Consumption of low technology fuel (i.e. the fossil fuels, mostly produced in the less developed countries) by industrialized countries contributes to enlarging the existing gap between countries and civilizations.

The purpose of the paper is to provide a view about the current status of the nuclear technology by streamlining the attention toward emblematic topics that are discussed in a provocative way. The scientific rigor, needed when energy problems are considered, may not be found from the performed analysis (references are avoided to stress the "spreading" nature of the memory). Nevertheless, the out-comings of the study are expected to be useful for open minds willing to support the nuclear energy option in the present civilization and to protect the environment from the products associated with the chemical fire when adopted for electricity production [1]

The fight against the conditions that prevented the suitable exploitation of the nuclear energy has been lost at the time being within the industrialized countries. In order to remove the established conditions, an attempt is made here to propose, in a merely academic way, a unique application for the nuclear technology [2].Producing substantial amounts of clean water from the ocean could help establish the conditions for an agriculture-based civilization in the desert areas of the planet. The analysis outlined in the paper has been carried out for a desert land as Algeria. The idea is to find a dry or seasonal river that originates from a dry lake and ends up into the sea. The sea water could be pumped towards a nuclear power desalination site, which in turns feeds the dry lake by desalinated water. Conditions are created in this way to develop a human civilization based upon the autonomous agriculture resources around the river and the lake. The benefit of using nuclear technology for providing water resources around a dry river is emphasized.

2. NUCLEAR TECHNOLOGY: THE MAIN FEATURES

The features of the nuclear technology constitute by themselves an emblematic topic: an impartial judge will state that these typical and unique characteristics are beneficial for the development of the human civilization [3] . In order to boast such positive peculiarities, nuclear technologists undertook an unprecedented effort and many branches of learning were explored: mathematics, physics, chemistry, thermal hydraulics, neutronics, material technology, computer science, electronics, robotics, safety engineering, risk analysis, quality control procedures, quality assurance and many other disciplines are concerned.

Thanks to this effort, many branches of learning have born and other existing branches have been developed, exploited and strengthened to such an extent to contribute to the technological development of the world. This is a unique, cannot-say-how-much-positive feature of the nuclear technology, as outlined by Figure 1.

Unless nuclear technology typical features are objectively positive, detractors usually use them to 'prove' that the technology is unsafe, non-competitive, non-economic, polluting, in a word, immoral. Probably they don't know what a proof is. Anyway, let's briefly analyze some of these features.

2.1. Energy Competition

Existing nuclear power plant technology has been compared with both renewable and non-renewable energy technologies, with both consolidated/state of the art or developing technologies. Such a characterization could involve a large number of aspects, but only a few have been provocatively considered. The addressed items do not allow a comprehensive or systematic or subjective comparison. Rather, significant aspects that distinguish the technologies are mentioned that are improperly used to run nuclear down.

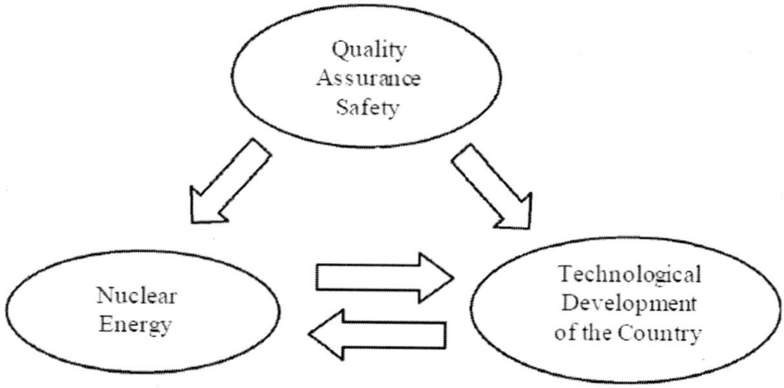

Figure 1. Role of nuclear technology in the development of the civilization.

2.1.1. Nuclear vs. Hydropower

The hydropower exploitation implies the construction of dams that unavoidably impact the surrounding animal and plant life. Huge territory extensions are needed to generate electricity from dams, as discussed in the section "Environment Protection". Alternatively, dams can be built in mountain regions creating the potential for enormous disaster. In addition, suitable natural conditions for the construction of dams may occur thousands kilometers far from the energy consumption location. As a consequence, substantial amounts of energy are lost and an additional impact on the territory is caused by the energy transportation.

2.1.2. Nuclear vs. Solar

Solar energy is generally made in desert areas that are located, again, away from the important energy consumption centers. In economic terms, both solutions are not (yet) competitive with other energy sources. If non-desert areas are used, these are allotted to become desert. Toxic materials must be manipulated when electronic cells are used. Energy storage (strictly needed to allow energy consumption when the sun is obscured by clouds or by the earth motion) also requires the use of pollutant materials [4].

2.1.3. Nuclear vs. Wind

Wind mills are suitable to generate electricity with reduced coasts. Their installation in the desert areas of the planet causes increase of cost that, in any case, is not (yet) competitive with other energy sources. Unfortunately, hundreds wind mills have been installed in beautiful areas of the planet, forgetting landscape impacts. The majority of these areas were not accessible to cars: the installation of the wind mills implied the construction of the roads and the consequent irreversible degradation of the nature. As in the case of sun, energy storage (strictly needed to allow energy consumption when the wind is not strong enough) also requires the use of pollutant materials [4].

2.1.4. Nuclear vs. Fossil (Coal, Oil, Gas)

From the economic point of view, fossil fuels constitute the only competitive energy source. Drawbacks, related to the nuclear energy source, which apply to a different extent for any of the coal, oil or gas fuels are recalled in the following.

The amount of fuel needed (either in volume or in mass terms) to produce the unit electric energy is million times larger.

- Hundreds peoples die per year just owing to transportation of suitable amounts of fossil fuels. Huge territories of the earth have been irreversibly contaminated owing to the transportation of fossil fuels.
- Hundreds peoples die per year just owing to the extraction of suitable amounts of fossil fuels. Explosions of coal mines cause fatalities that sum up with cancer deaths of miners and in cities close to the mines.
- Hundred thousand deaths per year are addicted to the pollution causes by burning of fossil fuels.
- Irreversible consumption of the fuels that were formed in chemical processes lasting million years happens in less than a century.

2.2. Safety and Residual Risk

Design of application and safety could only be developed in parallel. Any design concept was not considered if safety concern could be raised.

Authorities controlling the design were created in any nation involved with the technology. Their only purpose was to prevent any design that could have any reasonable possibility to endanger the environment or the population.

Without having the objective to go into detail into the meaning of safety, an overview is given hereafter of the safety issue as presently considered. Then, examples are provided in relation to the values of 'safety indicators' for the nuclear technology that are compared with safety indicators of other technologies that are fully accepted in the present civilization. As shown in Table 1, the energy system with by far the greatest amount of controversy about its risk is undoubtedly nuclear power [5] . Actually, Chernobyl and Fukushima accidents "killed" together less than one hundred persons, but possibly prolonged the life to million peoples.

2.2.1. Overview of the Safety Issue

The safety of nuclear power plants (NPP) is based on the defense in depth concept, which relies on subsequent physical barriers (fuel matrix, cladding, primary system pressure boundary, containment) and other provisions to control radioactive materials and on multiple levels of protection against damage to these barriers and against undue radiological impact on the NPP itself and on its surroundings. Demonstration that there is no undue risk caused by plant operation is obtained by means of safety assessment.

Safety assessment is a broad term dealing with the systematic processes aimed at ensuring that all relevant safety requirements are met. These cover the various aspects connected with the siting, the design, the construction, the operation including the maintenance and the qualification of the involved personnel and the decommissioning of any NPP.

The entire range of conditions for which an NPP is designed according to established design criteria, including all the national regulatory requirements, and for which the damage to the fuel and the release of radioactive material are kept within authorized limits, form the design basis of an NPP [3]. Within the design basis, a number of unintended events are considered, including operating errors and equipment failures, whose consequences or potential consequences are not negligible in terms of safety. According to the probability of its occurrence and potential consequences, an event may be classified as an anticipated operational occurrence or a Design Basis Accident (DBA).

An accident occurring outside the NPP design basis is called a Beyond Design Basis Accident (BDBA). A beyond design basis accident may or may not involve degradation of the reactor core (leading to significant core damage). An accident involving core degradation (typically core melt) is also called a severe accident. It must be noted that such event causes a huge capital lost and is not expected to cause harms to the population because the 'fourth barrier' is designed to protect the environment (including the human being) against such an event.

Safety assessment includes safety analysis as an essential component. By the term safety analysis, an analytical study is meant by which it is demonstrated how safety requirements, such as ensuring the integrity of barriers against radioactive releases and various other requirements, are met for initiating events (both internal and external) occurring in a suitable range of operating conditions, and in other circumstances, such as varying availability of the plant systems.

Accident analysis is an important tool for confirming the adequacy and efficiency of provisions in the defense in depth concept to cope with challenges to plant safety. It is used in a number of applications, such as licensing of a new plant, modification of an existing plant, periodic safety reviews, etc.

2.2.2. Comparison among Safety Indicators and Off-Nominal System Performance

All machines and systems designed by human brain are subject to fail and everything that surrounds us constitutes a machine or a system, i.e. from houses, to bridges, to TV screens, to cars, to guns, to airplanes. No system is designed to fail and "off-nominal" system performance implies a failure. The probability to fail and the damage produced to the environment owing to the failure distinguish the various possible failures.

Table 1. Risk per unit energy output (1969-2000) [5] .

Energy chain	Accidents	Fatalities
Coal	1119	20,276
Oil	375	20,218
Nuclear	2	31
Natural Gas	135	2043
Hydro	12	29,935

What distinguishes the various technologies, or even the individual machines and systems, in relation to risk is listed below:

- probability of occurrence of an environment harmful event,
- damage associated to the event expressed in terms of human deaths,

- knowledge and acceptance of the risk.

Typical results from that report is that risks coming from smoking, from travelling by car, by plane, by train, by ship etc. or due to natural events are much larger than any conceivable risk associated with the existence of whatever large number of nuclear power plants.

In the forthcoming years it was clear that the death owing to the consumption of tobacco, or of alcohol, or the pleasure to drive a car in a dangerous environment was among the freedom of individuals who could get a pleasure from the planned desire, while the death eventually caused by harmful radiation was to be addicted to benefits for the entire society and therefore "less acceptable".

In the case of Chernobyl and/or Fukushima, there is no study showing that the number of deaths owing to the accidents caused a number of late fatalities for any reasonable group of population outside the statistical tolerance range. Studies conducted by scientific institutions show the opposite.

2.3. Environment Protection

The evaluation of the impact upon the environment brings to the best evaluation of the nuclear technology compared with any other technology suitable for electrical energy production. The issue of environment protection is huge in terms of indicators, social impact, costs, global climate change and potential harmful consequences. Any energy source has an impact upon the environment. This must be evaluated considering the various stages of developing, exploiting and decommissioning the technology.

In the case of fossil fuel technology and making reference to coal at first, environmental impact includes, but is not limited to drilling a mine, diffusion to the atmosphere of gases and particulate from the mine, transportation of miners to the mine, cleaning of the coal from the mine, transportation of the coal from the mine to the steaming plant, production of gases and solid particulate at the chimney of the plant, production and transportation of ashes.

Energy sources, like geothermal, wind, hydro, solar are impact-free only to some mass-media or following ignorant institutions. Namely, radioactive pollution caused by a geothermal plant may be order of magnitudes larger than the pollution (in terms of activity) due to a nuclear power plant. In addition various acids are extracted from the underground together with steam, as already mentioned. Wind energy determines irreversible impact on the ground. In the South of Italy, thousands "wind-mills" are installed along several hundred miles on the top of beautiful Appennino's hills (more than 1000 m elevation on the sea level) that before were inaccessible to cars, and tens of miles apart from the nearest road. Now these lands are open to cars and to the human impact. Hydro power implies the flooding of hundreds or thousands square kilometers. On average, making reference to the Itaipu hydro power station, i.e. the biggest hydro-plant in the world installed at the border between Argentina, Brazil and Paraguay, the production of 1 MWe (on the site) requires more than 10^5 square meters. The production of the same power in a nuclear plant requires about 500 square meters (i.e. 200 times less). Normally the ground around dams is not a desert and is excellent for agriculture.

NPP do not emit any pollutant during the operation. Only water is emitted in the form of droplets from cooling towers or owing to the increased evaporation of water reservoirs adopted to cool the condenser. Radioactivity connected with noble gases is only emitted by the power plants (less than 1/5 of the total number of power plants) equipped with boiling water reactors with amounts about 100 times less than allowed by regulations and lower than released by some hospitals.

3. PROPOSAL FOR "CIVILIZATION-SUSTAINABLE" DEPLOYMENT

It is well known that most of the ancient civilizations that brought to the present development for the human societies developed along rivers: the ancient Egyptians around the Niles, the Babylonians between the Tigris and the Euphrates, the Romans along Tevere, the Etruscans along the Arno and the Chinese along the Yellow River, just to name some of the best known examples.

Developed societies tend to transfer to underdeveloped societies currency. The underdeveloped society supplements the received currency with any raw strategic matter that may offer and buy technological goods from the developed societies. In this way, at the end of the process (i.e., receiving currencies, delivering raw matters and buying goods for the underdeveloped countries) the developed societies have all the currency and all the strategic raw matters in their possession.

The effective help for these countries is to supply them with the tools necessary to exploit the resources which they dispose of and to supply the conditions which allow this exploitation. We must supply them with a hoe and teach them how to use it, not supply them with vegetables posing as magnanimous.

Authors of this paper do think that the pacific use of nuclear energy might be a smart hoe. More precisely, nuclear technology not only can help but constitutes the only technology that can be suited for the plan. It is proposed to transfer from developed to under-developed societies, the means to sustain the development i.e. not the apparent product of the development constituted by the currency.

In addition, the water needs and the fact that the poorest societies actually live in the desert areas of the planet are considered. Having this in mind the proposal is simple: to provide the condition to create a 'new' river suitable for the rise of plantation in the desert to such an extent to sustain the creation of a people and of a civilization.

An academic and provocative proposal is formulated and minor resources (in the sense of devoted effort not sufficient even for a preliminary technological project) have been invested so far. Many questions probably are lightening in the reader's brain: "why to create such a civilization" or "where to create such a river" or "who might do such a huge work" and, especially, "how to create a river in the desert" and "which technology is so powerful to make of a such strange and extravagant proposal a realistic scenario and an interesting issue, which might induce someone to reflect".

3.1. The Task

The task is to find out a suitable coastal site close to a desert area of Africa and to build-up a certain number of nuclear power plants (see Figure 2). Desalinated seawater should be produced and pumped several tens of kilometers away of to form a lake into the desert. A dry river originating from the lake should end up into the sea very far from the coastal nuclear site. In this way, conditions are created to develop a human civilization based upon the autonomous agriculture resources around the river and the lake.

The analysis outlined in the next section has been carried out having in mind a five million population that should occupy a suitable territory and have sufficient water resources for a self-sustained farming.

Algeria, has been chosen where all the aforementioned conditions exit. Figure 2 shows the location of the considered desert land around which a dry river that can be fertilized thanks to nuclear energy resources.

3.2. Technical Data to Exploit the Idea

Even though the proposal is provocative and academic, a limited scope and rough technical evaluation has been performed.

The first step leads to fix the specifications. This is outlined in Table 2. Agriculture products should be cultivated around the dry river. The civilization should grow at some distance from the energy production that should remain in a remote area for strategic and security reasons (easy to protect possible opponents to the civilization). In order to exploit the idea, a number of assumptions must be formulated that constitutes the second step for the study. The main assumptions are listed in Table 3.

In the third step of the study simplified calculations have been carried out considering the above assumptions. The main results are given in Table 4. The information in the last row can be further elaborated. If one assumes the cost of a reference standard NSSS unit equal to 2 billion US $, the overall pro-capita cost to sustain a new civilization of 5 million people results less than 3000 US $. The additional rough assumptions of negligible fuel costs and of design lifetime for the NSSS units equal to the average human life have been made.

4. CONCLUSIONS

All energy sources are needed for the "sustainable" development of the mankind. Competition should be intended in order to achieve a proper energy mixing. However, the current share of the nuclear energy source appears regrettably small. Occurred or potential nuclear accidents should be given no role in the reasonable planning of the share for energy sources. The number of fatalities and the environmental impact, unavoidably associated with any energy source, must be at the basis for designing a consistent share of energy sources. The actual cost of energy should have a role inside a broader context where the available natural resources and the level of technological development of individual countries are properly considered. Advantages of the nuclear energy applications over competing energy sources in the area of electricity generation have been pointed out. Overall technological development has been associated to the exploitation of the nuclear power. Quality assurance and safety constitute industry related discipline that has been largely developed following nuclear energy uses.

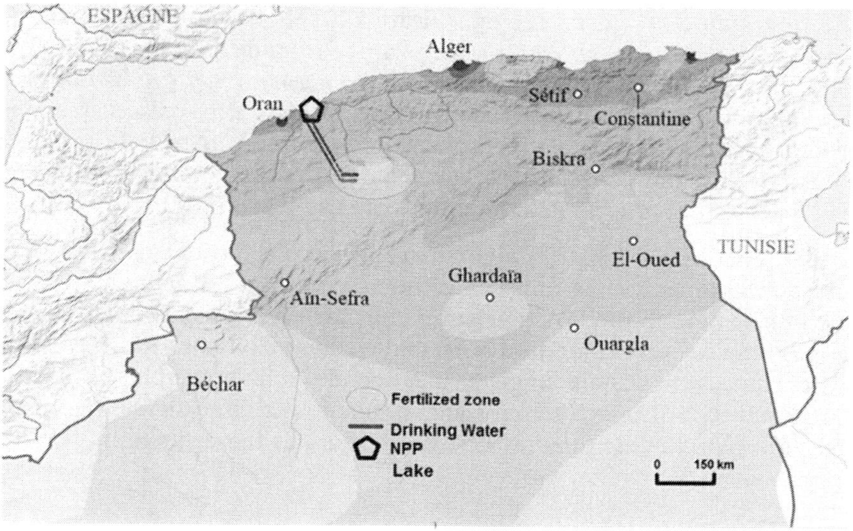

Figure 2. Dry desert lake fed by NPP desalinated sea water.

Table 2. Main targets for the analysis.

Geographical data	Quantity
Population that constitutes the civilization	≅ 5 million
Distance from the coastal site to the open air lake (suitable pipelines must be built along this distance)	>100 km
Lake surface	>100 km^2
Length of the river (*i.e.* between the lake and the ocean)	>200 km
Distance between the mouth of the river and the coastal site	>100 km
Percentage of African desert area involved	<0.1%

Table 3. Main assumptions adopted in the analysis.

Resource data	Quantity
Water needed to fertilize 104 m^2 of desert land per year	3×10^3 m^3
Number of peoples that can be fed by the food produced per each 104 m^2 of fertilized land	5
Unit reference Nuclear Steam Supply System power (Unit power per NSSS)	4500 MWth
Desalination plant Gained Output Ratio (GOR)*	17
Fraction of desalinate water lost by evaporation from the river and the lake	50%
Pro-capita thermal power consumption required for needs "other than agriculture"	1.7 kw
Reference plant thermodynamic efficiency	35%

*GOR = ratio between flow-rate of produced desalinated water and flow-rate of steam entering the desalination plant. GOR equal to 17 is representative of the current state of the art.

Table 4. Main results of the analysis.

Geographical data	Quantity
Desalinated water production rate per year	1.3×10^9 m^3
Area of desert land that can be cultivated	2300 km^2
Number of reference NSSS units needed for agriculture (desalination and desalinated water pumping power)	4.5
Number of reference NSSS units needed for energy "other than agriculture"	1.9
Total number of reference NSSS units	7

Therefore nuclear technology can be profitably exploited to create new civilizations. The paper illustrates a possibility to create a river in the desert, "powered" by a suitable number of remote coastal nuclear power plants. This will allow the creation of conditions for the growth of an agriculture self-sustained society.

ACKNOWLEDGEMENTS

This work was carried out for the "E. Fermi and Nuclear Energy" conference, held in Pisa (I), October 15-16, 2001.

REFERENCES

1. IAEA-TECDOC-574 (1990) Use of Nuclear Reactors for Seawater Desalination.
2. IAEA-TECDOC-1326 (2002) Status of Design Concepts of Nuclear Desalination Plants.
3. D'Auria, F., Glaeser, H., Kim, M. (2014) A "New" Vision for Nuclear Reactor Safety. 3rd International Scientific and Technical Conference "Innovative Designs and Technologies of Nuclear Power" (ISTC NIKIET-2014), Moscow, 7-10 October 2014.
4. Inhaber, H. (1977) Is Solar Power More Dangerous than Nuclear? IAEA Bulletin, 21, 11-17.
5. OECD/NEA N. 6861 (2010) Comparing Nuclear Accident Risks with Those from Other Energy Sources.

CHAPTER 2

Understanding Accelerator Driven System (ADS) Based Green Nuclear Energy: A Review

Mohammad Kamal Hossain*, Md. Abu Taher, Mithun Kumar Das

Department of Physics, Comilla University, Comilla, Bangladesh

ABSTRACT

This paper reflects the scopes of accelerator driven system (ADS) based nuclear energy, as a reliable source of electric energy generation, comparing to the other existing non-renewable and renewable sources. There are different limitations in the use of every source of electric energy but in consideration of minimum environmental impact, exclusively inherently low greenhouse gas (GHG) emission, and also, high life time with maximum power production efficiency, nuclear would be the best choice. From this study it was found that several difficulties involved in the ADS based energy production, more specifically, difficulties regarding the target parameters, coding system, waste management, etc. Hence suggestions from this study points out that if it is possible to ensure more energy efficient production of enriched uranium, improved nuclear fuels and reactors that allow greater utilization, extended life times for nuclear power plants (NPPs) that reduce the need to build new facilities, improved coding system capable of minimizing the discrepancy between theoretical and experimental calculation of spallation products, improved data library with sufficiently available high energy nuclear data to perform a better coding analysis, and finally, considering the environmental safety if the disposal of the radioactive wastes could manage more effectively, nuclear energy would then play a significant role in minimizing future energy crisis worldwide as well as to save our loving green earth.

Keywords: Green Nuclear Energy, Spallation, Accelerator Driven System, Transmutation of Wastes, Sustainable Energy Development

1. INTRODUCTION

Energy is central to economic growth. Higher living standards and increased energy consumption are intrinsically related. Fighting poverty worldwide will require a means of raising the world's living standards to levels the developed nations take for granted. This will involve the introduction of a basic energy infrastructure and consumption patterns [1].Worldwide demand for energy is rapidly increasing and it is expected that it could double by 2050. Therefore it is important to develop and deploy multiple energy sources in the context of a comprehensive energy plan i.e., a broad mix of energy technologies is essential to meet the growing demand. At the same time, the evidence is clear that CO_2 emissions must be reduced globally. Abundant, affordable, and environmentally responsible energy must be developed, both domestically and internationally, to meet that demand [2].

There is a range of energy resources, often called non-renewable and renewable energy resources, available to generate electric energy. Non-renewable sources of energy can be divided into two types: fossil fuels and nuclear fuel. Coal, oil and natural gas are fossil fuels. Among the renewable sources solar, wind, etc. play the dominating roles. Whatever the contribution to the total world energy is, all of them have some notable environmental limitations, individually. For instance, air quality is the primary environmental concern for coal-fired power plants, which emit a significant amount of air pollutants. Coal mining also can significantly affect the environment by polluting water and destroying wildlife habitats. The health effects from coal mining include death from cave-ins, coal dust explosions, methane explosions and methane inhalation. Many coal miners also have suffered lung ailments and other disabilities. Natural gas plants emit an amount of particulates, smog-forming pollutants and carbon dioxide. It is also responsible for the ground water pollution from the hydraulic fracturing process [3]. In addition to ecological disturbances from fossil fuel extraction, there are certain cultural consequences for communities around the Pacific Rim. These communities, many indigenous, are threatened by the depletion of specific resources they depend upon for their livelihoods and culture [4].

Most types of renewable energy sources create little or no air pollution or greenhouse gases. But the fact is that both wind and large-scale solar facilities can require significant land area. For example, a solar thermal plant uses between 5 (five) and 10 (ten) acres of land per megawatt of electricity generated. Water also can be a concern for certain solar thermal electric plants that, like conventional power plants, need water to generate electricity. Installing both solar and wind can disturb the natural environment, so wildlife studies are often conducted before projects proceed. Bird and bat migration and flight patterns are important considerations for wind turbines since poorly located turbines can cause excessive deaths [3]. Therefore, most of the environmental problems the world facing today—including climate change, air pollution, oil spills, and acid rain—result from our dependence on fossil fuels [4and a little bit ecological disturbance that arises from the use of renewable energy sources. Growing concerns over the consequences of climate change may severely limit future access to fossil fuels. A forced choice between energy and environment could precipitate a major economic crisis, an environmental crisis, or both [1].

Among the non-renewable sources nuclear power has been used to produce electricity for public distribution since 1954 [5]. This option of electric power generation should be retained, precisely because it is an important carbon free source of power that can potentially make a significant contribution to future electric power supply [6]. But now it appears that hesitations in further developments of nuclear power are primarily due to the environmental concerns arising from the nuclear waste disposal schemes involving long-term geological storage of commercial reactor waste [7]. The possible solution that overcomes the above problem and ensures the future growth of nuclear power as a reliable, safe and abundant source of electricity is the incineration of long-lived transuranic elements (TRU) and transmutation of fission products (FP) which can make its disposal in geological repository environmentally acceptable. With a view to do so, several projects and programs are going on worldwide, in finding the possibility of separating (or partitioning) the long-lived radioactive waste from the used fuel and transmuting it into shorter-lived radionuclides so that the management and eventual disposal of this waste are easier, less expensive and will be environmentally friendly [8]. The above is done in an assembly equipped with the spallation arrangement and an accelerator driven sub-critical nuclear reactor system (ADS). The aim of this study is to understand the ADS based green nuclear energy as a part of understanding sustainable energy development in order to meet future energy needs as well as to save our green earth.

2. SPALLATION AND ADS

Starting from the observations of particle cascades in cosmic rays interactions in 1930 scientist's attempt in investigating the spallation yields is growing day to day but not been established yet. A theoretical picture of an incident particle colliding successively with several nucleons inside target nucleus, losing a large fraction of its energy was proposed by Serber in 1947. Before, in 1937 Weisskopf considered possibility of emission of neutron from excited target nucleus. In the end of fifties, Metropolis and Dostrovsky (who used the ideas of Serber and Weisskopf) suggested description of spallation as two-step process involving energy deposition and subsequent evaporation. They formulated and performed first Monte Carlo calculations of the reactions. Such treatment of spallation reactions is used from that time up to now [9].As spallation applications require more precise knowledge, they are being investigated with the increasing interest in the last two decades while in recent years the world has registered obvious progress in the accelerator technique. The successful functioning of several high energy research accelerators (Berkeley, KEK), the construction of the LHC, ... brought new technologies also to mid- and low-energy accelerators. Also, advances like superconductivity for magnets and RF cavities, ion sources have led to the practical realization of high-power beams [10].

Next to fission and fusion, spallation is an efficient process for releasing neutrons from nuclei [11]. By definition it is a process in which a light projectile such as proton, neutron, or light nucleus with the kinetic energy from several hundreds of MeV to several GeV i.e., a high-energy hadron or a nucleus interacts with a heavy nucleus and causes the emission of a large number of hadrons mostly neutrons or fragments [12].Therefore spallation is by far the most attractive means of neutron production when it comes to energy deposition per neutron. The spallation mechanism initiates with intra-nuclear cascade (INC) reactions, followed by de-excitation of nuclei through evaporation, multi-fragmentation and fission [13].The first, so-called fast stage of the spallation is highly non-equilibrated process. High energy proton causes an intra-nuclear cascade on a time scale $\sim 10^{-22}$ s. The incident projectile goes through the target nucleus and deposits a significant amount of excitation energy and angular momentum, while ejecting only a few high energy nucleons and, with a minor yield, pions and light ions. The result of the first stage is excited residual nucleus in thermodynamical equilibrium (totally or partly equilibrated), with excitation energy ~a few MeV/nucleon. The second, so-called slow stage of the spallation, consists in de-excitation of the residuum by evaporation of particles. The isotropic emission, in the system of nucleus, of nucleons (mainly neutrons), light and heavy ions (d, t, He, Li, Be, B, ..., γ) takes place on a time scale ~10^{-18} - 10^{-16} s [9. Broadly, the different stages of spallation reactions are:

Fast Direct Process:

- Intra-Nuclear Cascade

(nucleon-nucleon collisions)

Pre-Compound Stage:

- Pre-Equilibrium
- Multi-Fragmentation
- Fermi Breakup

Compound Nuclei:

- Evaporation (mostly neutrons)
- High-Energy Fissions

Inter-Nuclear Cascade

Low-Energy Inelastic Reactions

- (n, xn)
- (n, nf), etc. [14]

Both the INC and de-excitation processes lead to the production of neutrons and spallation products as illustrated in Figure 1 [13]. Spallation products are of two types: spallation evaporation and spallation-fission products. The spallation products can be measured using two methods: direct or inverse kinematics. In direct kinematics, a relativistic light projectile hits a heavy target. The spallation products, which stop in the target, can be detected using γ-spectroscopy and mass spectrometry. This method has the possibility to measure the yields of the meta-stable states of residual nuclei, it can use radioactive targets, and it consumes less beam time. On the other hand, it is impossible to measure the yields of very long-lived, stable and very short-lived nuclei, and the off-line yields measurements are more time consuming. In the case of inverse kinematics, a relativistic heavy nucleus hit a light target. The spallation products leave the target in forward direction and can be immediately in flight using the appropriate experimental set-up [12].

The neutrons produced in spallation reactions can be characterized by their energy and spatial distributions and multiplicity. The neutron multiplicity will determine the current and beam energy of the proton-driver accelerator while their energy and spatial distribution shapes the geometry of the spallation target and the shielding to high-energy neutrons [15. The spallation neutrons have only a very small probability of causing additional fission events in the target [8]. Spallation reactions do not only produce neutrons but also residual nuclei. Most of these nuclei are radioactive. In addition, the residual nuclei will contribute to the corrosion and to the radiation damages in the target, accelerator window and structural materials [15].

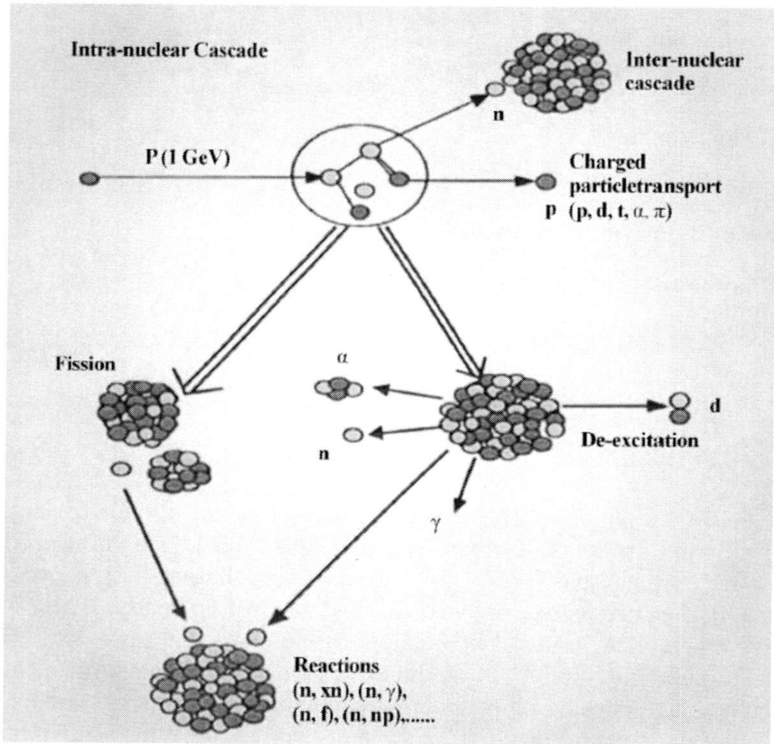

Figure 1. Spallation reaction and different stages with the products [14.

Neutrons from Spallation have huge potential in research in earth, in space, in medical industry, in energy production, etc. More clearly, they play a significant role in a wide domain of applications ranging from neutron sources for condensed matter, particle physics experiments, transmutation of nuclear waste, material studies, isotope production to astrophysics, simulation of detector set-ups in nuclear study and radiation protection around an accelerator and also in space, etc.

The long-term hazard of radioactive waste arising from nuclear energy production is a matter of continued discussion and public concern in many countries [10]. Transmutation of highly radioactive nuclear waste can be performed using an accelerator driven system (ADS), where high energy protons impinge on a spallation target to produce neutrons. These neutrons are multiplied in a sub-critical core, while simultaneously fissioning the minor actinides into short-lived or stable nuclides [13]. The accelerator driven system (ADS) has recently been receiving increased attention due to its potential to improve the flexibility and safety characteristics of transmutation systems [10].

The idea of the ADS was born in 90's with articles of C.D. Bowmam and C. Rubbia, who independently proposed a new approach to the problems of radioactive waste and limited uranium resources: to introduce extra neutrons produced in spallation reaction to the core of the subcritical reactor. A subcritical reactor is a nuclear fission reactor that produces fission without achieving criticality. Instead of sustaining a chain reaction, a subcritical reactor uses additional neutrons from an outside source. Such a reactor coupled to a particle accelerator to produce neutrons by spallation is called an ADS [10]. Therefore, it is said that, a sub-critical reactor is driven by an Accelerator-Driven System (ADS) to compensate for the loss of neutron economy which otherwise would have been self-sustaining in a critical reactor [13. Figure 2 shows how the electric energy is generated using a subcritical reactor.

In particular, as the spallation target is surrounded by a blanket assembly of nuclear fuel, such as fissile isotopes of uranium or plutonium (or thorium-232 which can breed to U-233), there is a possibility of sustaining a fission reaction. This is described as an accelerator-driven system (ADS). In this system, the neutrons produced by spallation would cause fission in the fuel, assisted by further neutrons arising from that fission. Up to 10% of the neutrons could come from the spallation, though it would normally be less, with the rest of the neutrons arising from fission events in the blanket assembly. An ADS can only run when neutrons are supplied to it because it burns material which does not have a high enough fission-to-capture ratio for neutrons to maintain a fission chain reaction. One then has a nuclear reactor which could be turned off simply by stopping the proton beam, rather than needing to insert control rods to absorb neutrons and make the fuel assembly subcritical. Because they stop when the input current is switched off, accelerator-driven systems are seen as safer than normal fission reactors.

An ADS can be used to destroy heavy isotopes contained in the used fuel from a conventional nuclear reactor—particularly actinides. Here the blanket assembly is actinide fuel and/or used nuclear fuel. One approach is to start with fresh used fuel from conventional reactors in the outer blanket region and progressively move it inwards. It is then removed and reprocessed, with the uranium recycled and most fission products separated as waste. The actinides are then placed back in the system for further incineration. Figure 3 shows the different aspects of ADS products based on their (ADSs) physical characteristics.

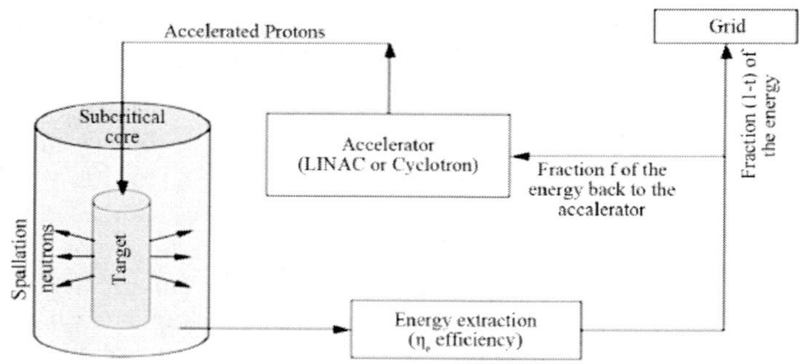

Figure 2. ADS based electric energy generation [14].

ADS could also be used to destroy longer-lived fission products contained in spent nuclear fuel, such as ^{99}Tc and ^{129}I (213,000 and 16 million years half-lives, respectively). These isotopes can acquire a neutron to become ^{100}Tc and ^{130}I respectively, which are very short-lived, and beta decay to ^{100}Ru and ^{130}Xe, which are stable [8].

It is therefore, can then be said that the term ADS comprehensively includes all non-self-sustaining fissioning neutron multiplying assemblies which are driven by an external neutron source provided by a charged particle accelerator and a neutron producing target. Worldwide, ADS systems under current study include both thermal and fast neutron multiplying media comprised of either liquid or solid (lattice) fuel and driven by either cyclotron or linear proton accelerators and spallation targets (liquid and solid) of various heavy metals. The underlying missions targeted for ADS systems span the range from nuclear waste incineration with ancillary power production through power production with integral waste self-incineration to finally, excess neutron production for the purpose of isotope production via neutron capture reactions on targets [16].

Another area of current interest in the use of ADS is in their potential to dispose of weapons-grade plutonium, as an alternative to burning it as mixed oxide fuel (MOX) in conventional reactors. Two alternative strategies are envisaged: the plutonium and minor actinides being managed separately, with the latter burned in ADS while plutonium is burned in fast reactors; and the plutonium and minor actinides being burned together in

ADS, providing better proliferation resistance but posing some technical challenges. Both can achieve major reduction in waste radiotoxicity, and the first would add only 10% - 20% to electricity costs (compared with the once- through fuel cycle) [8].

3. PARAMETRIC AND TECHNOLOGICAL CONSIDERATIONS IN SPALLATION ANDADS STUDY

The spallation target constitutes the physical and functional interface between the accelerator and the sub-critical reactor. For this reason it is probably the most innovative component of the ADS. The target design is a key issue to investigate in designing an ADS and its performances are characterized by the number of neutrons emitted for incident proton, the mean energy deposited in the target for neutron produced, the neutron spectrum and the spallation product distribution. The development and design of the target implies a detailed assessment of different aspects mutually interacting, from the physics of spallation reaction including neutron generation and distribution, spallation product yields and damage rates to technological issues, such as choice of the most suitable material, power density distribution, heat removal, thermo-mechanics and fabricability. In particular, accurate and rigorous assessment of nuclear parameters under different physical conditions is the prerequisite for an optimal design of the target and its interaction with the sub-critical core [17].

The physical study on the spallation target is one of the basic researches of the ADS design. From the point of view on the engineering application, the basic problems such as spallation neutron yield, energy spectrum and energy deposition will be firstly concerned. The neutron yield per incident proton is the most interesting data for the practical application in the ADS study, which is related with the incident proton energy and the target material [18. Also, the neutron yield is an important consideration in selecting the energy and current parameters of the proton beam. Many experimental data show that for a given neutron yield, lower beam current is required at higher beam energy. An advantage of higher beam energy is greater penetration depth in the target, leading to lower axial power densities [19].

The ADS system utilizes the neutrons produced in a spallation target by a high-energy proton beam to drive a sub-critical reactor containing both fissionable fuel and radioactive waste [18. The neutron fluxes produced in spallation reactions will depend strongly on the projectile-target combination. Specifically, it depends on the material properties, the kinetic energy of the projectile and the geometric configuration of the target [13]. In principle the heavier the target nucleus the larger the neutron excess and the larger the neutron yields [15]. The number of spallation neutrons per incident proton depends on the beam energy and on the mass of the target nuclei [19]. As because of the above, spallation target is ideally conceived to be high atomic number material and heavy density metals like lead, bismuth, mercury, uranium, tungsten, tantalum and lead-bismuth eutectic (LBE) to fit the requirement extremely well. The gain factor between heavy and light targets is around a factor of five. However, the radiotoxicity induced in the spallation target could be drastically reduced when using lighter targets [15].

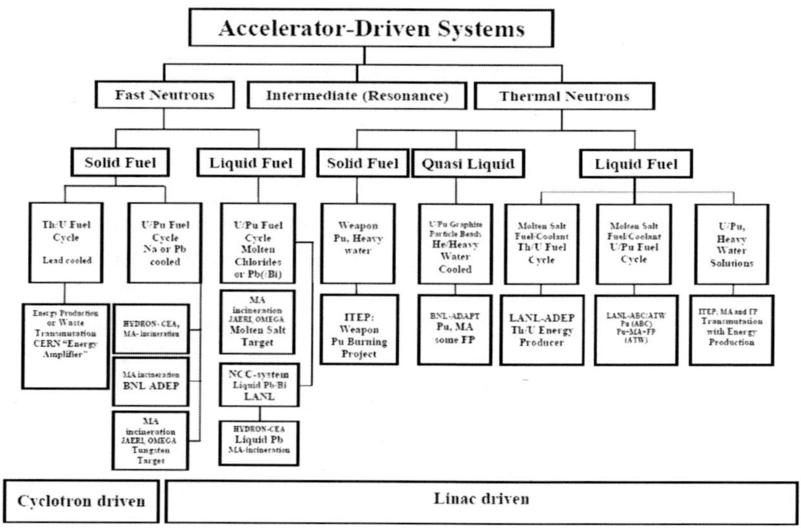

Figure 3. Classification of existing ADS concepts according to their physical features and final objectives [14].

The size of the target has a great influence in parameter consideration. Determination of the appropriate size of the target is so complicated. One must consider the design of the reactor, heat dissipation, the production process of the target, neutron yield and the amount of a combination of factors. The reaction mechanisms of different neutron spectra and reactor materials are different. Consequently, the target of different sizes results in different energy spectra of neutrons, and thus matches with a different reactor. Therefore, a comparison of the neutron spectrum in the different target sizes is necessary [18]. Moreover, because of the requirement to minimize the size of the target in relation to the core dimensions to provide good neutron economy; the target design is driven to maximum power density. For the very high power densities reached in a spallation target (several hundred kW per liter for proton beam powers of several MW), a flowing liquid metal target material provides the best option for practical removal of the heat by convection. The use of a heavy liquid metal (HLM) also significantly reduces the damage caused by intense radiation to the target itself and the structural materials. However, the use of HLM gives rise to problems of erosion, corrosion, and liquid metal embrittlement (LME) of structural materials that are in contact with the circulating metal at its characteristic high temperatures and high fluid velocities. In addition, difficulties concerning the confinement of volatile spallation products and the separation between the proton beam vacuum and the liquid spallation target must also be addressed [19].

Both window and windowless target concepts are currently being investigated in ongoing ADS design projects worldwide. However, the constraints introduced by a window in terms of the required subcritical-core working temperature, the need of frequent window replacement, and/or the need for an extra cooling loop for the spallation target have raised major design challenges for this approach. These concerns are reinforcing an R&D effort aimed at solving key design issues for the windowless target concept, namely compatibility of the beam line vacuum and target outgassing, and control of spallation products evolved from the target, and of HLM evaporated from the target surface. These R&D efforts are not dedicated exclusively to supporting the windowless target concept, but also address the safety issues arising from a possible window rupture in the window target case [19].

Two severe operating conditions of the spallation target are the very high volumetric heat deposition rate and irradiation by high fluence of energetic protons and neutrons. Regarding the former, the peak power dissipation density in the target volume can be of the order of a few kW/cm^3. In such a case the use of heavy liquid metals (HLM) like lead and bismuth may serve both as spallation target and heat removing media. Also the problems associated with irradiation damage like swelling, cracking/embrittlement and other degradations can easily be avoided using such a target. However, these and corrosion/erosion issues need to be resolved with new and better materials for liquid target container and process equipment [7].

In order to get the full pledge benefit from the spallation dependent ADS technology it is important to consider the parametric factors having direct or indirect influence on the system. Among these, the parameters which characterize the spallation module are extremely important since they will have several impacts on the design of the whole ADS. Some important notes regarding this are:

- The neutron angular and energetic distribution will determine the transmutation potential of the system.
- Target with high neutron source strength will drive an ADS with lower multiplication factor, thus improving safety conditions.
- Costs saving can be achieved by using a target with high proton to neutron conversion factor. In this way fewer particles current is required.

Therefore, for achieving the best from an ADS assembly, the spallation module design should then be based on a balanced optimization between neutronic efficiency, material properties and thermo-hydraulic performances [17]. Moreover, in considerations of different issues related to the target, technology, efficiency and environmental safety, an ADS should also have the following attractive features:

- extremely high level of inherent safety;
- minimum production of long lived wastes;
- ability to transmute long-lived nuclear wastes generated in normal reactors and;
- more efficient use of abundantly available natural fuel (^{232}Th) [18].

All these four are directly dependent on the above three. Hence a combined attempt would provide the path to reach the goal. Butin the practical realization of the ADS, the most challenging task is the development of a high energy (~1 GeV) and high current (≥10 - 15 mA) proton accelerator, which is reliable, rugged and stable in order to provide uninterrupted beam power to the spallation target over long periods of time [7].

In spallation neutron sources and Accelerator Driven Systems, along with all, radiation protection has mainly to deal with specific problems induced by high-energy reactions in the target. This is the case, for instance, of the radioactivity in the target in which the spallation reactions create a much larger variety of radioactive isotopes than activation by low energy neutron fluxes. The existence of a high-energy tail in the spectrum of neutrons generated in the spallation target is also of importance for the shielding of the facility [20].

Besides the above, it is better to add here the following that, research work on the spallation target design and ADS set up have been going on for a long periods of time since its concept given by Bowman and Rubbia. As a result, data available for different parameters obtained from different experimental observations may help in further parametric and technological research and development. Moreover, a large variety of existing nuclear data may provide the scope of reliable ADS design as it facilitates a detail study of the ADS assembly. For this reason some experimental findings are discussed here with a view to take it into considerations during future research and development so that it can then be made more efficient and reliable, also. For instance:

Both lead (Pb) and lead-bismuth (Pb-Bi) exhibit very low neutron capture making them good candidates from a neutronic standpoint. Lead (Pb) might at first appear a better target choice than LBE, because of its considerably lower production of ^{210}Po (a migratory α-emitter) under neutron bombardment. However, lead has the critical disadvantage of a higher melting point (327°C compared with 123.5°C), which presents severe thermo-mechanical design challenges for the target structural materials. The ^{210}Po release problem in LBE is mitigated by the fact that Po forms a chemical bond with Pb to form metastable Pb-Po, which has an emanation rate (vapour pressure) 1000 times lower than that of Po by itself. Therefore, Lead-bismuth eutectic (LBE) becomes today the reference target material for ADS applications [19].

Mercury (Hg) has also been considered as a spallation target material because of its low melting point (−38.9°C) and the absence of the ^{210}Po problem. However, its high volatility would seriously complicate the confinement of its radioactive isotopes, and its low boiling temperature would make its use in ADS difficult due to the high working temperature of the target and also of the subcritical multiplier. In addition its high neutron capture cross section degrades the leakage and therefore the neutron efficiency of the target [19].

The optimum proton energy for production of neutrons by spallation in a heavy metal target, in terms of costs, target heating, and system efficiency, lies in the range 600 to 1000 MeV. Two completely different kinds of machines can be considered for acceleration of high currents of protons to an energy of 600 - 1000 MeV: linear accelerators (linacs) and cyclotrons. CW linear accelerators will be the baseline technology for XADS demonstration facilities and full-scale industrial ADS plants, because of their high potential for achieving the required level of beam reliability and availability, and also because of their power upgrade capability. Linac beam theory and recent technology advances have confirmed that a linac capable of delivering up to 100 MW at 1 GeV is a relatively direct extension of existing technology. For ADS applications involving a few tens of megawatts, HLM is the target-material of choice at present. This selection is principally driven by the need for high-neutron yield per proton, the high power deposition densities reached within the target, and compatibility of the target material with structural materials [19].

4. MODELS AND CODES IN SPALLATION STUDY

The detailed engineering design of an ADS will require that the performance of the spallation target and all the problems related to the existence of high-energy particles can be predicted with sufficient accuracy. Some of the important quantities to assess are:

- intensity and spatial distribution of the spallation neutron flux outside the target;
- radiation damage and gas production in target, window and structural materials;
- radiotoxicity, activity and corrosion problems inside the target;
- required shielding to neutrons with much higher energies than in conventional reactors.

Provision of ADS related data is mainly done by means of theoretical development and computational implementation of nuclear models. For ADS, cross sections for the important materials need to be known for all possible reaction channels. This total amount of required information is so large that experiments alone can never cover the nuclear data needs. To fill this gap, the nuclear data must be simulated computationally, with the help of theoretical nuclear reaction models. The models can then be used in areas where no measurements exist [21]. Also, designing and operating a spallation neutron source requires reliable high-energy transport codes. In these codes, the elementary cross-sections and characteristics of all produced particles are either obtained from nuclear data libraries or calculated by nuclear-physics models. For several decades spallation models have been developed aiming at reproducing the particle and the residue production [22]. Moreover, application of high energy accelerators for production of energy and radioactive nuclides as well as transmutation of nuclear waste is based on large scale using of spallation reactions for neutron generation in thick targets and following multiplication of the neutrons in subcritical blankets [23]. As a result, neutron production calculations have to be as reliable as possible when designing a neutron spallation source. However, correct predictions of light charged particles (LCP), in particular tritium and helium, and residues in the spallation target and surrounding materials are important as well, most of the time for radioprotection and safety issues. As regards residues, β/γ or α-emitters, delayed neutron progenitors that can be produced in the target, in particular volatile elements in the case of liquid metals, are some examples of what must be properly estimated [22] and all of these are done by the help of different computer codes.

The development of a transmutation program based on the ADS requires accurate spallation simulation tools in order to design such systems. Correct description of such a complicated process as high energy particle interaction with matter can be performed by means of Monte Carlo method. It is the most suitable tool to take into account all details of nuclear reaction mechanism and particle transport both in high and low energy range [23. But there are some limitations and the main limitations of the Monte Carlo method are:

- The correctness of the neutron cross-sections, but this is common to all transport codes.
- The physical model used, but for low energy neutron transport this is mainly expressed by the partial reaction cross-sections, double differential cross-sections, etc.
- Its intrinsic imprecision, due to the random nature of the events generated, though it may be reduced by increasing the number of trial events N, now possible with the help of fast parallel computers which can generate many events simultaneously [10].

Several simulation codes and combinations of these codes exist. Now though there are some limitations in the existing codes there is a strong need of simulation codes for ADS assembly projection. Among the existing codes the two widely used transport codes used to simulate the spallation reactions are LCS (LAHET Code System) and FLUKA. The LCS code, developed at Los Alamos National Laboratory is a Monte Carlo code for treating the transport and interactions of nucleons, pions, muons, light ions and antinucleons in complex geometry. LAHET includes both the Bertini and the ISABEL intranuclear cascade model as user options. An evaporation model for the break-up of light nuclei is also included. An optional multistage pre-equilibrium model has been implemented as an intermediate stage between the intranuclear cascade and the evaporation phase of a nuclear reaction. Alternative level density parameterizations are also included. The FLUKA (FLUctuating KAscade simulation program) code, developed at CERN, is a Monte Carlo code able to simulate transport and interaction of electromagnetic and hadronic particles in any target material over a wide range of energies. FLUKA use the PEANUT (Pre-Equilibrium Approach to Nuclear Thermalization) model to describe an elastic nuclear interaction. This model consists of intranuclear cascade (INC), pre-equilibrium, evaporation and de- excitation. The current version of the code can simulate neutron interaction and transport down to thermal energies (multigroup below 20 MeV) and hadron-hadron and hadron-nucleus interactions up to 100 TeV [17].

The neutronics behaviour of an ADS is an important subject to be studied in the development phase. Detailed ADS design studies call for an accurate 3D geometry model. The code MCNP in its standard shape can only be used for static calculations. However, the time-dependent behaviour of an ADS requires full 3D burnup analyses. In order to make these analyses possible the NRG OCTOPUS code system was extended to include the possibility of performing 3D burnup analyses with MCNP and a burnup code (ORIGEN or FISPACT). The version of MCNP which is used at NRG was substantially extended with important modifications for ADS applications. A useful addition for ADS analyses is the possibility to generate 3D reaction rates in a calculation. This allows the calculation of power distributions and flux distributions in a 3D Monte Carlo calculation. Thus, the insight in the neutronics behaviour of an ADS may be substantially increased. Both the criticality analyses and the radiation shielding analyses of an ADS may be performed with MCNP. Finally, MCNP is used for the calculation of various nuclear data sensitivity coefficients. This is an important feature in the ADS design phase, as it allows parametric studies with a 3D Monte Carlo code. The results of sensitivity analyses may be combined with covariance data in order to calculate the uncertainty in calculated responses [21].

The Binary Cascade (BIC) is a new approach to cascade calculations, based on a detailed 3D model of the nucleus, and exclusively based on binary scattering between reaction participants and nucleons within this nuclear model. This feature makes it a hybrid between a classical cascade code, and a quantum molecular dynamics (QMD) model. In particular, BIC agrees better with the data for neutron production, low and medium energy protons, light ion production and isotropic distribution. A very good prediction by BIC is seen for the description of the forward peak spectra in several materials. Tritium production is also well predicted by the model [13].

The MCNPX code is a coupling of two previous calculations codes: LAHET and MCNP. MCNPX only needs one input file for both codes and avoids the transfer of large data files. It allows the treatment of transport problems in a large range of energies, from thermal energy, 25 meV to a few GeV. To treat the transport, MCNP uses data deduced of International cross-sections libraries such as ENDF, JEFF, JENDL, etc, after processing them with NJOY or ACER. After running an MCNPX-job, several evaluations can be performed with an auxiliary code, HTAPE3X, to obtain specific information e.g., neutron spectrum, energy deposition, residual nuclei, etc. Most options from older LAHET are used in MCNPX. They include the following models:

- the BERTINI and the ISABEL INC models as in LAHET, and a third INC model: the CEM model;
- the INCL4 model based on Cugnon INC approach;
- a multistage pre-equilibrium model;

- DRESNER and ABLA evaporation model;
- RAL and ORNL fission models;
- a nucleon elastic scattering model;
- a gamma production model [10].

The code SONET being developed for many years in NAS Belarus is dedicated for Monte Carlo simulation of intra- and inter-nuclear cascades in thick targets of arbitrary geometric configuration and material composition. One can calculate transport of nuclei (A < 16), nucleons and pions in the energy region up to 10 GeV. Electro- magnetic showers are simulated by means of the well-known EGS4 computer code, which is included in the package also. Ionization energy losses, range straggling, pion decay, multiple Coulomb scattering, inelastic interaction of particles (n, p, $\pi^{\pm}$, γ) and nuclei with nuclei as well as high energy fission have been taken into account. The code SONET describes nuclear collision products by means of a cascade-exciton model. The model treats a nuclear reaction as a three-stage process i.e., cascade, pre-equilibrium and equilibrium stages unlike well-known two-stage model. The code SONET provides complete description in energy as well as angular and spatial coordinates of the low energy neutrons produced in spallation reactions. Further, low energy neutron transport is described by means of methods developed for radiation shielding and nuclear reactor calculations. The first alternative is represented by the code SYNTES-Q. It allows to calculate neutron transport for subcritical systems in multi-group diffusional approximation in 2D geometry and is based on iterative synthesis method for solving the neutron transport equation with an external source. The second alternative is represented by the well-known code MCNP 4A. One of the versions of the code SONET allows to calculate the evolution of a nucleon-meson cascade in time, which is of interest when considering external pulsed sources. More generally, the code SONET allows calculating the following quantities:

- neutron energy-angle leakage spectra as well as neutron and nuclide yield;
- neutron fluence as well as specified reaction rates for specified geometric cells;
- time evolution of fuel nuclide composition;
- thermal power due to (n, f) and (n, γ) reactions;
- energy production in different geometric cells;
- distribution of fission and capture rates for different nuclides [23].

A Brazilian research group (IFUSP and CBPF) developed the MCMC/MCEF (Multi Collisional Monte Carlo plus Monte Carlo for Evaporation-Fission calculation) model to study nuclear reaction such as spallation. The MCMC and the MCEF model utilize the Monte Carlo approach to describe the intranuclear cascade and the evaporation/fission processes respectively. Their coupling originates the CRISP (Colabora ÇãoRIo-São Paulo) package. The code takes into account the possibility of neutron, proton and alpha particle evaporation and gives information about neutron and proton multiplicity, angular distribution and energy spectra [17].

The NRG Petten and CEA Bruyères-le-Châtel created a computer code system for the simulation and analysis of nuclear reactions named TALYS. The basic objective behind the construction of TALYS is the ability to give a complete description of nuclear reactions that involve neutrons, photons, protons, deuterons, tritons, ^{3}He- and alpha-particles, for target nuclides of mass 12 and heavier [24. TALYS is a deterministic code which implements various physical models and can quite reliably reproduce various cross-sections in the energy range 1 keV - 250 MeV [10].

The radiotoxicity of the spallation residue must be accurately known for safety reasons. The concentration limit of the radioactivity can be found based on the ORIGEN2 code. It is found that for the production of the radiotoxic isotopes, the CEM2K and the INCL4K models give the best results with a slight preference for the INCL4K model. Concerning the multiplicity (differential) of neutrons, the Bertini-preequilibrium-dresner model, the Isabel-preequilibrium-dresner model, the CEM2K model, and the INCL4-KHS model all reproduce rather well the experimental measurements. With these models, most of the ratios (theory to experiment) stay within 25% of the discrepancy [25].

Most shielding designs for high-energy accelerators have been performed by using a point kernel method, Moyer model. Moyer model is based on exponential attenuation of neutron dose equivalent for neutrons reaching the equilibrium state after thick shield. This method uses a single built-up factor and an attenuation length. However, for lower energies, the attenuation length depends on the neutron energy and the simple Moyer model is no longer applicable. In the practical neutron shield designs of spallation neutron sources there is a need for quick calculation method such as point kernel code. Neutron attenuation calculations are sometimes made by Monte Carlo method based on intra-nuclear cascade model. However this method is very time consuming and complicated [26].

Besides above there are some existing codes and models also. These are: HERMES [23, HETC (High Energy Transport Code.), NMTC (Nucleon Meson Transport Code), NUCLEUS, SHIELD, GEANT4 (Geometry and tracking), GEM (Generalized Evaporation Model), JAM (Jet AA Microscopic Transport Model), LAQGSM (Los Alamos Quark-Gluon String Model), MARS, TIERCE, BRIEFF [12, TUL (Tamura-Udagawa-Lenske model) [24, Exciton pre-equilibrium decay model and Fermi break-up model [23, KENO-Va (multi-group Monte Carlo code) [27, ALICE, LANCELOT, LILITA, PACE [28and all of these are used in ADS arrangement for various purposes.

5. SPALLATION WASTES AND ITS TRANSMUTATION IN ADS

5.1. Spallation Wastes

Continued development of nuclear fission as a source of energy is blocked by three main fears: potential reactor meltdown, proliferation of weaponizable fuel, and unsolved buildup of nuclear waste. As it turns out, the least progress has been made on addressing the waste issue. The concern with nuclear waste is its radioactivity— radioactive elements are unstable and can decay into other elements, releasing energy in the process. In some cases this energy can be threatening to humans or other life close enough to be exposed to the radiation. Waste is categorized according to two metrics—the danger it poses, and the lifetime over which it remains dangerous. Some radioactive waste is a byproduct of medical and research efforts, but the majority is produced through nuclear fission in reactors [29].

During the spallation process not only neutrons but also protons and other light nuclei (^{2}H, ^{3}H, He) are emitted from the excited nuclei. As a consequence, the residual nuclei are not only neutron-poor isotopes of the parent nucleus that decay (mainly by internal $p \rightarrow n$ conversion and β^{+} emission) into lower Z elements, but these lighter elements are also created directly in the spallation process. Furthermore, there is always a certain probability for a highly excited nucleus to fission and therefore it becomes usual to find medium-heavy and light nuclei in the spectrum of spallation products even for classically "non-fissionable" materials. As well as fission products, the process generates spallation products from the target material, in direct proportion to the energy of the proton beam. Some of these are volatile and will find their way into the cover gas system above the coolant, posing a major maintenance challenge. Their radiotoxicity is likely to exceed that of the fission products in the short term, which is relevant to operation and storage rather than final disposal. Finally, capture reactions can even lead to heavier nuclei than the parent nucleus. It can be seen that, although in small quantities, most elements of the periodic table are generated in a heavy spallation target. The consequences this has on materials properties and possible reactions are not yet well explored [11].

Generally, the Spent Nuclear Fuel (SNF) is the source of nuclear waste in the ADS arrangement (Figure 4). Hazard from the SNF is mainly due to a few chemical elements—plutonium, neptunium, americium, curium; out of which the latter three are classified as minor actinides—MAs and some long-lived fission products such as iodine and technetium at concentration levels of grams per ton and extended burn-ups or use of mixed oxides (MOX) fuel increases the amount of MAs produced [30].

Though the concentration of these nuclides are very minute in comparison to the target material, the hazardous nature stems from the fact that these nuclides have very long half-life and relatively higher radiotoxicity measured in Sv/Bq [13. In addition, the residual nuclei will contribute to the corrosion and to the radiation damages in the target, accelerator window and structural materials.

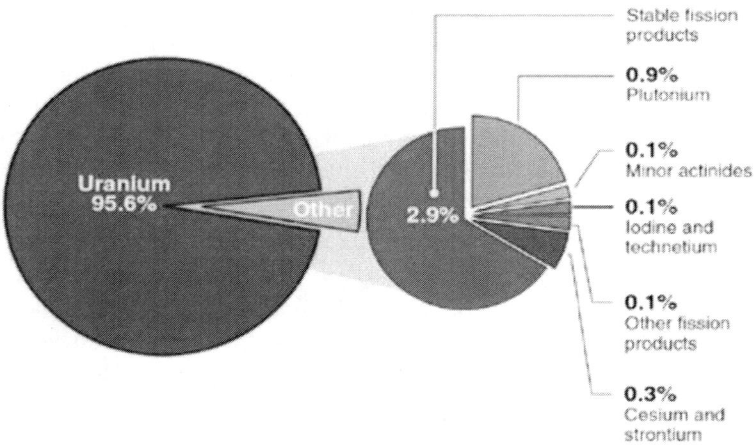

Figure 4. Nuclide distribution of spent nuclear fuel [31].

5.2. Transmutation in ADS

The radioactive waste is generated at different stages of the fuel cycle, and can arise in the form of radioactive liquids, gases or solids and with a large spectrum of activity levels. Depending on its activity level and its future management and disposal, it is classified as low, intermediate or high level waste [5. Management of radioactive waste in an environmentally safe manner is an important issue being addressed by all countries developing a nuclear industry. In many countries it has become a serious political issue attracting intense critical attention of the general public [32].

Treatment, conditioning and long term storage of all kinds of waste are mature technologies and are normally performed at the nuclear facilities where the waste is generated. Storage periods of 50 years or more are not unusual. This allows decay of most radioactive nuclides and flexibility for decisions on disposal. It is the widely held view of technical experts that the method of final disposal for high level wastes (HLW) and spent nuclear fuel (SNF) is likely to be in deep geological repositories. Experience indicates that the time needed to site and develop geological repositories is several decades. Most of the world's spent fuel continues to be stored in reactor pools or dry storage. However, storage represents an interim stage in all spent fuel management strategies, and the final disposal of spent fuel or HLW from spent fuel reprocessing can take decades. Spent fuel continues to accumulate in larger quantities and needs to be stored for longer time periods than initially envisaged (over 100 years). Furthermore, fuel designs are developing to allow much higher burn ups than initially considered in the design basis of many types of storage. Therefore, many different physical, chemical and thermal processes, for example, need to be researched and tested for continued operability, reliability, safety and security of the storage and the spent fuel, and to ensure that the spent fuel can ultimately be safely and securely transported from storage to reprocessing or disposal [5].

The disposal of the wastes arising from nuclear power production is a talkative issue in today's energy world which also discourages the further growth of nuclear power. Research and development (R&D) work has been going on for a long period of time to find out an efficient and a reliable solution to this problem. As a result of this, today, a technology is working namely ADS which stands for Accelerator Driven System. The concept and design of an accelerator driven system (ADS) is based on the burning up of the wastes produces at the time of power production. The capability of the ADS concept can then be applied to simultaneous burning of the long-lived fission products and actinides in the waste from nuclear power reactors. The goal of burning these fission products is to destroy those long-lived isotopes that, because of their mobility, would be the greatest contributor to radioactive doses from a repository, namely ^{99}Tc, ^{129}I, and ^{135}Cs. These three isotopes constitute about 5% by weight of the total fission products generated during power production and are the principal contributors to possible long-term risks associated with geologic disposition. If these three isotopes are destroyed internally, the remaining short-lived and stable fission product waste can be confined in engineered storage, perhaps even at the power production site. Maximum burn-up of the fission products requires that they be processed on site. By optimizing the accelerator-driven target or blanket assembly for this burning, the volume and half-life of the waste stream can be reduced considerably, thus simplifying long-term storage and secure disposal requirements. Even in this system, where the primary function is to burn radioactive wastes, it is expected that sufficient electrical energy can be produced to pay for the capital and operational costs of the facility [32].

An ADS destroys heavy isotopes contained in the spent fuel from a conventional nuclear reactor—particularly actinides. Here the blanket assembly is actinide fuel and/or spent nuclear fuel. One approach is to start with fresh spent fuel from conventional reactors in the outer blanket region and progressively move it inwards. It is then removed and reprocessed, with the uranium recycled and most fission products separated as waste. The actinides are then placed back in the system for further incineration. The whole process is done following a flow chart given in Figure 5.

Commercial application of partitioning and transmutation (P&T), which is attractive particularly for actinides, is still a long way off, since reliable separation is needed to ensure that stable isotopes are not transmuted into radioactive ones. New reprocessing methods would be required, including electrometallurgical ones e.g., pyroprocessing. The cost and technology of the partitioning together with the need to develop the necessary high-intensity accelerators seems to rule out early use. An NEA study showed that multiple recycling of the fuel would be necessary to achieve major (e.g. 100-fold) reductions in radiotoxicity, and also that the full potential of a transmutation system can be exploited only with commitment to it for 100 years or more.

Another area of current interest in the use of ADSs is in their potential to dispose of weapons-grade plutonium, as an alternative to burning it as mixed oxide fuel (MOX) in conventional reactors. Two alternative strategies are envisaged: the plutonium and minor actinides being managed separately, with the latter burned in ADSs while plutonium is burned in fast reactors; and the plutonium and minor actinides being burned together in ADSs, providing better proliferation resistance but posing some technical challenges. Both can achieve major reduction in waste radiotoxicity, and the first would add only 10% - 20% to electricity costs (compared with the once-through fuel cycle) [8].

Figure 6 illustrates the effectiveness of transmutation in reducing the overall radiotoxicity of the system. For example, ^{245}Cm with half-life of 8500 years is fissioned using a neutron. The fission products ^{110}Pd is stable and ^{134}Cs has a comparatively shorter half-life of two years.

Besides the above, another problem is the reactor itself. As power reactors reach the end of their life cycles they need to be decommissioned. As some parts of the reactors are radioactively contaminated, they will need to be dismantled in a controlled way and the radioactive waste taken care of. The timing of the dismantling is dependent on several factors, for example, radiation protection considerations, availability of funding and availability of disposal facilities. The radioactive waste from decommissioning is mostly low and intermediate level and can be handled and disposed of accordingly. For some of the components that are very large, special approaches, such as intact disposal, have been successfully used [5].

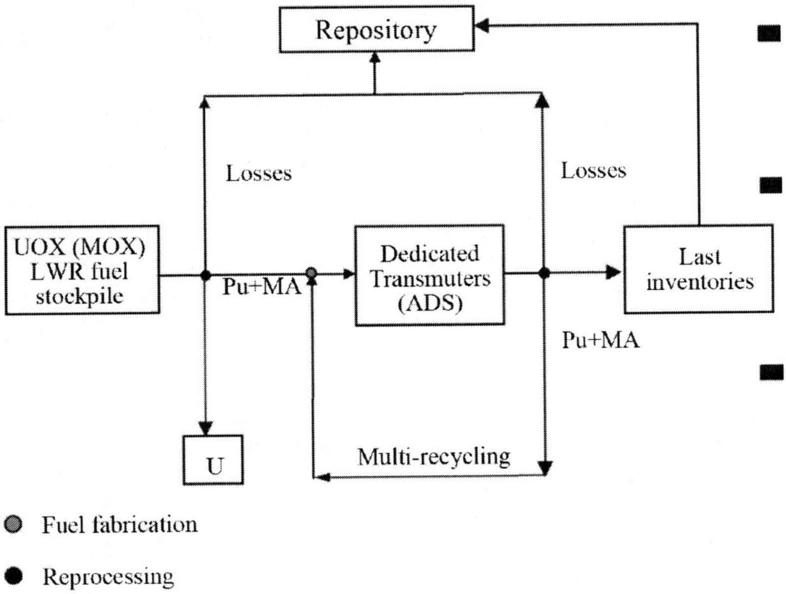

Figure 5. Partitioning and transmutation of nuclear wastes in ADS [30].

6. CONCLUSIONS

Energy is vital to human civilization and underpins national security, economic prosperity, and global stability. The first line is more than appropriate if it is of green in nature. Therefore, standing on different limitations of the existing non-renewable and renewable sources of electric energy generation as the basis, this paper includes the scopes of nuclear source as a green, efficient and reliable one.

Although nuclear power already has inherently low GHG emissions, future emissions would be even lower due to more energy efficient production of enriched uranium, improved nuclear fuels and reactors that allow greater utilization, and extended life times for NPPs that reduce the need to build new facilities. Moreover, it offers unique benefits as it is the only existing technology with capability for major expansion that can simultaneously provide stability for base-load electricity, security through reliable fuel supply, and environmental stewardship by avoiding emissions of greenhouse gases and other pollutants. Furthermore, it has proven reliability (greater than 90 percent capacity factor), exemplary safety, and operational economy through improved performance [2]. Despite all of the above, the public acceptability of nuclear power is adversely affected by negative perceptions concerning its safety, links with proliferation, and the radiotoxicity of its waste. Therefore for future efficient growth of nuclear power, it will be necessary to satisfactorily address the troublesome issue of disposal of the nuclear waste. Therefore, nuclear waste disposal is certainly an urgent and important issue to be tackled to ensure that future growth of this sector is an important finding of this study and an ADS is, therefore, quite relevant for further growth of nuclear power. In many respects, accelerator-driven systems (ADS) are worth pursuing. By producing electricity, they can contribute to the world's growing energy needs, and by incinerating plutonium and highly radioactive wastes, they can contribute to the goals of environmental protection and safe waste management. Beyond this, there is the promise of systems with the goal to burn weapons plutonium and to incinerate spent nuclear fuel, including its major fission products, from commercial nuclear power plants [32].

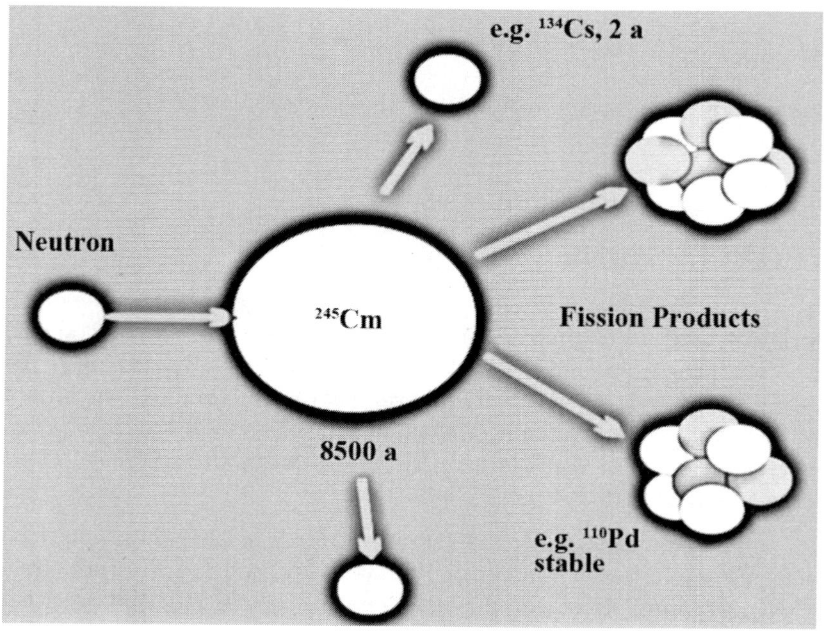

Figure 6. Transmutation of ^{245}Cm isotope using neutron in ADS [13.

But from this study it is found that all of the above can be achieved only if it is possible to ensure a reliable and successful ADS design. However a successful ADS design requires an efficient spallation technique with all its peripherals. Moreover, the efficiency of spallation reactions basically depends on the beam energy and on the target characteristics. Available information regarding the beam energy is more or less sufficient to carry out spallation reactions effectively but most of the problems lie in the target module. The outcome of the spallation reactions sharply varies with different parameters, for example, target material and its composition, atomic mass, size, thickness, etc., involved in the target. Also, appropriate coding systems play a vital role in theoretical analysis. It is found that most of the existing codes used for analyzing data show a discrepancy of about 20% - 25% with the experimental one. Also, the codes take their inputs from data libraries whereas the existing high energy data are not sufficiently available for ADS applications. Hence this study says that there is a strong need of further and specific research in designing the target characteristics, improving the coding systems (highly capable of minimizing the discrepancy with experimental results), and high energy reactions (to enrich nuclear data library) so that the ADS would be able to show its potential, in a greater domain, to provide a more sustainable, cost-effective and safer form of nuclear energy in the future.

REFERENCES

1. Lackner, K.S. (2010) Comparative Impacts of Fossil Fuels and Alternative Energy Sources, Issues in Environmental Science and Technology 29, Carbon Capture: Sequestration and Storage, Royal Society of Chemistry. www.rsc.org
2. (2008) A Sustainable Energy Future: The Essential Role of Nuclear Energy. https://inlportal.inl.gov/.../NLDC_ltr_Future_of_Nuclear-Final.pdf
3. (2010) Meeting the Energy Challenges of the Future—A Guide for Policymakers. National Conference of State Legislatures, a Forum for America's Ideas. www.ncsl.org http://www.renewable energysources.com/
4. (2011) International Status and Prospects of Nuclear Power, 2010 Edition, International Atomic Energy Agency Vienna. www.iaea.org/Publications/Booklets/NuclearPower/np10.pdf
5. MIT Study on the Future of Nuclear Power, the Future o f Nuclear Power—Over View and Conclusions. Chapter #1. http://web.mit.edu/nuclearpower/pdf/nuclearpower-ch1-3.pdf
6. Kapoor, S.S. (2002) Accelerator-Driven Sub-Critical Reactor System (ADS) for Nuclear Energy Generation. Pramana-Journal of Physics, 59, 941-950. http://dx.doi.org/10.1007/s12043-002-0143-z
7. http://www.world-nuclear.org/info/Current-and-Future-Generation/Accelerator-driven-Nuclear-energy/
8. Kowalczyk, A. (2007) Proton Induced Spallation Reactions in the Energy Range 0.1 - 10 GeV. A Doctoral Dissertation Prepared at the Institute of Nuclear Physics of the Jagiellonian University, Submitted to the Faculty of Physics, Astronomy and Applied Computer Science at the Jagiellonian University, Cracow, 1-162.

9. Majerle, M. (2008) Monte Carlo Methods in Spallation Experiments. Report for the PhD State Exam, Czech Technical University in Prague, 1-55.
10. Bauer, G.S. (1998) Physics and Technology of Spallation Neutron Sources. Lecture Notes of a Course Given at the Frederic Jolliot Summer School in Cadarache, France; Intended for Publication, in Revised Form, in a Special Issue of Nuclear Instruments and Methods A, 1-44.
11. Krása, A. (2010) Spallation Reaction Physics. Manuscript for the Lecture "Neutron Sources for ADS" for Students of the Faculty of Nuclear Sciences and Physical Engineering at Czech Technical University, Prague, 1-61.
12. Shetty, N.V. (2013) Study of Particle Transport in a High Power Spallation Target for an Accelerator-Driven Transmutation System. Master's Thesis, RWTH Aachen University, Aachen, 1-138.
13. Kadi, Y. and Revol, J.P. (2005) IAEA/ICTP Workshop on Technology and Applications of Accelerator Driven Systems (ADS). ICTP, Trieste.
14. Armbruster, P. and Benlliure, J. (2001) Basic Nuclear Data at High and Intermediate Energy for Accelerator-Driven Systems. http://www.nupecc.org/iai2001/report/A6.pdf
15. Wade, D.C. (2001) Safety Considerations in Design of Fast Spectrum ADS For Transuranic or Minor Actinide Burning: A Status Report on Activities of the OECD/NEA Expert Group. Overview Paper, 95-118. www.researchgate.net/.../237451934
16. Mongelli, S.T., Maiorino, J.R., Anéfalos, S., Deppman, A. and Carluccio, T. (2005) Spallation Physics and the ADS Target Design. Brazilian Journal of Physics, 35, 894-897. http://dx.doi.org/10.1590/S0103-97332005000500048
17. Zhang, Y.L., Zhang, X.C., Qi, J., Wu, Z. and Yang, L. (2013) Study on the Parameters of the ADS Spallation Target. Journal of Physics: Conference Series, 420, Article ID: 012064. http://dx.doi.org/10.1088/1742-6596/420/1/012064
18. OECD (2005) Accelerator and Spallation Target Technologies for ADS Applications, Nuclear Energy Agency, Organization for Economic Co-Operation and Development. A Status Report, NEA No. 5421.
19. Leray, S., Boudard, A., David, J.C., Donadille, L., Villagrasa, C. and Volant, C. (2005) Impact of High Energy Nuclear Data on the Radio-Protection in Spallation Sources. Radiation Protection Dosimetry, 115, 242-246.http://dx.doi.org/10.1093/rpd/nci086 Klippel, H.T. (2003) ADS Activities in the Netherlands. IAEA-TECDOC-1365, 166-175.
20. Khandaker, M., Mank, G., Mengoni, A., Otuka, N., David, J.-C., Leray, S., et al. (2010) Codes and Data for Spallation Sources, Benchmark of Nuclear Spallation Models. http://scholar.google.com
21. Chigrinov, S.E., Kievitskaia, A.I., Rakhno, I.L., Serafimovich, I.G., Rutkovskaia, C.K., et al. (2003) Research on Accelerator Driven Systems in the National Academy of Sciences of Belarus. IAEA-Tecdoc-1365, 78-91.
22. Koning, A., Beijers, H., Benlliure, J., Bersillon, O., et al. (2002) A European Nuclear Data Programme for Accelerator-Driven Systems. Seventh Information Exchange Meeting on Actinide and Fission Product Partitioning and Transmutation, 723-734. https://www.oecd-nea.org/pt/docs/iem/jeju02/session5/SessionV-05.pdf
23. Aoust, T., Malambu, E.M. and AïtAbderrahim, H. (2005) Importance of Nuclear Data for Windowless Spallation Target Design. In: Proceedings of the International Conference on Nuclear Data for Science and Technology, American Institute of Physics, College Park, 1572-1575.
24. Fragopoulou, M. (2006) Shielding around Spallation Neutron Sources. Journal of Physics: Conference Series, 41, 514-518.http://dx.doi.org/10.1088/1742-6596/41/1/058
25. Magill, J. and Peerani, P. (1999) (Non-) Proliferation Aspects of Accelerator Driven Systems. Journal de Physique IV, 9, Pr7-167.
26. Benlliure, J. (2006) Spallation Reactions in Applied and Fundamental Research. Lecture Notes in Physics, 700, 191-238. http://dx.doi.org/10.1007/3-540-33787-3_5
27. Ray, J. (2011) Accelerator Transmutation of Waste. Submitted as Coursework for Physics 241, Stanford University, Stanford.

28. OECD (2006) Physics and Safety of Transmutation Systems. Technical Report, NEA No. 6090.
29. GAO (2011) Nuclear Fuel Cycle Options. Technical Report GAO-12-70, United States Government Accountability Office, Washington DC.
30. Arkhipov, V. (1997) Future Nuclear Energy Systems: Generating Electricity, Burning Wastes. IAEA Bulletin, 39, 30-33.

CHAPTER 3

The global nuclear liability regime post Fukushima Daiichi

Raphael J. Heffron[1,*] ,Stephen F. Ashley[2], William J. Nuttall[2]

1 Plato, The Laws [of] Plato; Translated with an Introduction by Trevor J. Saunders.

2 Steven Shavell, Economic Analysis of Accident Law (Harvard University Press

ABSTRACT

Nuclear liability regimes are important as they ensure that potential victims will be compensated promptly and efficiently after a nuclear accident. The accident at Fukushima Daiichi in Japan in 2011 prompted a review of the global nuclear liability regime that remains on-going. Progress has been slow, but over the next few years the European Union is set to announce its new proposals. Meanwhile, in 2015, another global nuclear liability regime, the Convention on Supplementary Compensation for Nuclear Damage, has entered into force. This paper aims to move the debate in the literature on nuclear liability and focuses on the four following major issues: (1) reviews third-party nuclear liability regimes currently in operation around the world; (2) analyses the international nuclear liability regime following the accident at Fukushima Daiichi; (3) comparatively assesses the liability regimes for nuclear energy and the non-nuclear energy sector; and (4) presents the future outlook for possible developments in the global nuclear liability regime.

Keywords: Nuclear law; Nuclear liability; Fukushima; India; UK; US

1. INTRODUCTION

"Accidents and calamities...are the universal legislators of the world." Plato, The Laws.[1]

The accident at Fukushima Daiichi was a watershed moment for the global nuclear energy industry that has since led to an industry-wide revision of practices and regulations within the international nuclear energy community. The estimated costs of the incident at Fukushima Daiichi vary according to different calculations (as detailed in Section 4) but in every case, the incurred costs of the accident so far have been significantly higher than existing capped liability limits that are in operation around the world, as detailed further in Section 2. This has prompted the international nuclear energy community to scrutinise and revise what 'liability' means in the sector.

'Nuclear Liability' has been an area of legal scholarship for many decades, and has its roots especially in the area of tort law and more specifically in relation to the examination of law and accidents.[2] This article aims to advance the legal literature with not just a focus on nuclear accidents themselves, but also we intend to extend the discussion to: (1) a comparative review of the issues surrounding nuclear liability internationally; and (2) a comparative analysis of nuclear and non-nuclear energy accidents.

Nuclear accidents are transboundary in nature especially where there is an off-site release of radioactive material. Hence, various international liability regimes have been developed for the nuclear energy sector to which countries can be members. The nuclear energy sector demonstrates a level of international cooperation on liability for transboundary pollution and damage that is not observed in other parts of the energy industry. Nuclear liability regimes are of vital importance for two main reasons: (1) the ability for the nuclear energy industry to function; and (2) for the protection of the public as potential victims of a nuclear accident. For example, India is relying on significant growth of its nuclear energy sector to meet future electricity demands. However, as will be described in Section 3, uncertainty surrounding the extent of their liability for suppliers in its national liability regime has deterred potential non-domestic investors. Furthermore, nuclear liability regimes enshrine commonality that is important because of the potential cross-boundary effects that a nuclear accident may have on other countries.

Actions for damages and for injuries suffered are major areas of law. In general, the issue of accidents and insurance is approached from the perspective of law and economics, or through an economic analysis of the law. Here there are two questions that will be asked: (1) what is the effect of the legal rules (descriptive); and (2) is the current system desirable and what should it look like (normative)?[3,4] This paper examines the issue of accidents and insurance in the energy sector, and in particular, after a major accident. In this context, this paper provides four contributions: (1) it examines the third-party nuclear liability regimes currently in operation around the world and maps out the divergence between liability limits in the event of an accident in the European Union; (2) it analyses the international nuclear liability regime post the Fukushima Daiichi accident; (3) it comparatively assesses the liability regimes for nuclear energy and the non-nuclear energy sector; and (4) it presents the future outlook for possible developments in the global nuclear liability regime. More specifically, Section 2 details the current global nuclear liability regime while also analysing what it means if it is applied, i.e. if there were to be an accident. Section 3 focuses on global nuclear liability regime in light of the accident at Fukushima Daiichi, and Section 4 explores the difference between third party liability regimes for different energy sources. Finally, in Section 5, the paper advances the debate on whether the current global nuclear liability regime is desirable and points to what it should look like.

2. CURRENT NUCLEAR LIABILITY REGIMES

Many international institutions play a role in the energy sector. One of the most influential in terms of the international energy sector is the International Atomic Energy Agency (IAEA) which plays a lead role in the international nuclear energy industry. There are no similar UN-affiliated international institutions for other energy sources. The IAEA is located in Vienna, Austria, and 162 countries are currently Member States. It acts as an international driver of the nuclear energy sector, and it has a particular focus on maintaining the safe operation of nuclear energy plants across the world.

The IAEA is responsible for ensuring what are, in essence, global safety and security standards for the industry. It also gives law and policy advice, to assist countries in implementing laws to safely develop nuclear energy infrastructures. The IAEA also plays a role in the event of an accident. The main action by the IAEA in this context is its promotion of a liability regime for nuclear accidents. This encompasses a number of international treaties on nuclear liability.

There are three main international liability agreements, and the members of IAEA are all signed up to one or more of these three agreements. There are two main international third party nuclear energy liability regimes operating at present:

1) the "Paris and Brussels" Conventions, that are under the auspices of the Organisation for Economic Co-operation and Development (OECD); and

2) the "Vienna" Convention that is under the auspices of the IAEA.

There also exists a third nuclear energy liability regime, which is the Convention on Supplementary Compensation for Nuclear Damage (CSC), also under the auspices of the IAEA but which is championed by the USA and while only a few countries have ratified it, the CSC recently came into force, as further detailed in Section 5.

These liability regimes are an example of international cooperation in one area of the energy sector. Such liability regimes are not present for other energy sources and this highlights the value of research in this area in that other energy sources and global industrial sectors can learn from the nuclear energy sector and the liability regimes in place for accidents. A brief overview of the international nuclear liability regimes currently in force are detailed in Sections 2A–C.

2.1. The USA - Price–Anderson Act of 1957

The first nuclear liability regime was developed by the USA. The USA's third-party liability convention operates under the Price–Anderson Act of 1957. This Act inserted the new Section 170 into the Atomic Energy Act of 1954[5] and has been revised several times, most recently by the Energy Policy Act of 2005[6] that extends its application to the civil nuclear energy sector for a further 20 years.

The purpose of the Price–Anderson Act is to provide a Federal compensation fund should there be a nuclear accident. The Price–Anderson Act, in essence, requires each reactor site to pay an annual premium. Insurance protection consists of two tiers: the first tier provides $375 million in liability insurance per incident based on maximum level of private insurance available; the second tier (if the first tier is exhausted), allows each nuclear reactor site to pay up to a maximum of $111.9 million per nuclear incident and is classed as a retrospective premium equal to its proportionate share of the excess loss (i.e. the loss beyond the first tier amount of $375 million).[7] Estimates of the US insurance pool suggest that $12.6 billion in insurance cover is available.[7] Hence, The Act does limit liability and does not guarantee payment should possible funds already have been exhausted. Instead, in the instance where the damages are likely to exceed the applicable amount of aggregate public liability, the Nuclear Regulatory Commission will seek and give recommendations for additional sources of funds from the US government after they have conducted an assessment of potential costs.[8]

2.2. The Paris and Vienna Conventions

Thereafter, the OECD established a nuclear liability regime with the Paris Convention (PC) in 1960, which has since been updated in 1964 and 1982.[9] It was signed first by Western European states.[10] It requires ratification by national legislation of a contracted state. It is based on seven principles[11] outlined in Table 1 below.

Table 1. The Seven Principles of the Paris Convention, as equally applicable to the Vienna Convention.

	Principle
1	Strict liability
2	Channelling liability to the operator
3	Limitation of liability in time
4	Limitation of liability in amount
5	Insurance or other financial security
6	Jurisdiction
7	Applicable law

The Vienna Convention (VC), established by the IAEA, is similar to the PC in its aims and was signed in 1963 mainly by non-OECD countries in Eastern Europe and Latin America.[12] The two conventions are similar in terms of the aforementioned seven principles but differ in number of other aspects such as financial liability limits, territorial scope and dispute resolution.[13]

The Brussels Supplementary Convention (BSC) was the next development, and it was signed in 1963 by the majority of the same countries that signed the PC.[14] In essence, it is complimentary to the PC and allowed additional funds to be made available to victims of a nuclear accident.

The accident at Chernobyl in 1986 prompted the provision of additional funds to victims of a nuclear accident. Notably, The Joint Protocol that entered into force in 1992, which offers linkages between countries that adopted either the PC or the VC.[15]

2.3. The revised conventions and the CSC (new nuclear liability limits)

The revised Vienna Convention (RVC) was introduced in 1997 and increased the minimum financial security required from US$5 million to

300 million Special Drawing Rights (SDRs), broadened the definition and geographical scope of nuclear damage and increased the time-period for certain claims to be raised. [16,17] The CSC (also introduced in 1997) in many ways is an alternative nuclear liability convention. It can act as supplementary to the PC and VC/RVC, providing an additional international fund that allows second-tier compensation. [18] Countries not party to the PC or VC/RVC can join the CSC so long as their national legislation is in agreement with the provisions of the Annex to the CSC with the first-tier of compensation fixed at 300 million SDRs. The CSC is based on seven principles that are similar to that of the PC, as shown in Table 1. The CSC required ratification from at least five countries that contain 400,000 units of installed capacity[19] for it to enter into force.

The other amendments take the form of the 2004 Protocol to Amend the Paris Convention and the 2004 Protocol to Amend the Brussels Supplementary Convention (hereafter referred to as the revised Paris Convention (RPC) and revised Brussels Supplementary Convention (RBSC)). The aim of the RPC and RBSC was to increase the scope of, limit of and time to claim compensation for instances of loss of life and personal injury as a result of a nuclear accident; with the Protocols having factored in the losses experienced as a result of Chernobyl.[20] The RPC plus RBSC now cover a lower liability limit of €1.5 billion, and like the RVC broadened the definition, geographical scope and timeliness for certain claims to be raised.[21] A summary of the losses that can be claimed is shown below; it is noted that the PC only covers items (i)–(iii):

i. loss of life or personal injury;

ii. loss of or damage to property;

iii. economic loss arising from (i) and (ii) above;

iv. the costs of measures of reinstatement of impaired environment;

v. loss of income deriving from a direct economic interest in any use or enjoyment of the environment; and

vi. the cost of preventative measures including loss or damage caused by such measures.

vii. any other economic loss, other than any caused by the impairment of the environment if permitted by general law on civil liability of a competent court [22,23]

2.4. The principle of channelling

One of the key principles of present nuclear liability regimes is the channelling principle, where all liability is channelled to the operator.[24]

Table 2. IAEA Model provisions on Liability of the Operator provisions on Liability of the Operator.Stoiber and others (n 22).

Article##: liability of the operator
1. Subject to the provisions of this Law, only the operator of a nuclear installation shall be liable for nuclear damage wherever suffered, upon proof that such damage has been caused by a nuclear incident at the operator's nuclear installation. 2. Liability for nuclear damage caused by nuclear material which was stolen, lost, jettisoned or abandoned lies with the operator who was last authorized to possess such material 3. Liability for nuclear damage shall apply to nuclear damage wherever suffered.

The channelling principle is expressed best in the phrase from the first paragraph of IAEA Implementing Legislation drafting notes (see Table 2 below) and along with the accompanying phrase from both the Vienna and Paris Conventions "Except as otherwise provided in this Convention, no person other than the operator shall be liable for nuclear damage".[25]

Through the nuclear liability regimes, liability is legally channelled (i.e. it can be considered as exclusive liability). In essence this follows that a nuclear operator will be solely liable for the accident. This means that all victims will be compensated by the operator and reduces the cost burden for victims and the speed to which they can receive compensation. The responsibility of suppliers to nuclear operators falls on the operator themselves but in some cases there may be specific clauses within contracts between the operator and supplier in relation to the supplier being liable to a certain degree.[26]

3. AN INTERNATIONAL REVIEW OF THE NUCLEAR LIABILITY REGIME POST-FUKUSHIMA

One issue surrounding the conventions is that not all countries have signed and ratified (i.e., enter into force) these conventions, especially the most recent revisions to the conventions. To illustrate this, the map in Fig. 1 below highlights the 'patchwork' of conventions that are in force in Europe. A synopsis from the most recent OECD review of Liability Amounts and Financial Security Limits[27] is presented. At present, there are 21 countries that have signed the RVC, of which 12 have ratified these. Of the 28 EU Member States, only Latvia, Poland, and Romania have signed and ratified the RVC; whilst the Czech Republic, Hungary, Italy, and Lithuania have signed the RVC. Under the non-legislative EU Act 2013/434/EU[28]: Bulgaria, the Czech Republic, Estonia, Lithuania, Hungary, Poland, and Slovakia are authorised to ratify the RVC.

The RPC has only been ratified by two countries: Norway and Switzerland. It has been signed by 13 of the EU member states and Turkey. One reason why there have been few ratifications of this convention across Europe is that under EU directive 2004/294/EC[29]all Member States (with exception of Austria, Denmark, Ireland, and Luxembourg)[30]should simultaneously ratify, i.e. enter into force the revised Convention at the moment they sign the RPC. Interestingly, all of the states that were members of the EU prior to 2004 signed the convention (albeit with reservations from Germany, Austria, Greece, and Sweden). Of the post-2004 countries, only Slovenia has signed the Convention (under Directive 2007/727/EC[31]). It is suggested that the requirement of simultaneous ratification may act as a barrier to the majority of states ratifying the RPC.[32]

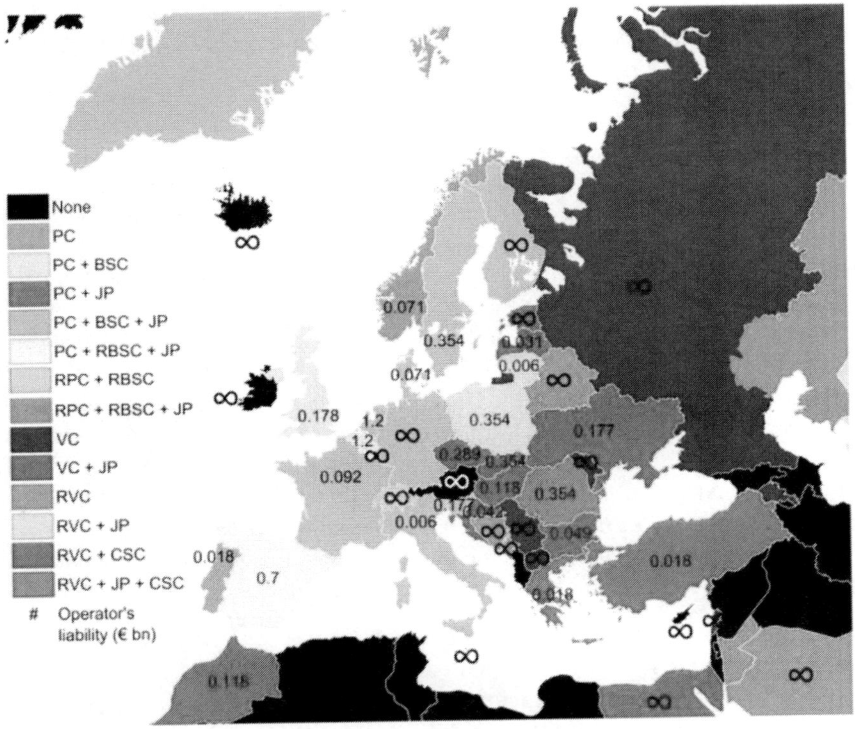

Figure. 1. Nuclear Liability Regime in EU countries including Operator's liability. Constructed by authors from OECD data contained in n 27. Acronyms: Paris Convention (PC); Brussels Supplementary Convention (BSC); Joint Protocol (JP); Revised Brussels Supplementary Convention (RBSC); Revised Paris Covention (RPC); Vienna Convention (VC); Revised Vienna Convention (RVC); Convention on Supplementary Compensation for Nuclear Damage (CSC).

With the ongoing uncertainty around which countries have signed and ratified which conventions, the nuclear liability regime was in a state of relative confusion prior to the accident at Fukushima Daiichi. Burns (2012)[33] has highlighted interesting statistics that further demonstrate the level of uncertainty on the potential effectiveness of existing nuclear liability regimes. His conclusion is the majority (74 per cent) of the world's reactors in operation or under construction are in eight states: Canada, China, France, India, Japan, South Korea, Russia, and the US. More importantly however, six of these countries — Canada, China, Japan, South Korea, India and the US — which account for 57 *per cent* of the world's reactors that are either in operation or under construction were until recently outside the scope of an international third party nuclear liability regime. Both Japan and the US are now as of April 2015 under the CSC liability regime.

There are renewed efforts post-Fukushima Daiichi to reduce the uncertainty around nuclear liability regimes in Europe. The European Union, as stated earlier, is aiming to revise the nuclear liability regime. In addition, the IAEA has called for a move towards developing a global nuclear liability regime. This was agreed by members of the IAEA in the IAEA's Action Plan on Nuclear Safety that was developed in 2011 by the Board of Governors and endorsed by all Member States at the IAEA General Conference in 2012.[34]A functional and transparent nuclear liability regime is important from the context of the number of countries that aim to build nuclear energy plants for the first time. In addition, there are new developments in the nuclear industry that also add to the case for reform. For example, Chinese nuclear energy firms are planning to enter the European market in the UK and Romania (a cooperation agreement was signed in July 2014 and a contract is on schedule to be signed).[35] Russian and South Korean nuclear energy companies are also planning international operations, in Turkey and UAE, respectively.[36]

The approach of India has been in contrast to the rest of the international nuclear industry. India has its own national nuclear liability that is set out in The Civil Liability for Nuclear Damage Act (2010).[37] One significant difference compared to other liability regimes is that liability is not exclusively channelled to the operator. Nuclear suppliers are also prospectively liable and this is specified in Section 17 of the Act (as detailed in Table 3). India has also signed the CSC (see Section 5). However, Section 17 contravenes the Annex of the CSC.[38] The intention in this paper is not to discuss in depth the clause (b) in Section 17, but to highlight more its ambiguity or vagueness.

The intent behind the direction by India in terms of civil nuclear liability is unclear. If anything it effectively prevents foreign suppliers from operating in India and it has stalled industry progress. This it seems was contrary to one of the intentions of The Civil Liability for Nuclear Damage Act in India, which was to attract foreign nuclear technology investment.[39] Compensation for major industrial accidents (especially from foreign companies) is a public issue in India after the Bhopal disaster in 1984 where compensation was extremely slow in been received by victims and that was if it was received at all.[40] There may be an attempt to avoid similar problems, but if anything the new law creates more confusion.[41] For example in its current state, a victim of a nuclear accident may have to take a multitude of court cases against several suppliers and the operator—especially in the instance where the operator invokes Section 17 clause (b) "the nuclear incident has resulted as a consequence of an act of supplier or his employee" (see Table 3). In this context, it is also unclear in terms of administrative law if the Indian court systems has improved in terms of speed to compensate claimants; and especially in light of this been an issue after the Bhopal accident. It has been noted in the literature that the Indian legal system is far from being in a situation to manage the legal consequences of a nuclear accident.[42]

Overall, many authors are positive towards the Indian approach to nuclear liability regime and its imposition of liability on suppliers though it is not exactly clear why.[43] There are several themes that emerge. One of these is that the international nuclear liability regime is in need of reform and the Indian approach is the way forward. This is despite one commentator noting that "In fact, in India, the operator companies will also be wholly government-owned, which could explain in part why the Indian government chose to introduce an expanded concept of supplier liability".[44] Yet this issue is never addressed. There also seems to be a distrust of foreign companies in the sector and also perhaps distrust in the state nuclear company — Nuclear Power Corporation of India Limited (NPCIL) — which will operate the nuclear reactors in the country. Indeed, a case that was before the Indian courts in 2004 highlighted this problem — *People's Union for Civil Liberties v Union of India* (2004) 2 SCC 476 — and this case concerned the disclosure of 130 issues of safety concern by the Atomic Energy Regulatory Board. [45] The disclosure was held not to be in the Indian State's interest and to be an issue of national security. This case highlighted the widespread view that the Indian nuclear industry operated under a veil of secrecy.[46,47]

A significant oversight in the academic literature concerns the role of the nuclear safety regulator in India (Atomic Energy Regulatory Board, AERB) and its effect on India's relationship with the international nuclear liability regimes. There has just been one notable review of the AERB and other government agencies in India such as the Department of Atomic Energy and the Atomic Energy Commission. In this case the roles of the Indian government, the AERB and the other latter agencies have been heavily criticised with the authors pointing three key problems which have served to reduce the importance of the need for India to sign an international nuclear liability convention. These are: (1) the attitude that is demonstrated by comments made by the Chairman of the Atomic Energy Commission who stated that the probability of an accident was 'one in infinity'; (2) the lack of independence of the AERB from the influence of Government; and (3) the comment by the Secretary of the Department of Atomic Energy who stated that Indian reactors were 100 *per cent* safe. [48]

Table 3. Section 17 of The Civil Liability for Nuclear Damage Act (2010). Civil Liability for Nuclear Damages Act, 2010 (n 37); The Civil Liability For Nuclear Damage Rules, 2011 (n 52).

Operator's right of recourse
17. The operator of the nuclear installation, after paying the compensation for nuclear damage in accordance with section 6, shall have a right of recourse where- (a) such right is expressly provided for in a contract in writing; (b) the nuclear incident has resulted as a consequence of an act of supplier or his employee, which includes supply of equipment or material with patent or latent defects or sub-standard services; (c) the nuclear incident has resulted from the act of commission or omission of an individual done with the intent to cause nuclear damage.

The regulator has a role to ensure safety in the activities of the suppliers, as has the nuclear operator. Not allowing liability to be passed on to suppliers does not reduce the role of the AERB to ensure that suppliers provide safe, reliable and non-defective equipment and services. Even NPCIL has commented that the effect of the Civil Liability for Nuclear Damage Act (2010) in India will be to reduce business activity in the Indian nuclear energy industry to a standstill.[49]

Nevertheless, it is clear that despite the introduction of the Civil Liability for Nuclear Damage Bill in India, there remains a realisation that the issue of supplier liability is important in the nuclear energy industry. Further clarification will happen in this area, and in particular after the recent meeting in January 2015 between the US and India between their Presidents, Barack Obama and NarendraModi, respectively. The outcome of this meeting will ensure there is a limit to the amount a supplier will be liable for in the event of a nuclear accident – though exact specifics still have yet to emerge.[50] In June 2015, aRs 1500 crore ($234 million) Indian nuclear insurance pool was announced - this insurance pool comprises domestic insurers plus Nuclear Risk Insurers which manages the UK's nuclear insurance pool.[51]

In India, it seems the operator will not have strict liability until there is a major shift in Indian legislation. In particular, the supplier needs to consider the Civil Liability for Nuclear Damage Rules (2011) that apply from the same day the Civil Liability for Nuclear Damage Act went into force (11 November 2011).[52] The three main issues as identified by a Committee on Subordinate Legislation of the LokSabha, which produced a Report on the Civil Liability for Nuclear Damage Rules, 2011 (on 28 August 2012)[53] which are outlined in Table 4 below.

In essence Rule 24 aims to limit the wide scope of Section 17 (b) of the Civil Liability for Nuclear Damage Act (see Table 3) both limiting the right of recourse of an operator on a supplier in time and the financial amount. However, it is perhaps not sufficiently unambiguous in asserting these limits. Also, the question as to whether Rule 24 is enforceable over Section 17 (b) of the Civil Liability for Nuclear Damage Act remains open. Indeed, it has been asserted that Rule 24 could not be relied upon. It is not only inconsistent with Section 17 (b) but would also be prohibited by the Indian Supreme Court.[54] Hence, despite the introduction of Rule 24, the situation remains the same and is why the US nuclear industry have sought, through the intervention of President Obama, further clarification of the issue of supplier liability in India.

Table 4. Issues concerning Rule 24 of the Civil Liability for Nuclear Damage (CLND) Act, 2011.OECD, 'Annual Report on India' (2012) 90 OECD-NEA Nuclear Law Bulletin 119.

" - Rule 24(1) provides that any contract under S. 17(a) of the Civil Liability for Nuclear Damage Act shall not be for an amount less than the amount of the operator liability specified under S. 6(2) of the CLND Act, or the value of the contract itself, whichever is less. - Rule 24(2) states that the right of recourse referred to in Rule 24(1) shall be for the duration of the initial licence issued under the Atomic Energy (Radiation Protection) Rules, 2004, or the product liability period, whichever is longer. Such initial licence is granted for five years. - Section 6(1) of the Civil Liability for Nuclear Damage Act specifies that the maximum amount of liability with respect to each nuclear incident shall be SDRs 300 million (approximately USD 450 million) or such higher amount as may be notified by the government of India. Section 6(2) further subdivides the liability of an operator for each nuclear incident, and with respect to nuclear reactors having thermal power equal to or above ten MW, it shall amount to Rupees 1500 Crores (approximately USD 292.4 million)."

4. A COMPARATIVE FOCUS BETWEEN NUCLEAR AND NON-NUCLEAR ENERGY SECTORS

In many analyses of the nuclear liability regime, it has been held that a major nuclear accident will be extremely costly, with some studies suggesting that it could cost as much as £5000 billion.[55] In general these studies allude that the existing financial limits of liability are insufficient and therefore it is considered that the nuclear energy industry is in receipt of a subsidy.[56] The various approaches to calculating the damages from a nuclear accident by these studies can be excessive. Recent cost estimates on nuclear energy accidents from the Institute for Radiological Protection and Nuclear Safety (IRSN) in France estimates the cost of ~€120 billion for a severe nuclear accident and ~€430 billion for a major nuclear accident, such as Chernobyl or Fukushima, occurring in France.[57] As of 2014, ~$21 billion has been claimed from the accident at Fukushima Daiichi.[58] The current projected cost of Fukushima is estimated at ~€36.5 billion which is significantly below the IRSN figures[59] but much greater than the $12.6 billion U.S. insurance pool and the €1.5 billion limit of the RPC + RBSC. Hence, discussion is presently focussed on what the lower limit of third-party nuclear liability should be and how the insurance sector can increase its coverage.

As a comparison, accidents in offshore oil and gas can be very costly in terms of human life and financially. For example, the major accident at the Piper Alpha rig in the North Sea in the UK on 6 July 1988 caused 167 fatalities.[60] Piper Alpha cost $1.4 billion ($3.6 billion in 2014 US$ values) in insurance claims and was unprecedented for the sector in realising the potential scale and spread of insurable claims across geographical and business scope.[61] The operating company, Occidental, paid £66 million to families of the deceased, but this compensation was not awarded through the UK criminal or civil courts.[62]

More recently, the BP Deepwater Horizon Oil Spill in the Gulf of Mexico in 2010 (which began on the 20 April 2010) caused so far has cost BP near $43 billion, as outlined in Table 5.[63] This may be set to increase further with BP having been found grossly negligent in its activities that may have led to a potential fine of $18 billion under the Clean Water Act by the US courts.[64] That figure was then reduced to a $13.7 billion in the US courts but the expectation was that the Judge would seek a fine close to that maximum amount that will take the cost of the accident close to $55 billion — with a final decision expected in the summer of 2015.[65]

Table 5. Cost to BP of BP deepwater horizon oil spill.

Issue	Amount BP has spent/set aside, in 2010 US$
Litigation and settlement cots	$25.87 billion
Spill response costs	$14.30 billion
Clean Water Act penalties	$3.51 billion
Environmental costs	$3.03 billion
Other costs	$1.94 billion
Sub-total	$48.65 billion
Less recoveries	$5.68 billion
Total	$42.97 billion

The final fine is estimated to be $5.5 billion under the Clean Water Act penalty, with an additional $7.3 billion for natural resource damages, $4.9 billion in economic claims to five states and $1 billion in 400 local government claims – however, it is expected that this will be stretched out over 18 years (at $1.1 billion a year) and the expectation is BP will deduct the majority of these fines out of taxes.[66]

Similar to the nuclear industry, oil industry liability rests with the operator. This is mainly achieved through Knock-for-Knock indemnity contracts.[67] The offshore oil and gas industry has no similar third-party agreements such as the PC, the VC, or the CSC — with the exception of transportation. The marine oil sector for oil tankers has an international liability regime governed by the International Marine Organization to which 133 countries are signatory.[68] Oil tanker accidents are insured for between €7 million to €136 million depending on units carried and two mechanisms are in place if the cost of the accident exceeds this to pay further compensation: (1) the International Oil Pollution Compensation Fund (IOPC) 1992 increases the amount up to 203 million SDRs; and (2) the IOPC 2003 Supplementary Fund Protocol raises this to 750 million SDRs.[69] Even still this compensation fund may not be sufficient, as highlighted by the Exxon-Valdez oil spill that yielded punitive damages of $5 billion (in 1995 US$).[70]

A key issue for the operator, BP, as a result of the Deepwater Horizon disaster, was again the broadening of scope of damages beyond that experienced with Piper Alpha. Timeliness was also a key issue for BP in settling claims; even though they have since argued they settled too early in some cases,[71] due in part, to the division of geographical areas into settlement zones according to proximity to the spill.

There are two major lessons for the nuclear insurance industry from the BP Deepwater Horizon case and these are:

1) timeliness is very important in the claims management process; and

2) each major accident in the oil and gas sector will mean an increase in the scope and scale of those individuals and entities that seek, and can claim damages.

These are similar to what has happened in the nuclear energy sector after its major accidents. Chernobyl, as stated earlier, broadened the scope of the categories of damage that could be claimed and prompted updates to the nuclear liability regime. Fukushima Daiichi has shown that timeliness is very important and that again the scope for claiming by those affected to some degree by the accident has also increased. The logic dictates that as a result of a future accident in the nuclear energy sector, the scope of claims will increase. In essence, the definition of loss may expand and become more inclusive and expansive. There maybe other reasons for this in that, public trust in the industry is considered of vital importance and a return to 'business as usual' as soon as possible is more important for the industry than the amount of compensation finally claimed.[72]

5. CONCLUSION AND FUTURE OUTLOOK

After Fukushima Daiichi, as after Chernobyl, the accident prompted a desire to review and possibly to reform the nuclear liability regime to ensure victims are adequately compensated.[73] To-date discussion around the global nuclear liability regime post Fukushima Daiichi has received little attention in academic literature and this article aims to address this. Over the next few years, the global nuclear liability regime is set to come under scrutiny at an international institutional level, and in particular, because of developments in the EU and also due to the CSC coming into force.

The EU has so far played a lead role in reviewing the international global nuclear liability regime. The EU held a meeting on this subject matter in January 2014.[74] The meeting was preceded by an EC Public consultation on nuclear liability[75] and the ambition of this meeting was to build on this consultation and also begin to draft a new approach for nuclear liability within the EU. The necessity of the consultation by the EU Commission and the need for a revision of the liability regime had been previously questioned in 2013 by the nuclear industry.[76] Observations from the January 2014 meeting were no clearer, though it was highlighted that there were lessons to be learned from the accident at Fukushima Daiichi. In particular, these related to the claims management process, an issue that Japanese national authorities struggled with initially.[77]

In contrast, the US is pursuing a different agenda, and is pushing the CSC as the global solution towards third-party nuclear liability,[78] especially recommending that a country's national law needs be in line with the Annex of the CSC. At present six countries have ratified the CSC, these are: Argentina, Morocco, Romania, Japan, UAE, and the USA (which from Table 6 totals 152 reactors in operation and 11 reactors under construction) and came into force on 15 April 2015.[79] There are 13 other signatories, four of which have installed nuclear capacity, including: Canada, Czech Republic, India, and Ukraine.[80] Japan was the most recent country to sign and ratify the CSC, and with 48 nuclear reactors of installed operation, and brought the CSC into force on 15 April 2015.[81] However, it is unlikely that all of Japan's capacity will come back online, with recent estimates from Reuters,[82] coupled with data from the IAEA's Power Reactor Information System,[83] suggesting that circa a third of the 48 nuclear reactors are likely to come back online; a third and their return to operation remains uncertain; and the final third unlikely to come back into operation. Nevertheless, the CSC is now in force.

Table 6. Estimated Nuclear Reactors under the Scope of the CSC International Nuclear Liability Regime.Constructed by the authors using data contained in n 79 and n 83.

Country/region	Number of reactors			Capacity (MW_{th})		
	In operation	Under construction	Total	In operation	Under construction	Total
CSC contracting states						
Argentina	3	1	4	5354	100	5454
Japan	48	2	50	131,073	7852	138,925
Morocco	0	0	0	0	0	0
Romania	2	0	2	4360	0	4360
UAE	0	3	3	0	11,949	11,949
United States	99	5	104	306,274	17,025	323,299
	152	11	163	447,061	36,926	483,987
CSC signatories						
India	21	6	27	19,771	12,819	32,590
Canada	19	0	19	45,487	0	45,487
Czech Rep.	6	0	6	12,016	0	12,016
Ukraine	15	2	17	41,750	6400	48,150
	61	8	69	119,024	19,219	138,243
Noted CSC non-signatories						
South Korea	24	4	28	62,703	15,752	78,455
China	24	25	49	59,459	73,912	133,461
	48	29	77	122,252	89,664	211,916

This paper has assessed what the international nuclear regimes are (in Section 2). In Section 3 the effect of these regimes is highlighted through Fig. 1 where the map of Europe shows the effect of the 'patchwork' of conventions that are in force which results in varying levels of liability amounts and financial security limits. The absence of a common coherent liability regime that is accessible to victims raises many issues, but principally the question of whether there is merit in the range of international nuclear liability regimes themselves. Section 3 also discusses the nuclear liability regime in India where it is evident that India has to engage in reform of its own domestic liability laws before further development will happen in relation to being a signatory to an international nuclear liability regime. [84] In the penultimate section, it is clear that from a comparative focus between nuclear and non-nuclear energy sectors that the international nuclear industry is more advanced in its approach to accidents. Non-nuclear energy sectors have lower liability limits and yet (based on the BP Deepwater Horizon Oil Spill) result in similar if not higher amounts of compensation. Further, it will in more likelihood be necessary to pursue action in the courts after an accident for a non-nuclear energy sector as there is not the same cooperation on liability as there is in the nuclear energy sector.

The effect of the Fukushima Daiichi accident is set to continue on the nuclear energy sector in relation to liability and compensation. Indeed, issues are around compensation and accidents are a contentious area and as one legal scholar noted: "One thing seems clear. If we are concerned... with accident compensation, the tort law system is unsatisfactory."[85] The debate will continue and in relation to the nuclear energy sector and liability, Burns (2012) highlights the essence of the future direction that research in the area should take:

"The Fukushima Daiichi accident brings a new focus to the international liability debate, one that is not merely about theoretical problems and impacts of a nuclear accident. In my own view, the accident does not so much newly reveal heretofore "unknowable" or "unthinkable" events and impacts as it does the need for conscientious re-dedication to the essential tenets of nuclear safety and the courage to face the "what ifs" and consequences of an accident."

The area of nuclear liability identifies with previous literature on the issue of torts and accidents and the law, in that it is an area that has been recognised as complex and where there remains a lack of a definitive way forward. It is certain that after a nuclear accident there is a need to restore public trust in the industry and in essence, there needs to be less uncertainty as to how to achieve this. There is no immediate solution but certainly the possibility of the harmonisation of the current regimes needs further examination, however, new departures such as what has happened in India do not indicate a way forward.[86] In particular, developments between Rosatom and its plans to build in Finland, India and Turkey point towards further complexity with international nuclear liability regimes. All three countries have signed separate and different bilateral agreements with Russia relating to nuclear liability.[87]

Finally, while an accident within the nuclear sector may inevitably exceed the amount stated in the legal liability cap though to what extent will be determined by the characteristics of the accident and the response to the accident. Nevertheless, it is evident that the issue of liability in the nuclear energy sector is not a unique question when accidents and the energy sector are considered. If anything, the nuclear energy sector is further along the in the process of resolving this issue than in other parts of the energy sector, for example, in the offshore oil and gas industry.

ACKNOWLEDGEMENTS

We acknowledge support from the UK Engineering and Physical Sciences Research Council under grant no. EP/K007580/1, as part of the research project "Management of Nuclear Risks: Environmental, Financial and Safety (NREFS)" led by Professor Philip J. Thomas at City University London. The views expressed herein are those of the authors alone and the usual disclaimers apply. Regarding data availability: data provided in full within this article.

REFERENCES

1. Plato, The Laws [of] Plato; Translated with an Introduction by Trevor J. Saunders. (Penguin 1970).
2. Steven Shavell, *Economic Analysis of Accident Law* (Harvard University Press 1987).
3. ibid
4. Steven Shavell, Foundations of Economic Analysis of Law (The Belknap Press of Harvard University Press 2004).
5. Atomic Energy Act of 1954, Pub. L. No. 83–703, 42 U.S.C. § 2011 et seq. (1954), as amended by Pub. L. 85–256 (the "Price–Anderson Act") 42 U.S.C. § 2210.
6. Energy Policy Act of 2005, Pub. L. No. 105-98, 42 U.S.C. § 19 (2005).
7. Helen Cook, *The Law of Nuclear Energy* (Sweet & Maxwell: Thomson Reuters 2013) 86–87.
8. The Price–Anderson Act (n 5) ss (i).
9. Paris Convention on Third Party Liability in the Field of Nuclear Energy (29 July 1960), as amended by the Additional Protocol of 28 January 1964, by the Protocol of 16 November 1982 and by the Protocol of 12 February 2004 (Paris Convention); the most recent revision is not yet in force.
10. The countries that have ratified the Paris Convention are: Belgium, Denmark, Finland, France, Germany, Greece, Italy, Netherlands, Norway, Portugal, Slovenia, Spain, Sweden, Turkey and United Kingdom. Austria and Luxembourg have signed the Paris Convention but not ratified it. Switzerland has ratified the Paris Convention but it does not take effect until the 2004 Protocol comes into force. It is open to all OECD member countries and non-OECD countries if all the contracting countries agree.
11. For more discussion on these and more detailed review of the nuclear liability regime prior to the accident at Fukushima Daiichi please see: Anthony Thomas and Raphael J Heffron, 'Third Party Nuclear Liability: The Case of a Supplier in the United Kingdom' (Faculty of Economics, University of Cambridge, UK 2012) Working Paper <http://www.repository.cam.ac.uk/handle/1810/242219> accessed 9 January 2016.
12. The full title is the: Vienna Convention on Civil Liability for Nuclear Damage (21 May 1963) (Vienna Convention); Vienna Convention Countries are Argentina, Armenia, Belarus, Bolivia, Bosnia and Herzegovina, Brazil, Bulgaria, Cameroon, Chile, Croatia, Cuba, Czech Republic, Egypt, Estonia, Hungary, Latvia, Lebanon, Lithuania, Mexico, Montenegro, Niger, Nigeria, Peru, Philippines, Poland, Republic of Moldova, Romania, Russian

Federation, Saint Vincent & the Grenadines, Senegal, Serbia, Slovak Republic, The former Yugoslav Republic of Macedonia, Trinidad and Tobago, Ukraine and Uruguay.

13. DG Tren, European Commission, 2005. 'TREN/CC/01–2005 Legal Study for the Accession of Euratom to the Paris Convention on Third Party Liability in the Field of Nuclear Energy.' Report < http://www.mng.org.uk/gh/private/2009_12_accession_euratom.pdf> accessed 9 January 2016.
14. Convention of 31 January 1963 Supplementary to the Paris Convention of 29 July 1960, as amended by the Additional Protocol of 28 January 1964 and by the Protocol of 16 November 1982 (Brussels Supplementary Convention); it entered into force in 1974 and the following countries are a party to the Brussels Convention: Belgium, Denmark, Finland, France, Germany, Italy, the Netherlands, Norway, Slovenia, Spain, Sweden and the United Kingdom. Austria, Luxembourg and Switzerland have signed it but it is not in force.
15. 1988 Joint Protocol on the Application of the Vienna Convention and the Paris Convention (Joint Protocol).
16. 1997 Protocol to amend the Vienna Convention on Civil Liability for Nuclear Damage (Revised Vienna Convention).
17. The SDR is an international reserve asset, created by the IMF in 1969 to supplement its member countries' official reserves <http://www.imf.org/external/np/exr/facts/sdr.htm> accessed 9 January 2016. On July 26 2015, 1SDR = 1.393 US$.
18. With regard to the CSC, first-tier compensation corresponds to the amount made available by a nation state; whereas second-tier compensation corresponds to an international fund which Contracting Parties contribute toward that can be used when first-tier compensation is exhausted.
19. Units of installed capacity refers to the thermal MW rating of the plant (referred to as MWt). One unit is one MWt.
20. Julia A Schwartz, 'Liability and Compensation for Third Party Damage Resulting from a Nuclear Incident', *International Nuclear Law: History, Evolution and Outlook* (OECD 2010) <http://www.oecd-nea.org/law/isnl/10th/isnl-10th-anniversary.pdf> accessed 9 January 2016.
21. This is a combined total from under the Paris–Brussels regime and under a three-tier compensation system with: the operator's financial security will be at least €700 million; the state in whose territory the liable operator's installation is situated and will have liability up to €500 million; all of the contracting parties may have a liability up to €300 million.
22. Carlton Stoiber and others, 'Handbook on Nuclear Law: Implementing Legislation' (2010) 104–106 <http://www-pub.iaea.org/MTCD/publications/PDF/Pub1456_web.pdf> accessed 9 January 2016.
23. Paris Convention (n 9).
24. Cook (n 7) 177.
25. Revised Vienna Convention (n 16) art II para 5; Paris Convention (n 9) art 6b
26. Karine Fiore, 'No-Fault Compensation Systems' in Michael Faure (ed), *Tort Law and Economics* (Edward Elgar Publishing 2009).
27. Organisation for Economic Co-operation and Development, 'Nuclear Operator Liability Amounts and Financial Security Limits, July 2014' <https://www.oecd-nea.org/law/2014-table-liability-coverage-limits.pdf> accessed 9 January 2016.
28. Council Decision (EU) 2013/434/EU of 15 July 2013 authorising certain Member States to ratify, or to accede to, the Protocol amending the Vienna Convention on Civil Liability for Nuclear Damage of 21 May 1963, in the interest of the European Union, and to make a declaration on the application of the relevant internal rules of Union law [2013] OJ L220/1.
29. Council Decision (EC) 2004/294/EC of 8 March 2004 authorising the Member States which are Contracting Parties to the Paris Convention of 29 July 1960 on Third Party Liability in the Field of Nuclear Energy to ratify, in the interest of the European Community, the Protocol amending that Convention, or to accede to it [2004] OJ L97/53.

30. The exception of Austria, Ireland, and Luxembourg stems from Council Decision 2004/294/EC (n 29) only applying to Member States that are Contracting Parties to the Paris Convention. Denmark is excepted due to Articles 1 and 2 of the Protocol on the position of Denmark annexed to the Treaty on European Union and to the Treaty establishing the European Community.
31. Council Decision (EC) 2007/727/EC of 8 November 2007 authorising the Republic of Slovenia to ratify, in the interest of the European Community, the Protocol of 12 February 2004 amending the Paris Convention of 29 July 1960 on Third-Party Liability in the Field of Nuclear Energy [2007] OJ L294/23.
32. Maxine Symington, 'Nuclear Energy Policy Update: Key Recommendations from the Conference on Nuclear Third Party Liability' (*Wragge Lawrence Graham & Co*, 28 January 2014) <http://www.wragge-law.com/insights/nuclear-energy-policy-update-key-recommendations- f/> accessed 9 January 2016.
33. Stephen G Burns, 'A Global Nuclear Liability Regime: A Journey or a Destination?' (2012) <http://www.burges-salmon.com/inla_2012/10156.pdf> accessed 9 January 2016.
34. International Atomic Energy Agency, 'IAEA Action Plan on Nuclear Safety' (2014) <https://www.iaea.org/sites/default/files/actionplanns.pdf> accessed 9 January 2016.
35. World Nuclear Association, 'Nuclear Power in Romania' (November 2014) <http://www.world-nuclear.org/info/Country-Profiles/Countries-O-S/Romania/> accessed 9 January 2016.
36. World Nuclear Association, 'Nuclear Power in Turkey' (October 2015) < http://www.world-nuclear.org/info/Country-Profiles/Countries-T-Z/Turkey/> accessed 9 January 2016; World Nuclear Association, 'Nuclear Power in the United Arab Emirates' (September 2015) <http://www.world-nuclear.org/info/Country-Profiles/Countries-T-Z/United-Arab-Emirates/> accessed 9 January 2016.
37. Civil Liability for Nuclear Damage Act 2010 <http://dae.nic.in/sites/ default/files/civilnucliab.pdf> accessed 9 January 2016.
38. ArghyaSengupta and SanhitaAmbast, 'A Dangerous Recourse? A Critical Relook at Section 17 of the Civil Liability for Nuclear Damage Act, 2010' (2012) 3 International Journal of Nuclear Law 292.
39. MP Ram Mohan, 'Nuclear Liability Law of India: An Appraisal of Extent of Liability, Right of Recourse and Transboundary Applicability' (2014) 17 Journal of Risk Research 115.
40. A similar argument is made by Ted Jones, 'NEI Nuclear Notes: Why India's Nuclear Liability Law Is Harming Indian Interests' <http://neinuclearnotes. blogspot. co.uk/2013/10/why-indias-nuclear-liability-law-is.html> accessed 9 January 2015.
41. AyushiSutaria, 'Placing the Indian Civil Nuclear Liability Regime in Context: The Extent of Supplier's Liability' (2014) 17 Journal of Risk Research 97.
42. AnupamJha, 'Dynamics of Legal Regime on Safety of Nuclear Power Plants in India after Fukushima Disaster' (2014) 17 Journal of Risk Research 145.
43. AryaHariharan, 'India's Nuclear Civil Liability Bill and Supplier's Liability: One Step Towards Modernizing the Outdated International Nuclear Liability Regime' (2011) 36 Wm. & Mary Envtl. L. &Pol'y Rev. 223; Mohit Abraham, *Nuclear Liability: A Key Component of the Public Policy Decision to Deploy Nuclear Energy in Southeast Asia - American Academy of Arts & Sciences* (American Academy of Arts and Sciences 2014) <https://www.amacad.org/content/publications/publication.aspx?d=1499> accessed 9 January 2016; MV Ramana and SuvratRaju, 'Nuclear Safety before Vendor Interests' *The Hindu* (30 October 2012) <http://www.thehindu.com/opinion/op-ed/nuclear-safety-before-vendor-interests/article4044438.ece> accessed 9 January 2016; MV Ramana and SuvratRaju, 'Profitability without Accountability' *The Hindu* (16 February 2015) <http://www.thehindu.com/opinion/op-ed/comment-profitability-without-accountability/article6898851.ece> accessed 9 January 2016.
44. Abraham (n 43).
45. Jha (n 42).
46. ibid.
47. Yash Thomas Mannully, 'Law Relating to Nuclear Liability and Compensation in India' (2010) 3 International Journal of Nuclear Law 112.

48. MV Ramana and Ashwin Kumar, '"One in Infinity": Failing to Learn from Accidents and Implications for Nuclear Safety in India' (2014) 17 Journal of Risk Research 23; ibid 27..
49. Zee News, 'N-Liability Bill Changes Tough, No Supplier Will Come: NPCIL' (*Zee News*, 24 August 2010) <http://zeenews.india.com/news/nation/n-liability-bill-changes-tough-no-supplier-will-come-npcil_650351.html > accessed 9 January 2016.
50. Dan Roberts, 'Obama and Modi Agree to Limit US Liability in Case of Nuclear Disaster' *The Guardian* (Delhi, India, 25 January 2015) <http://www.theguardian.com/ world/2015/jan/25/obama-modi-limit-us-liability-nuclear-disaster> accessed 9 January 2016.; Ministry of External Affairs (Government of India), 'Frequently Asked Questions and Answers on Civil Liability for Nuclear Damage Act 2010 and related issues (28 February 2015) <http://www.mea.gov.in/press-releases.htm?dtl/24766/Frequently+ Asked+Questions+and+Answers+on+Civil+Liability+for+Nuclear+Damage+Act+2010+and+related+issues > accessed 27 July 2015, which notably states that "Section 17 states that the operator shall have a right of recourse. While it provides a substantive right to the operator, it is not a mandatory but an enabling provision. In other words it permits but does not require an operator to include in the contract or exercise a right of recourse."
51. World Nuclear Association, 'Nuclear Power in India' (January 2016) <http://www.world-nuclear.org/info/Country-Profiles/Countries-G-N/India/#NuclearLiability> accessed 9 January 2016.
52. The Civil Liability For Nuclear Damage Rules, 2011 <http://dae.nic.in/ writereaddata/liab_rules.pdf> accessed 27 July 2015.
53. Committee on Subordinate Legislation, LokSabha Secretariat, *The Civil Liability For Nuclear Damage Rules, 2011* <http://164.100.47.134/lsscommittee/ Subordinate%20Legislation/27%20Report%20nuclear.pdf> accessed 9 January 2016.
54. Sengupta and Ambast (n 38).
55. Hans-Jürgen Ewers and Klaus Rennings, 'Economics of Nuclear Risks — A German Study' in Olav Hohmeyer and Richard L Ottinger (eds), *Social Costs of Energy* (Springer Berlin Heidelberg 1994) <http://link.springer.com/chapter/10.1007/978-3-642-85120-9_11> accessed 27 July 2015; Duncan EJ Currie, 'Problems and Gaps in the Nuclear Liability Conventions and an Analysis of How an Actual Claim Would Be Brought under the Current Existing Treaty Regime in the Event of a Nuclear Accident, The' (2006) 35 Denv. J. Int'l L. &Pol'y 85.
56. Jeffrey A Dubin and Geoffrey S Rothwell, 'Subsidy to Nuclear Power Through Price–Anderson Liability Limit' (1990) 8 Contemporary Economic Policy 73; Anthony G Heyes and Catherine Liston-Heyes, 'Subsidy to Nuclear Power Through Price–Anderson Liability Limit: Comment' (1998) 16 Contemporary Economic Policy 122.
57. Patrick Momal, 'The Cost of Nuclear Accidents', NEA Workshop on Approaches to Estimation of the Costs of Nuclear Accidents, May 2013 <http://www.oecd-nea.org/ndd/workshops/aecna/presentations/documents/PatrickMomal-Costestimationmethodology.pdf> accessed 9 January 2016.
58. Toyohiro Nomura, 'The Japanese experience on claims management after the Fukushima Daiichi accident', Stakeholder Conference on Nuclear Third Party Liability and Insurance, January 21 and 22, 2014 <http://ec.europa.eu/energy/en/events/international-workshop-nuclear-liability> accessed 9 January 2016.
59. Japan Atomic Energy Commission (Yoshihiro Nagaoki), 'Estimation of Accident Risk Cost of Nuclear Power Plants', NEA Workshop on Approaches to Estimation of the Costs of Nuclear Accidents, May 2013 <http://www.oecd-nea.org/ndd/ workshops/ aecna/presentations/documents/YoshihiroNAGAOKI-EstimationofAccidentRiskCostofNPP.pdf> accessed 9 January 2016.
60. Cullen, W. D. 1990.The public inquiry into the Piper Alpha Disaster. Department of Energy: London, HMSO, London, UK.
61. Lloyd's, '1988 The Piper Alpha Explosion.' (28 May 2013) <http://www. lloyds.com/lloyds/about-us/history/catastrophes-and-claims/piper-alpha> accessed 9 January 2016; Lloyd's, 'Piper Alpha after the Fire' (5 July 2013) <http://www.lloyds.com/news-and-insight/news-and-features/market-news/industry-news-2013/piper-alpha-after-the-fire> accessed 9 January 2016.

62. Terry Macalister, 'Piper Alpha Disaster: How 167 Oil Rig Workers Died' *The Guardian* (4 July 2013) <http://www.theguardian.com/business/2013/jul/04/piper-alpha-disaster-167-oil-rig> accessed 9 January 2016.
63. Margaret Cronin Fisk, Brian Swint and Laurel Brubaker Calkins, 'BP's Oil Spill Deal Sours as Claims Add Billions to Cost - Bloomberg Business' *Bloomberg* (4 June 2013) <http://www.bloomberg.com/news/articles/2013-06-04/bp-s-oil-spill-settlement-sours-as-claims-add-billions> accessed 9 January 2016.
64. Margaret Cronin Fisk and JefFeeley, 'BP Found Grossly Negligent in 2010 Gulf of Mexico Spill' *Bloomberg* (4 September 2014) <http://www.bloomberg.com/news/articles/2014-09-04/bp-found-grossly-negligent-in-2010-gulf-of-mexico-spill> accessed 9 January 2016.
65. For a recent discussion on the penultimate stage of the BP Ruling, see Dominic Rushe, 'Deepwater Horizon Oil Spill Legal Saga Enters Final Chapter' *The Guardian* (19 January 2015) <http://www.theguardian.com/business/2015/jan/19/bp-oil-spill-deepwater-horizon-final-fine-clean-water-act> accessed 9 January2016.
66. BP, 'BP to Settle Federal, State and Local Deepwater Horizon Claims for up to $18.7 Billion With Payments to Be Spread Over 18 Years' (2 July 2015) <http://www.bp.com/en/global/corporate/press/press-releases/bp-to-settle-federal-state-local-deepwater-horizon-claims.html> accessed 9 January 2016; Daniel Gilbert and Sarah Kent, 'BP Agrees to Pay $18.7 Billion to Settle Deepwater Horizon Oil Spill Claims' *Wall Street Journal* (2 July 2015) <http://www.wsj.com/articles/bp-agrees-to-pay-18-7-billion-to-settle-deepwater- horizon-oil-spill-claims-1435842739> accessed 9 January 2016.
67. For more on Knock-for-Knock contracts see: ChidiEgbchue, 'Reviewing "Knock for Knock" Indemnities Following the Macondo Well Blowout' (2013) 7 Construction Law International 7.
68. International Convention on Civil Liability for Oil Pollution Damage (CLC), <http://www.imo.org/About/Conventions/ListOfConventions/Pages/International-Convention-on-Civil-Liability-for-Oil-Pollution-Damage-%28CLC%29.aspx> accessed 9 January 2016.
69. The International Oil Pollution Compensation Funds, <http://www.iopcfunds.org/about-us/legal-framework/1992-fund-convention-and-supplementary-fund-protocol/> accessed 9 January 2016.
70. RT Paine and others, 'TROUBLE ON OILED WATERS: Lessons from the Exxon Valdez Oil Spill' (1996) 27 Annual Review of Ecology and Systematics 197.
71. Associated Press in New Orleans, 'Judge Orders BP to Stick by Deepwater Horizon Payouts Agreement' *The Guardian* <http://www.theguardian.com/ environment/ 2014/sep/24/bp-ordered-deepwater-horizon-payouts-agreement> accessed 9 January 2016.
72. This point is addressed further in Section 5 though is an area in need of further research.
73. This does prompt the argument that the development of nuclear law needs to be more proactive and is too reactive at the moment. It should not wait until there is an accident to reform. See the following for more discussion on this though a different reason is asserted for why nuclear law should be more proactive in its formulation: RJ Heffron, 'The Need for Proactive over Reactive Nuclear Law for the International Nuclear Industry' (2013) 11 Oil, Gas & Energy Law Journal (OGEL) <https://www.ogel.org/article.asp?key=3335> accessed 9 January 2016.
74. Stakeholder Conference on Nuclear Third Party Liability and Insurance, January 20 and 21, 2014 <http://ec.europa.eu/energy/en/events/international-workshop-nuclear-liability> accessed 9 January 2016.
75. EC Public consultation on nuclear liability <https://ec.europa.eu/energy/ en/consultations/consultation-nuclear-liability> accessed 9 January 2016.
76. Carolina Parra-Serrano, 'Necessity of New EU Nuclear Liability Proposals Questioned by Industry' [2013] *Post Magazine* <http://m.postonline.co.uk/post/ news/2302292/necessity-of-new-eu-nuclear-liability-proposals-questioned-by-industry> accessed 9 January 2016.
77. n 74.

78. Ben McRae, 'The Convention on Supplementary Compensation for Nuclear Damage: Catalyst for a Global Nuclear Liability Regime' (2007) 79 OECD-NEA Nuclear Law Bulletin 17.
79. International Atomic Energy Agency, 'Convention on Supplementary Compensation for Nuclear Damage' (Latest Status: 17 April 2015) <https://www.iaea.org/Publications/Documents/Conventions/supcomp_status.pdf> accessed 9 January 2016.
80. ibid
81. As this bring the total installed capacity to more than 400,000 MW_{th}, a condition needed to bring the CSC into force as mentioned in Section 2C.
82. Reuters, 'FACTBOX-Outlook for Japan Nuclear Reactor Re-Starts' *Reuters* (1 April 2014) <http://www.reuters.com/article/2014/04/01/japan-nuclear-restarts-idINL4N0MT0K020140401> accessed 9 January 2016.
83. All units of installed capacity taken from International Atomic Energy Agency, 'Power Reactor Information System' <http://www.iaea.org/PRIS/home.aspx> accessed 9 January 2016.
84. Mannully (n 47), who stated that India can be characterized as having an "inadequate development of tort law and lack of legal and legislative developments in the area of complex liability cases."; World Nuclear Association, 'Civil Liability for Nuclear Damage' (June 2015) <http://www.world-nuclear.org/info/Safety-and-Security/Safety-of-Plants/Liability- for-Nuclear-Damage/> accessed 9 January 2016.
85. R Ian McEwin, 'No-Fault Compensation Systems' in BoudewijnBouckaert and Gerrit De Geest (eds), *Encyclopedia of Law and Economics, Volume II. Civil Law and Economics.* (Edward Elgar 2000) 745 <http://encyclo.findlaw.com/3600book.pdf> accessed 9 January 2016.
86. This is similar to the view expressed by Robert J Gruendel and ElsReynaersKini, 'Through the Looking Glass' (2012) 2012 Nuclear Law Bulletin 45,who stated in relation to the Indian "variations among national liability regimes may exacerbate the phenomenon of fragmentation of international conventions pertaining to the nuclear sector".
87. In Finland's (Paris Convention) case it has signed an agreement with Russia (Vienna Convention) that in effect agrees that the Joint Protocol will be in force for that project, World Nuclear Association, 'Nuclear Power in Finland' (17 July 2015) <http://www.world-nuclear.org/info/Country-Profiles/Countries-A-F/Finland/> accessed 9 January 2016; In India there is an agreement that India will determine the insurance premium for each component and Russia will pay this premium for 20 years, Sonal Patel, 'India's Nuclear Liability Law: Breakthrough for Russia, Stalemate Endures for U.S.' (*POWER Magazine*, 5 January 2014) <http://www.powermag.com/indias-nuclear-liability-law-breakthrough-for-russia-stalemate-endures-for-u-s/> accessed 9 January 2016; In Turkey, where Rosatom is building a nuclear plant at Sinop, Russia and Turkey have signed a bilateral agreement where the project will be subject to the "international agreements and instruments to which the Republic of Turkey is or will be a party and national laws and regulations of the Turkish Party". Please see: Agreement Between the Government of the Republic of Turkey and the Government of the Russian Federation on Cooperation in Relation to the Construction and Operation of a Nuclear Power Plant at the Akkuyu Site in the Republic of Turkey, Article 16, OG No.27721 of 6 October 2010 (also in English), as quoted in ErinçErcan and Horst Schneider, 'Turkish Nuclear Legislation: Developments for a Nuclear Newcomer' (2013) 92 OECD-NEA Nuclear Law Bulletin 29.

CHAPTER 4

Reliability of Passive Systems in Nuclear Power Plants

Luciano Burgazzi

Reactor Safety and Fuel Cycle Methods Technical Unit, ENEA, Italian National Agency for New Technologies, Energy and Sustainable Economic Development, Bologna,, Italy

1. INTRODUCTION

In order to tackle the development of advanced nuclear technologies, the reliability of passive systems has become an important subject and area under discussion, for their extensive use in new and advanced nuclear power plants, (NEA, 2002), in combination with active safety or operational systems.

Following the IAEA definitions, [1], a passive component does not need any external input or energy to operate and it relies only upon natural physical laws (e.g. gravity, natural convection, conduction, etc.) and/or on inherent characteristics (properties of materials, internally stored energy, etc.) and/or 'intelligent' use of the energy that is inherently available in the system (e.g. decay heat, chemical reactions etc.).

The term "passive" identifies a system which is composed entirely of passive components and structures or a system which uses active components in a very limited way to initiate subsequent passive operation. That is why passive systems are expected to combine among others, the advantages of simplicity, a decrease in the need for human interaction and a reduction or avoidance of external electrical power or signals. These attractions may lead to increased safety and acceptability of nuclear power generation if the detractions can be reduced.

Besides the open feedback on economic competitiveness, special aspects like lack of data on some phenomena, missing operating experience over the wide range of conditions, and driving forces which are smaller - in most cases - than in active safety systems, must be taken into account: the less effective performance as compared to active safety systems has a strong impact on the reliability assessment of passive safety systems.

A categorisation has been developed by the IAEA in [1] distinguishing:

a. physical barriers and static structures (e.g. pipe wall, concrete building).

This category is characterized by:

- no signal inputs of "intelligence", no external power sources or forces,
- no moving mechanical parts,
- no moving working fluid.

Examples of safety features included in this category are physical barriers against the release of fission products, such as nuclear fuel cladding and pressure boundary systems; hardened building structures for the protection of a plant against seismic and or other external events; core cooling systems relying only on heat radiation and/or conduction from nuclear fuel to outer structural parts, with the reactor in hot shutdown; and static components of safety related passive systems (e.g., tubes, pressurizers, accumulators, surge tanks), as well as structural parts (e.g., supports, shields).

b. moving working fluids (e.g. cooling by free convection).

This category is characterized by:

- no signal inputs of "intelligence", no external power sources or forces,
- no moving mechanical parts, but
- moving working fluids.

Examples of safety features included in this category are reactor shutdown/emergency cooling systems based on injection of borated water produced by the disturbance of a hydrostatic equilibrium between the pressure boundary and an external water pool; reactor emergency cooling systems based on air or water natural circulation in heat exchangers immersed in water pools (inside containment) to which the decay heat is directly transferred; containment cooling systems based on natural circulation of air flowing around the containment walls, with intake and exhaust through a stack or in tubes covering the inner walls of silos of underground reactors; and fluidic gates between process systems, such as "surge lines" of Pressurized Water Reactors (PWRs).

c.moving mechanical parts (e.g. check valves).

This category is characterized by:

- no signal inputs of "intelligence", no external power sources or forces; but
- moving mechanical parts, whether or not moving working fluids are also present.

Examples of safety features included in this category are emergency injection systems consisting of accumulators or storage tanks and discharge lines equipped with check valves; overpressure protection and/or emergency cooling devices of pressure boundary systems based on fluid release through relief valves; filtered venting systems of containments activated by rupture disks; and mechanical actuators, such as check valves and spring-loaded relief valves, as well as some trip mechanisms (e.g., temperature, pressure and level actuators).

d.external signals and stored energy (passive execution/active actuation, e.g. scram systems).

This category addresses the intermediary zone between active and passive where the execution of the safety function is made through passive methods as described in the previous categories except that internal intelligence is not available to initiate the process. In these cases an external signal is permitted to trigger the passive process. To recognize this departure, this category is referred to as "passive execution/active initiation".

Examples of safety features included in this category are emergency core cooling and injections systems based on gravity that initiate by battery-powered electric or electro-pneumatic valves; emergency reactor shutdown systems based on gravity or static pressure driven control rods.

According to this classification, safety systems are classified into the higher categories of passivity when all their components needed for safety are passive. Systems relying on no external power supply but using a dedicated, internal power source (e.g., a battery) to supply an active component are not subject to normal, externally caused failures and are included in the lowest category of passivity. This kind of system has active and passive characteristics at different times, for example, the active opening of a valve initiates subsequent passive operation by natural convection.

Inclusion of failure modes and reliability estimates of passive components for all systems is recommended in probabilistic safety assessment (PSA)[1] - studies. Consequently the reliability assessment of passive safety systems, defined as the probability to perform the requested mission to achieve the generic safety function, becomes an essential step.

Notwithstanding that passive systems are credited a higher reliability with respect to active ones, – because of the smaller unavailability due to hardware failure and human error -, there is always a nonzero likelihood of the occurrence of physical phenomena leading to pertinent failure modes, once the system comes into operation. In fact the deviations of the natural forces or physical principles, upon which they rely, from the expected conditions can impair the performance of the system itself. This remark is especially applicable to type B passive systems (i.e. implementing moving working fluids) named thermal-hydraulic passive systems, due to the small engaged driving forces and the thermal-hydraulic phenomena affecting the system performance.

Indeed, while in the case of passive A systems the development of the structural reliability analysis methodology can be carried out with the application of the principles of the probabilistic structural mechanics theory, and operating experience data can be inferred for the reliability assessment of passive C and D components, there is yet no agreed approach as far as passive B systems are concerned.

In fact, such passive safety systems in their designs rely on natural forces, such as gravity or natural convection, to perform their accident prevention and mitigation functions once actuated and started: these driving forces are not generated by external power sources (e.g., pumped systems), as is the case in operating reactor designs. Because the magnitude of the natural forces, which drive the operation of passive systems, is relatively small, counter-forces (e.g. friction) can be of comparable magnitude and cannot be ignored as it is generally the case of systems including pumps. Moreover, there are considerable uncertainties associated with factors on which the magnitude of these forces and counter forces depends (e.g. values of heat transfer coefficients and pressure losses). In addition, the magnitude of such natural driving forces depends on specific plant conditions and configurations which could exist at the time a system is called upon to perform its safety function. All these aspects affect the thermal-hydraulic (T-H) performance of the passive system.

Consequently, a lot of efforts have been devoted mostly to the development of consistent approaches and methodologies aimed at the reliability assessment of the T-H passive systems, with reference to the evaluation of the implemented physical principles (gravity, conduction, etc.). For example, the system fault tree in case of passive systems would consist of basic events, representing failure of the physical phenomena and failure of activating devices: the use of thermal-hydraulic analysis related information for modeling the passive systems should be considered in the assessment process.

The efforts conducted so far to deal with the passive safety systems reliability, have raised an amount of open issues to be addressed in a consistent way, in order to endorse the proposed approaches and to add credit to the underlying models and the eventual reliability figures, resulting from their application. In fact the applications of the proposed methodologies are to a large extent dependent upon the assumptions underlying the methods themselves. At the international level, for instance, IAEA recently coordinated a research project, denoted as *"Natural Circulation Phenomena, Modelling and Reliability of Passive Systems"* (2004-2008), [2,3], while another coordinated research project on *"Development of Methodologies for the Assessment of Passive Safety System Performance in Advanced Reactors"* (2008-2011) is currently underway: while focus of the former project has been the natural circulation and related phenomena, the objective of the latter program is to determine a common analysis-and-test method for reliability assessment of passive safety system performance. This chapter provides the insights resulting from the analysis on the technical issues associated with assessing the reliability of passive systems in the context of nuclear safety and probabilistic safety analysis, and a viable path towards the implementation of the research efforts in the related areas is delineated as well. Focus on these issues is very important since it is the major goal of the international research activities (e.g. IAEA) to strive to reach a common consensus about the different proposed approaches. The chapter is organized as follows: after an overview on passive safety systems being implemented in the design of innovative reactors and an introduction on the main components of Probabilistic Safety Assessment approach, at first the current available methodologies are illustrated and compared, the open issues coming out from their analysis are identified and for which one of them the state of the art and the outlook is presented; the relative importance of each of them within the evaluation process is presented as well.

2. PASSIVE SYSTEMS IMPLEMENTATION IN ADVANCED REACTOR DESIGNS

Several advanced water cooled reactor designs incorporate passive safety systems based on natural circulation, as described in [2,3]: some of the most relevant design concepts for natural circulation systems are described hereafter and namely as regards AP600/AP1000, ESBWR and ABWR designs.

It is important to note that the incorporation of systems based on natural circulation to achieve plant safety and economic goals is being extended also to Generation-IV reactor concepts: however due to the early stage of the design - many systems are not yet established - they are not explicitly addressed.

2.1. Ap600/ap1000 passive residual heat removal systems (prhr)

Figure 1 presents a schematic that describes the connections of the primary system passive safety systems.

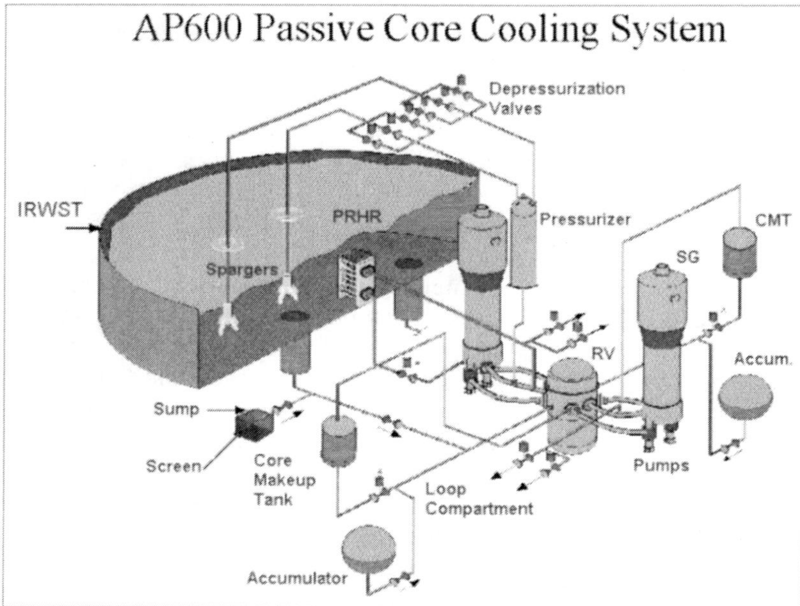

Figure 1. Passive Safety Systems used in the AP600/AP1000 Designs

The AP600/AP1000 passive safety systems consist of:

- A Passive Residual Heat Removal (PRHR) System
- Two Core Make-up Tanks (CMTs)
- A Four Stage Automatic Depressurization System (ADS)
- Two Accumulator Tanks (ACC)
- An In-containment Refueling Water Storage Tank, (IRWST)
- A Lower Containment Sump (CS)
- Passive Containment Cooling System (PCS)

The PRHR implemented in the Westinghouse AP1000 design consists of a C-Tube type heat exchanger in the water-filled In-containment Refuelling

Water Storage Tank (IRWST) as shown in the schematic given in Figure 2. The PRHR provides primary coolant heat removal via a natural circulation loop. Hot water rises through the PRHR inlet line attached to one of the hot legs. The hot water enters the tube sheet in the top header of the PRHR heat exchanger at full system pressure and temperature. The IRWST is filled with cold borated water and is open to containment heat removal from the PRHR heat exchanger occurs by boiling on the outside surface of the tubes. The cold primary coolant returns to the primary loop via the PRHR outline line that is connected to the steam generator lower head.

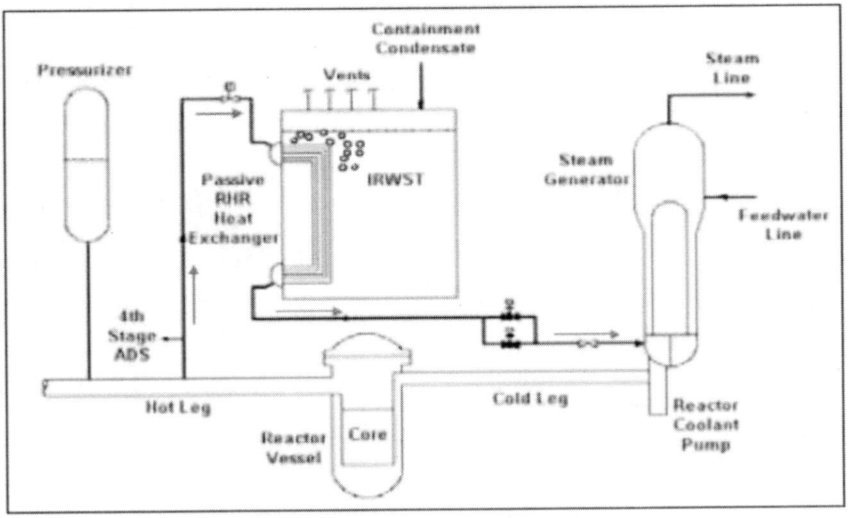

Figure 2. AP1000 passive residual heat removal systems (PRHR)

2.2. Esbwr (economic simplified boiling water reactor) isolation condenser system (ics)

During a Loss of Coolant Accident (LOCA), the reactor shuts down and the Reactor Pressure Vessel (RPV) is isolated by closing the main steam line isolation valves. The ICS removes decay heat after any reactor isolation. In other words, the ICS passively removes sensible and core decay heat from the reactor when the normal heat removal system is unavailable. Decay heat removal limits further increases in steam pressure and keeps the RPV pressure below the safety set point. The arrangement of the IC heat exchanger is shown in Figure 3.

The ICS consists of four independent loops, each containing two heat exchanger modules that condense steam inside the tube and transfers heat by heating/evaporating water in the IC pool, which is vented to the atmosphere. This transferring mechanism from IC tubes to the surrounding IC pool water is accomplished by natural convection, and no forced circulation equipment is required.

The ICS is initiated automatically by any of the following signals: high reactor pressure, main steam line isolation valve (MSIV) closure, or an RPV water level signal. To operate the ICS, the IC condensate return valve is opened whereupon the standing condensate drains into the reactor and the steam water interface in the IC tube bundle moves downward below the lower headers.

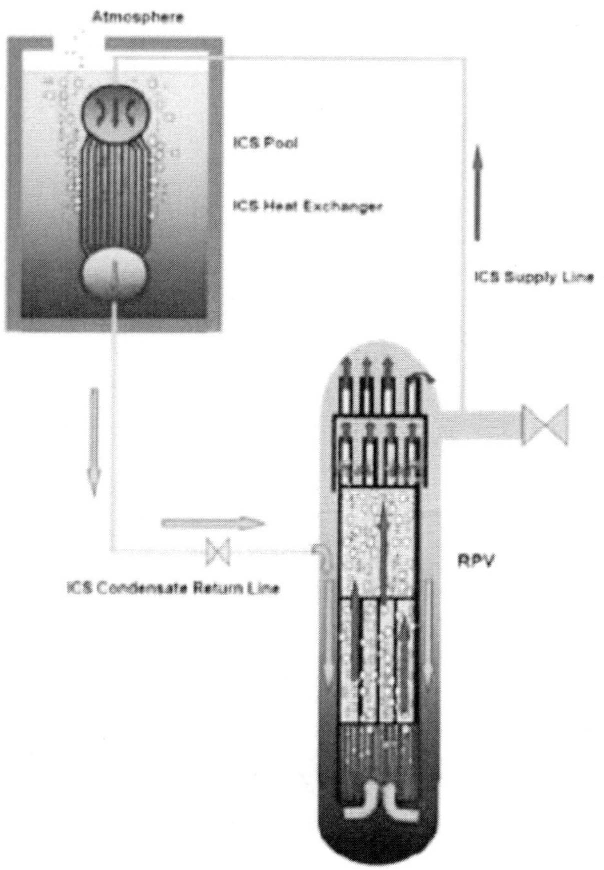

Figure 3. Isolation condenser arrangement

2.3. Esbwr passive containment cooling system (pccs)

The PCCS is a passive system which removes the decay heat released to the containment and maintains the containment within its pressure limits for design basis accidents such as a LOCA. The schematic of the PCCS is shown in Figure 4. The PCC heat exchangers receive a steam-gas mixture from the Dry Well (DW), condense the steam and return the condensate to the RPV via the Gravity Driven Cooling System GDCS pools. The non condensable gas is vented to the Wet Well (WW) gas space through a vent line submerged in the Suppression Pool (SP). The venting of the non condensable gas is driven by the differential pressure between the DW and WW. The PCCS condenser, which is open to the containment, receives a steam-gas mixture supply directly from the DW. Therefore, the PCCS operation requires no sensing, control, logic or power actuated devices for operation. The PCCS consists of six PCCS condensers. Each PCCS condenser is made of two identical modules and each entire PCCS condenser two-module assembly is designed for 11 MWt capacity. The condenser condenses steam on the tube side and transfers heat to the water in the IC/PCC pool. The evaporated steam in the IC/PCC pool is vented to the atmosphere. PCCS condensers are located in the large open IC/PCC pool, which are designed to allow full use of the collective water inventory.

2.4. Abwr (advanced boiling water reactor) passive reactor cooling system and passive containment cooling system

The passive heat removal system (PHRS) consists of two dedicated systems (Figure 5, right) namely the passive reactor cooling system (PRCS: the same as Isolation condenser) and the passive containment cooling system (PCCS), that use a common heat sink pool above the containment allowing a one-day grace period, with a 4*50% redundancy (Figure 5, left). These passive systems not only cover beyond DBA condition, but also provide in-depth heat removal backup for the RHR.

In addition, they provide the overpressure protection safety function, practically excluding the necessity of containment venting before and after core damage. Figure 6 shows PCCS functional schematic and an example of containment pressure transient following typical low pressure core melt scenario.

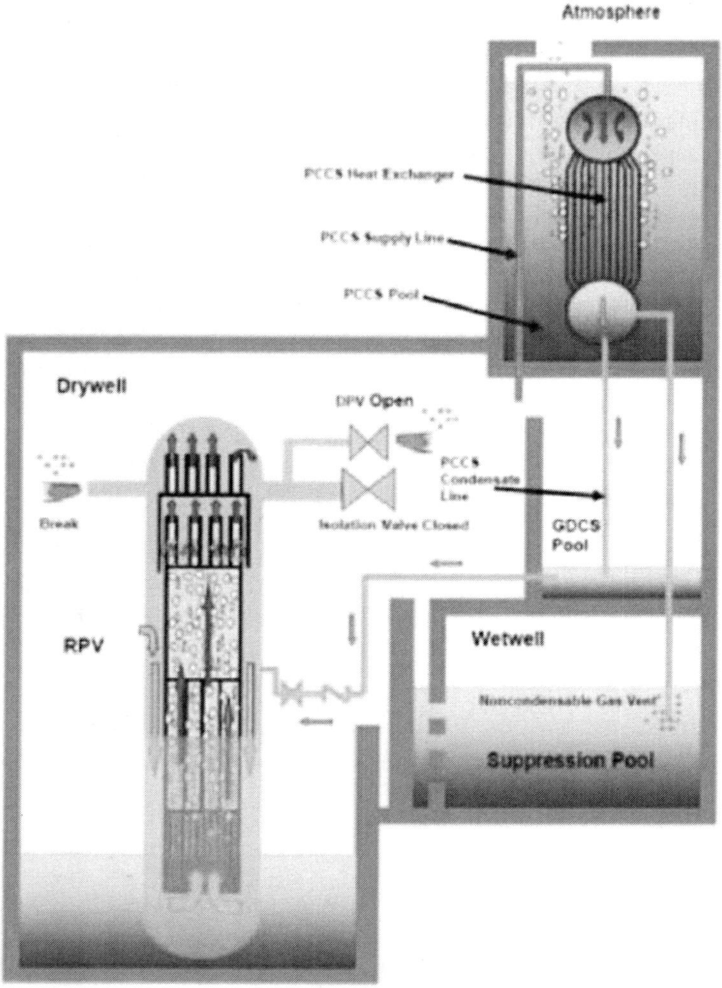

Figure 4. Passive containment cooling condenser arrangement

3. OVERVIEW OF PSA

PSA methodology widely used in the nuclear power industry is deemed helpful to the safety assessment of the facility and along the correspondent licensing process: probabilistic safety assessment can provide insights into safety and identify measures for informing designers of the safety of the plant.

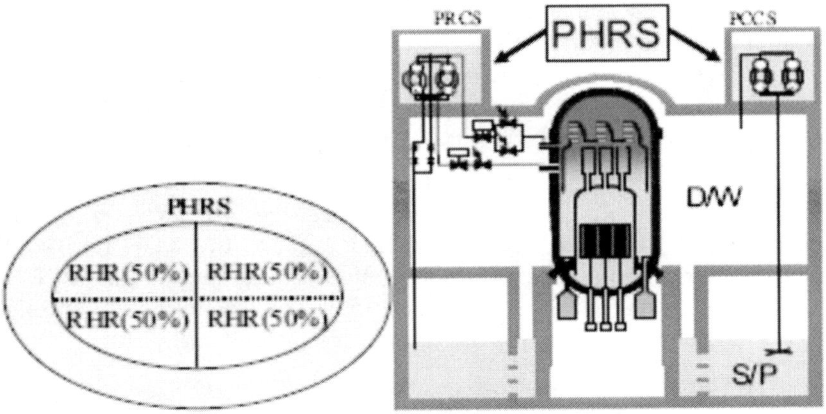

Figure 5. ABWR Passive heat removal system

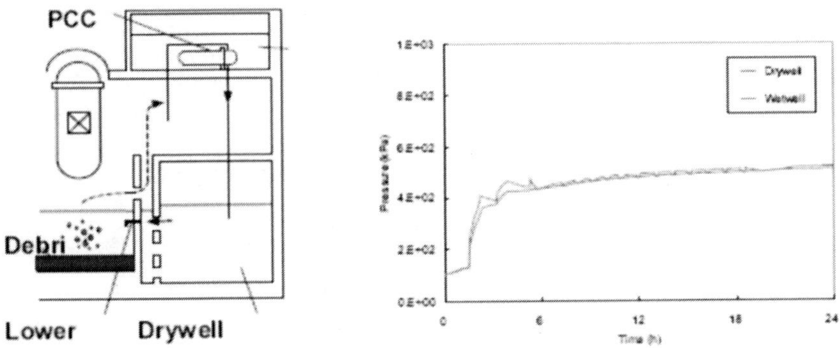

Figure 6. Example of containment pressure transient following typical low pressure core melt scenario.

The first comprehensive application of the PSA dates back to 1975, to the United States Nuclear Regulatory Commission's (U.S. NRC) Reactor Safety Study [4]. Since that pioneering study, there has been substantial methodological development, and PSA techniques have become a standard tool in the safety evaluation of the nuclear power plants (NPPs) and industrial installations in general. Due to historical reasons, the PSA sometimes is called PRA.

As the most important area of PSA projects remains nuclear power plants, mainly due to the specific features of the nuclear installations, three levels of PSA have evolved:

Level 1:The assessment of plant failures leading to core damage and the estimation of core damage frequency. A Level 1 PSA provides insights into design weaknesses and ways of preventing core damage. In the case of other industrial assessments, Level 1 PSA provides estimates of the accidents frequency and the main contributors.

Level 2: As possible releases are additionally protected by containment in most NPPs, PSA at this response and severe accident management possibilities. The results obtained in Level 1 are the basis for Level 2 quantification. In the case of other industrial assessments, Level 2 PSA might be fully covered by Level 1, as containment function is rather unique feature and is not common in other industries.

Level 3: The assessment of off-site consequences leading to estimates of risks to the public. Level 3 incorporates results om both previous levels.

Level1 PSA is the most important level and creates the background for further risk assessment, therefore it will be presented in detail. The structure of the other levels is much more application specific, and will be discussed only in general.

The methodology is based on systematically: 1) postulating potential accident scenarios triggered by an initiating event (IE), 2) identifying the systems acting as "defences" against these scenarios, 3) decomposing the systems into components, associating the failure modes and relative probabilities, 4) assessing the frequency of the accident scenarios. Two elements of the PSA methodology typically stand out:

- The event tree (ET) which is used to model the accident scenarios: it represents the main sequences of functional success and failure of safety systems appointed to cope with the initiating events and the consequences of each sequence. These consequences, denoted also as end states, are identified either as a safe end state or an accident end state.

- The fault tree (FT) which documents the systematic, deductive analysis of all the possible causes for the failure of the required function within an accident scenario modelled by the ET. A FT analysis is performed for each of the safety systems, required in response to the IE.

Assigning the safe end state to a sequence means that the scenario has been successfully terminated and undesired consequences have not occurred. In contrast the accident end state means that the sequence has resulted in undesired consequences.

Synthetically, the methodology embraced for the analysis consists of the following major tasks:

- identification of initiating events or initiating event groups of accident sequences: each initiator is defined by a frequency of occurrence;
- systems analysis: identification of functions to be performed in response to each initiating events to successfully prevent plant damage or to mitigate the consequences and identification of the correspondent plant systems that perform these functions (termed front-line systems): for each system the probability of failure is assessed, by fault tree model;
- accident sequences development by constructing event trees for each initiating event or initiating event groups;
- accident sequences analysis to assess the frequencies of all relevant accident sequences;
- identification of dominant sequences on a frequency-consequence base, i.e. the ones presenting the most severe consequences to the personnel, the plant, the public and the environment and definition of the reference accident scenarios to be further analysed through deterministic transient analysis (for instance by t-h code simulation), in order to verify the fulfilment of the safety criteria. Consequences in the case of Level 1 PSA of NPPs are usually defined as degrees of reactor core damage, including 'safe' state and 'severe' accident state.

One of the main issues encountered in probabilistic analysis concerns the availability of pertinent data for the quantification of the risk, which eventually raises a large uncertainty in the results achieved. Usually these data are accessible from consolidated data bases (e.g. IAEA), resulting from the operational experience of the plants.

They pertain, for instance, to component failure rates, component probability on demand, initiating event frequency: for this reason within a PSA study usually an uncertainty analysis, in addition to a sensitivity analysis, is required in order to add credit to the model and to assess if sequences have been correctly evaluated on the probabilistic standpoint.

Event trees are used for the graphical and logical presentation of the accident sequences. An example of an event tree is shown in Figure 7. The logical combinations of success/failure conditions of functions or systems (usually safety systems, also called front-line systems) in the event tree are modelled by the fault tree.

A fault tree logically combines the top event (e.g. complete failure of a support system) and the causes for that event (e.g. equipment failure, operator error etc.). An example of the fault tree is shown in Figure 8. The fault tree mainly consists of the basic events (all possible causes of the top event that are consistent with the level of detail of the study) and logical gates (OR, AND, M out of N and other logical operations). Other modelling tools, like common cause failures, house or area events are also used in the fault trees. All front-line and support systems are modelled by the fault trees and then combined in the event trees depending on the initiating event.

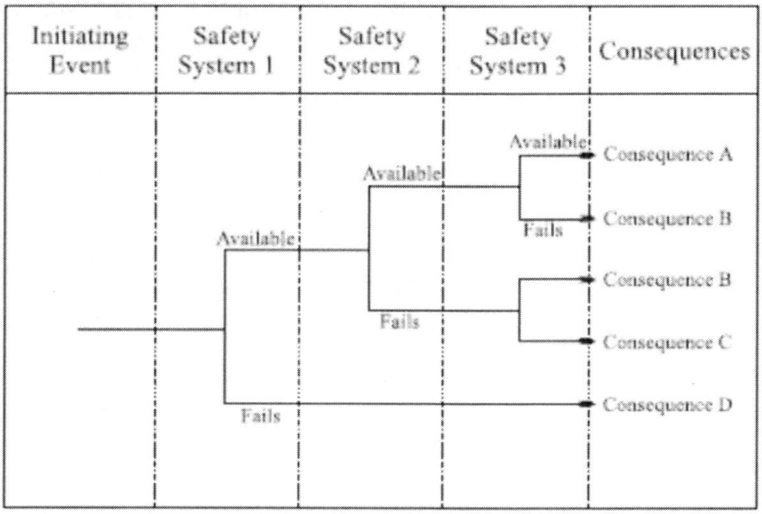

Figure 7. Example of an event tree

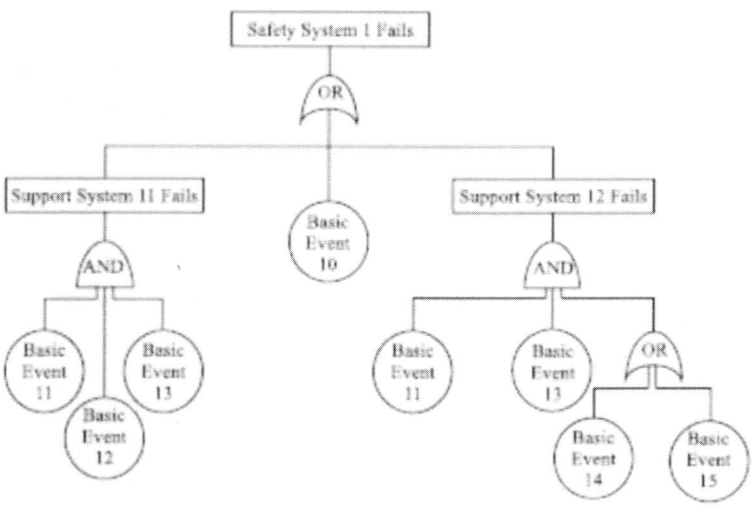

Figure 8. Example of a fault tree

A fault tree is capable to include rather special cases, usually identified in complex systems. These include system and components dependencies, called common cause failures (simultaneous failures of several components due to the same reason), area events (usually fire, flood etc., which damages groups of components in certain rooms), human actions (operator errors or mitigation actions).

The PSA is a powerful tool that can be used in many different ways to assess, understand and manage risk. Its primarily objectives are the following:

- estimate risk level of the facility,
- identify dominant event sequences affecting safety of the facility,
- identify systems, components and human actions important for safety,
- assess important dependencies (among systems or man-machine interactions),
- provide decision support in various application areas.

The growing area of PSA use is extensive support of probabilistic results in risk management and decision-making processes. The main areas of the PSA applications are assessment of design modifications and back-fitting, risk informed optimization of the Technical Specifications, accident management, emergency planning and others. Several modern tools of risk management are also based on the PSA model, such as risk monitoring, precursor analysis and others.

Despite its popularity among the risk assessment tools, the PSA has a number of imitations and drawbacks. The main limitations of the PSA model are the following:

Binary representation of the component state. Only two states are analyzed: failed state or fully functioning state. However, this is not always realistic, as intermediate states are also possible. The same limitation exists for the redundant systems with certain success criteria - system is in failed state (success criteria is not satisfied) or in full power. The intermediate states for redundant systems are even more important.

Independence. In most cases, the components are assumed to be independent (except modelled by CCF), however there are many sources of dependencies, not treated by the model.

Aging effect. The aging effect is ignored because of the constant failure rate assumption. The only conservative possibility to treat the aging impact is to perform sensitivity study.

Time treatment. The FT/ET model is not capable to treat time explicitly during the accident progression. This is one of the major drawbacks of the methodology. In realistic systems, many parameters and functions depend on time and this is not encountered in the model and only approximate chronological order is assumed.

Uncertainty of the calculations. Uncertainties are inevitable in the PSA results and calculations and therefore direct treatment of the quantitative PSA estimates might be misleading. Due to the fact of uncertainties, the qualitative PSA results (identification of dominant accident sequences, comparison of different safety modifications) are of greater importance than quantitative.

4. PASSIVE SYSTEM UNAVAILABILITY MODEL

The reliability of a passive system refers to the ability of the system to carry out a safety function under the prevailing conditions when required and addresses mainly the related performance stability.

In general the reliability of passive systems should be seen from two main aspects:

- systems/components reliability (e.g. piping, valves), as, for instance, the failure to start-up the system operation (e.g. drain valve failure to open)
- physical phenomena reliability, which addresses mainly the natural circulation stability, and the proneness of the system to the failure is dependent on the boundary conditions and the mechanisms needed for maintaining the intrinsic phenomena rather than on component malfunctions.

These two kinds of system malfunction are to to be considered as ET headings, to be assessed by specific FT components, as shown in figures 9 and 10.

The first facet calls for well-engineered safety components with at least the same level of reliability of the active ones.

The second aspect is concerned with the way the physical principle (gravity and density difference) operate and depends on the surrounding conditions related to accident development in terms of thermal hydraulic parameters evolution (i.e. characteristic parameters as flow rate and exchanged heat flux). This could require not a unique unreliability figure, but the unreliability to be re evaluated for each sequence following an accident initiator, or at least for a small group of bounding accident sequences, enveloping the ones chosen upon similarity of accident progress and expected consequences: with this respect thermal hydraulic analysis of the accident is helpful to estimate the evolution of the parameters during the accident progress.

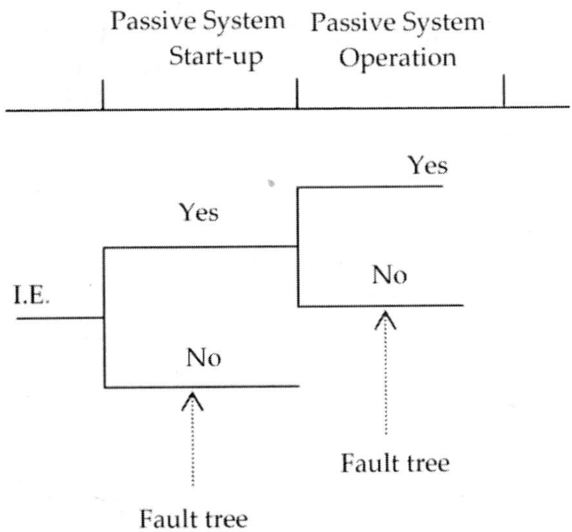

Figure 9. Event tree development

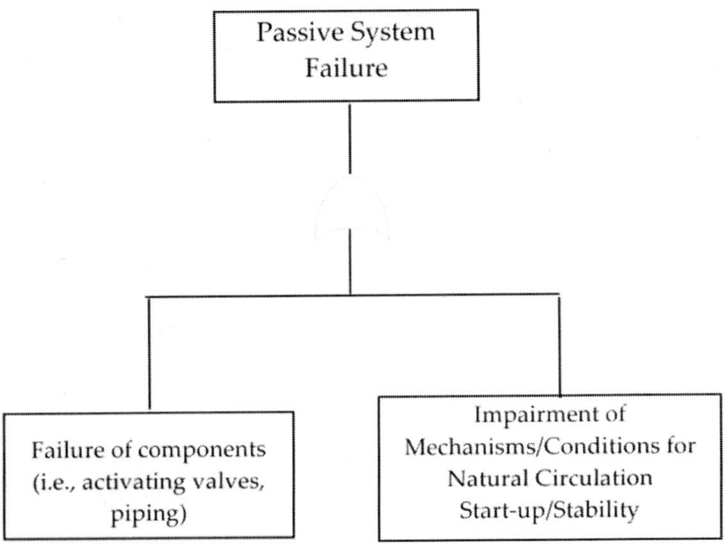

Figure 10. Fault tree model

First step of the analysis is the identification of the failure modes affecting the natural circulation: for this scope two well structured commonly used qualitative hazard analysis, as Failure Mode and Effect Analysis (FMEA) and HAZard and OPerability analysis (HAZOP), specifically tailored on the topic, by considering the phenomenology typical of natural circulation, are adopted.

This analysis concerns both mechanical components (e.g. valve, piping, heat exchanger) of the system and the natural circulation itself, as "virtual" component and the system under investigation is the aforementioned Isolation Condenser.

FMEA is a bottom-up procedure conducted at component level by which each failure mode in a system is investigated in terms of failure causes, preventive actions on causes, consequences on the system, corrective/preventive actions to mitigate the effects on the system, while the HAZOP procedure considers any parameters characteristic of the system (among pressure, temperature, flow rate, heat exchanged through the HX, opening of the drain valve) and by applying a set of "guide" words, which imply a deviation from the nominal conditions as for instance undesired decrease or increase, determines the consequences of operating conditions outside the design intentions. FMEA and HAZOP analysis are shown in Table 1 and 2 respectively.

The analysis points out several factors leading to disturbances in the Isolation Condenser system; the list of these includes:

- Unexpected mechanical and thermal loads, challenging the primary boundary integrity
- HX plugging
- Mechanical component malfunction, i.e. drain valve
- Non-condensable gas build-up
- Heat exchange process reduction: surface oxidation, thermal stratification, piping layout, etc.

Finally a set of critical parameters direct indicators of the failure of the system is identified; these include:

- Non-condensable fraction
- Undetected leakage
- Valve closure area in the discharge line
- Heat loss

- Piping layout
- HX plugged pipes

Table 1. FMEA Table for the Isolation Condenser System

Component	Failure Mode	Causes	Prev. Actions on Causes	Consequences	Corrective/Preventive Action on Consequences	Comment
System piping	Rupture	Material defects and aging; Corrosion; Abnormal operation conditions; Vibrations; Local. Stresses; Impact of heavy loads (missile)	Adequate welding process quality; Water chemistry control; In Service inspect; Design against missile generation	LOCA in the Drywell; Instantaneous loss of natural circulation; Emptying of the circuit; Loss of heat removal capability; Loss of reactor coolant inventory	Isolate the breached loop; Safety relief valves actuation; Automatic reactor depressurisation; Gravity Driven Cooling System actuation;	Includes both steam line and drain line Critical Parameter: Undetected Leakage
	Leak	Material defects and aging; Corrosion; Abnormal operation conditions; Vibrations; Local. stresses	Adequate welding process quality; Water chemistry control; In Service inspect.	Small LOCA in the Drywell; Slow emptying of the circuit and natural circulation arrest for long periods of operation; Reduced heat removal capability	Leak monitoring; Isolate the breached loop; Safety relief valves actuation	Critical Parameter: Undetected Leakage
Tube Bundle of the heat exchangers of the IC	Single pipe rupture	Wearing due to vibration and corrosion	Preventive maintenance; Water chemistry control; Leak monitoring	Release of primary water to the pool; Slow emptying of the circuit and natural circulation arrest for long periods of operation; Reduced heat removal capability	Flow monitoring; Isolate the breached loop; Safety relief valves actuation	Critical Parameter: Undetected Leakage
	Multiple pipe rupture	Wearing due to corrosion, vibration and pressure transient	Preventive maintenance; Water chemistry control; Leak monitoring	Release of primary water to the pool; Natural circulation stop; Emptying of the circuit; Loss of reactor coolant inventory; Loss of heat removal capability	Isolate the breached loop; Safety relief valves actuation; Automatic reactor depressurisation; Gravity Driven Cooling System actuation	Critical Parameter: Undetected Leakage
	Single pipe plugging	Crud in the cooling loop; Foreign object in the cooling loop	Water chemistry control; Yearly test of pipes flow; Preventive maintenance	No consequences		Critical Parameter: HX Plugged Pipes
Tube Bundle of the heat exchangers of the IC	Multiple pipe plugging	Violent pressure and vibration transient detaching large amount of crud from pipes walls.	Water chemistry control; Use of suitable materials for cooling loop pipes; Preventive maintenance	Natural circulation stop; Loss of heat removal capability; Reactor pressure and temperature increase	Safety relief valves actuation	Critical Parameter: HX Plugged Pipes

Component	Failure Mode	Causes	Prev. Actions on Causes	Consequences	Corrective/Preventive Action on Consequences	Comment
Drain valve on the return condensate line	Valve fails to open	Control circuit failure; Loss of electric power to motor; Electric motor failure	Redundancy of control devices; Signal to the operator; In Service inspect.	Non triggering of Isolation Condenser if bypass valve does not operate; Loss of heat removal capability; Reactor pressure and temperature increase	Reactor pressure and temperature control; Safety relief valves actuation; Realignment by the operator; Corrective maintenance	Critical Parameter: Partially Open Valve
	Inadvertent valve closing	Spurious signal; Control circuit failure; Human error	Redundancy of control devices; Signal to the operator; Procedured actions	Natural circulation stop in case bypass valve does not operate; Loss of heat removal capability; Reactor pressure and temperature increase	Reactor pressure and temperature control; Safety relief valves actuation; Realignment by the operator; Corrective maintenance	Critical Parameter: Partially Open Valve
Natural Circulation	Envelope failure	Material defects and aging; Corrosion; Abnormal operation conditions; Vibrations; Local. Stresses; Impact of heavy loads (missile)	Adequate welding process quality; Water chemistry control; In Service inspect; Design against missile generation	LOCA in the Drywell; Instantaneous loss of natural circulation; Emptying of the circuit; Loss of heat removal capability; Loss of reactor coolant inventory	Isolate the breached loop; Safety relief valves actuation; Automatic reactor depressurisation; Gravity Driven Cooling System actuation	Includes both steam line and drain line Critical Parameter: Undetected Leakage
	Cracking	Material defects and aging; Corrosion; Abnormal operation conditions; Vibrations; Local. stresses	Adequate welding process quality; Water chemistry control; In Service inspect.	Small LOCA in the Drywell; Slow emptying of the circuit and natural circulation arrest for long periods of operation; Reduced heat removal capability	Leak monitoring; Isolate the breached loop; Safety relief valves actuation	Critical Parameter: Undetected Leakage
	Modification of surface characteristics	Oxidation; Aerosol deposits	Water chemistry control	Reduction in heat exchange efficiency; Reduced heat removal capability	Flow monitoring	Critical Parameter: Oxide Layer
Natural Circulation	Thermal stratification	Temperature dishomogeneity; Density variations; Onset of local thermal hydraulic phenomena	Process control (pressure, flow, temperature)	Reduction of heat convection; Natural circulation blockage; Loss of heat removal capability; Reactor pressure and temperature increase	Flow monitoring; Reactor pressure and temperature control; Safety relief valves actuation	Critical Parameter: Piping Layout, Heat Loss

Component	Failure Mode	Causes	Prev. Actions on Causes	Consequences	Corrective/Preventive Action on Consequences	Comment
	Non condensable build-up	Onset of chemical phenomena; Radiolysis products; Impurities	Water chemistry control (PH, O2, H2)	Reduction in heat exchange efficiency; Reduction of heat convection; Natural circulation blockage; Loss of heat removal capability; Reactor pressure and temperature increase	Flow monitoring; Reactor pressure and temperature control; Purging through vent lines; Safety relief valves actuation	Critical Parameter: Non-Condensable Fraction
	Heat dissipation	Thermal insulation degradation; Inaccurate material assembly	In Service inspect.	Reduction of heat convection; Natural circulation impairment	Flow monitoring;	Critical Parameter: Heat Loss

Table 2. HAZOP Table for the Isolation Condenser System

Guide Word	Deviation	Possible Causes	Consequences	Safeguards/Interlocks	Actions Required
PARAMETER: Flow rate					
More of[1]	High Flow	N/A			
Less of	Low Flow	Modifications of surface characteristics (crud deposition, oxidation); Non-condensable build-up; Thermal stratification; Pipe partial plugging; Pipe leak; HX single pipe plugging; HX single pipe rupture Drain valve partial opening	Natural circulation degradation and reduced heat transfer capability; T and P increase	Safety relief valve actuation; Vent line valve actuation	Corrective maintenance; Operator action
No/None	No Flow	Non-condensable build-up; Thermal stratification; Pipe plugging; Pipe rupture; HX Multiple pipe plugging; HX Multiple pipe rupture; Drain valve closed	Natural circulation stop and loss of heat transfer capability; T and P increase	Safety relief valve actuation; Vent line valve actuation; Automatic Depressurisation System actuation; Gravity Driven Cooling System actuation	Corrective maintenance; Operator action
PARAMETER: Pressure					
More of	High Pressure	Non-condensable build-up; Surface modifications (crud, oxidation); HX tube plugging; HX tube rupture Partial valve opening	Natural circulation degradation and reduced heat transfer capability; T increase	Safety relief valve actuation; Vent line valve actuation	Corrective maintenance; Operator action
Less of[1]	Low Pressure	N/A			
No/None	No Pressure	N/A			
PARAMETER: Drain valve opening					
More of	N/A				
Less of	Reduced Opening	Partial blockage	Natural circulation degradation and reduced heat transfer capability; T and P increase	Safety relief valve actuation;	Corrective maintenance; Operator action
No/None	No Opening	Loss of electrical power; Circuit control failure; Electrical motor failure; Valve stuck	Natural circulation stop and loss of heat transfer capability; T and P increase	Safety relief valve actuation; Vent line valve actuation; Automatic Depressurisation System actuation; Gravity Driven Cooling System actuation	Corrective maintenance; Operator action
PARAMETER: Exchanged heat flux					
More of[1]	High flux	N/A			
Less of	Low Flux	Non-condensable build-up; Surface modifications (crud, oxidation); HX single tube plugging; HX single tube rupture	Natural circulation degradation and reduced heat transfer capability; T and P increase	Safety relief valve actuation; Vent line valve actuation	Corrective maintenance; Operator action
No/None	No Flux	Non-condensable build-up; HX multiple tube plugging; HX multiple tube rupture	Natural circulation stop and loss of heat transfer capability; T and P increase	Safety relief valve actuation; Vent line valve actuation; Automatic Depressurisation System actuation; Gravity Driven Cooling System actuation	Corrective maintenance; Operator action

[1] This deviation is not evaluated, even if it implies an overcooling of the system that could potentially induce to thermal stresses on core structures and reactor components, like the heat exchanger.

Each of these failure mode driving parameters is examined to determine the expected failure probability by defining the range and the probability distribution function pertaining to the parameter. These failure characteristics are then used to develop a probabilistic model to predict the natural circulation failure.

As stated before FT technique seems to be the most suitable mean to quantify the passive system unavailability, once introduced the failure modes in the form of critical parameters elementary basic events, linked following the Boolean algebra rules (AND et OR), or in the form of sub-fault trees. However the introduction of passive safety systems into an accident scenario, in the fashion of a safety or front line system, deserves particular attention. The reason is that its reliability figure depends more on the phenomenological nature of occurrence of the failure modes rather than on the classical component mechanical and electrical faults. This makes the relative assessment process different as regards the system model commonly adopted in the fault tree approach as depicted before.

In fact, since the failure of the physical process is addressed, the conventional failure model associated with the basic events (i.e. exponential, $e^{-\lambda t}$, λ failure rate, t mission time), commonly used for component failure model, is not applicable: each pertinent basic event will be characterized by defined parameters driving the failure mechanisms - e.g. non-condensable fraction, leak rate, partial opening of the isolation valve, heat exchanger plugged pipes, etc. - and the associated failure criterion. Thus each basic event model pertaining to the relevant failure mode requires the assignment of both the probability distribution and range of the correspondent parameter and the definition of the critical interval defining the failure (for example failure for non-condensable fraction >x%, leak rate > x gr./sec or crack size > x cm^2 and so on).In order to evaluate the overall probability of failure of the system, the single failure probabilities are combined according to:

$$Pe_t = 1.0 - ((1.0 - Pe_1)*(1.0 - Pe_2)*...*(1.0 - Pe_n)) \quad (1)$$

where:

Pe_t overall probability of failure

Pe_1 through Pe_n individual probabilities of failure pertaining to each failure mode, assuming mutually non-exclusive independent events

The failure model relative to each single basic event is given by:

$$Pe_i = \int p_i(x)\, dx \quad x > x_o \quad (2)$$

$p_i(x)$probability distribution function of the parameter x

$x_0$threshold value according to the failure criterion

It's worth noting that the assumed failure criterion, based on the failure threshold for each path, implies the neglecting of the "intermediate" modes of operation of the system or equivalently the degraded performance of the system (up to the failure point): this gives credit for a passive system that "partially works" and has failed for its intended function but provides some operation. This operation could be sufficient to prolong the window for opportunity to recover a failed system, for instance through redundancy configuration, and ultimately prevent or arrest core degradation.

Once the probabilistic distributions of the parameters are assigned, the reliability of the system can be directly obtained from (1) once a failure criterion is assigned and the single failure probabilities are evaluated through (2): this point is being satisfied by assigning both the range and the probability distributions, basing on expert judgment and engineering assessment. In fact, as further illustrated, difficulties arise in assigning both the range and the probability density functions relative to the critical parameters defining the failure modes, in addition to the definition of a proper failure criterion, because of the lack of operational experience and data.

5. METHODOLOGIES CHARACTERIZATION AND COMPARATIVE ASSESSMENT

A very good description of the various methodologies proposed so far and currently available in the open literature is given in [5].

The earliest significant effort to quantify the reliability of such systems is represented by a methodology known as REPAS (Reliability Evaluation of Passive Systems), [6], which has been developed in late 1990s, cooperatively by ENEA, the University of Pisa, the Polytechnic of Milan and the University of Rome, that was later incorporated in the EU (European Union) RMPS (Reliability Methods for Passive Systems) project. This methodology is based on the evaluation of a failure probability of a system to carry out the desired function from the epistemic uncertainties of those physical and geometric parameters which can cause a failure of the system.

The RMPS methodology, described in [7], was developed to address the following problems: 1) Identification and quantification of the sources of uncertainties and determination of the important variables, 2) Propagation of the uncertainties through thermal-hydraulic (T-H) models and assessment of passive system unreliability and 3) Introduction of passive system unreliability in accident sequence analyses. In this approach, the passive system is modelled by a qualified T-H code (e.g. CATHARE, RELAP) and the reliability evaluation is based on results of code runs, whose inputs are sampled by Monte-Carlo (M-C) simulation. This approach provides realistic assessment of the passive system reliability, thanks to the flexibility of the M-C simulation, which adapts to T-H model complexity without resort to simplifying approximation. In order to limit the number of T-H code runs required by M-C simulation, alternative methods have been proposed such as variance reduction techniques, first and second order reliability methods and response surface methods. The RMPS methodology has been successfully applied to passive systems utilizing natural circulation in different types of reactors (BWR, PWR, and VVER). A complete example of application concerning the passive residual heat removal system of a CAREM reactor is presented in [8]. The RMPS methodology tackles also an important problem, which is the integration of passive system reliability in a PSA study. So far, in existing innovative nuclear reactor projects PSA's, only passive system components failure probabilities are taken into account, disregarding the physical phenomena on which the system is based, such as the natural circulation. The first attempts performed within the framework of RMPS have taken into account the failures of the components of the passive system as well as the impairment of the physical process involved like basic events in static event tree as exposed in [7]. Two other steps have been identified after the development of the RMPS methodology where an improvement was desirable: the inclusion of a formal expert judgment (EJ) protocol to estimate distributions for parameters whose values are either sparse on not available, and the use of efficient sensitivity analysis techniques to estimate the impact of changes in the input parameter distributions on the reliability estimates.

R&D in the United States on the reliability of passive safety systems has not been as active at least until mid 2000. A few published papers from the Massachusetts Institute of Technology (MIT) have demonstrated their development of approaches to the issue. Their technique has examined TH uncertainties in passive cooling systems for Generation IV-type gas-cooled reactors. The MIT research on the reliability of passive safety systems has taken a similar approach but has focused on a different set of reactor technologies. Their research has examined thermal hydraulic uncertainties in passive cooling systems for Generation IV gas-cooled reactors, as described in [9,10]. Instead of post-design probabilistic risk analysis for regulatory purposes, the MIT research seeks to leverage the capabilities of probabilistic risk assessment (PRA) to improve the design of the reactor systems early in their development life cycle.

In addition to the RMPS approach, a number of alternative methodologies have been investigated for the reliability assessment of T-H passive systems.

Three different methodologies have been proposed by ENEA (Italian National Agency for New Technologies, Energy and Sustainable Economic Development). In the first methodology [11], the failure probability is evaluated as the probability of occurrence of different independent failure modes, a priori identified as leading to the violation of the boundary conditions or physical mechanisms needed for successful passive system operation.

This approach based on independent failure modes introduces a high level of conservatism as it appears that the probability of failure of the system is relevantly high, because of the combination of various modes of failure as in a series system, where a single fault is sufficient to challenge the system performance. The correspondent value of probability of failure can be conservatively assumed as the upper bound for the unavailability of the system, within a sort of "parts-count" reliability estimation.

In the second, [12], modelling of the passive system is simplified by linking to the modelling of the unreliability of the hardware components of the system: this is achieved by identifying the hardware failures that degrade the natural mechanisms upon which the passive system relies and associating the unreliability of the components designed to assure the best conditions for passive function performance.

Thus, the probabilities of degraded physical mechanisms are reduced to unreliability figures of the components whose failures challenge the successful passive system operation. If, on the one hand, this approach may in theory represent a viable way to address the matter, on the other hand, some critical issues arise with respect to the effectiveness and completeness of the performance assessment over the entire range of possible failure modes that the system may potentially undergo and their association to corresponding hardware failures. In this simplified methodology, degradation of the natural circulation process is always related to failures of active and passive components, not acknowledging, for instance, any possibility of failure just because of unfavourable initial or boundary conditions. In addition, the fault tree model adopted to represent the physical process decomposition is used as a surrogate model to replace the complex T-H code that models the system behaviour. This decomposition is not appropriate to predict interactions among physical phenomena and makes it extremely difficult to realistically assess the impact of parametric uncertainty on the performance of the system.

The third approach is based on the concept of functional failure, within the reliability physics framework of load-capacity exceedance [7,13,14]. The functional reliability concept is defined as the probability of the passive system failing to achieve its safety function as specified in terms of a given safety variable crossing a fixed safety threshold, leading the load imposed on the system to overcome its capacity. In this framework, probability distributions are assigned to both safety functional requirement on a safety physical parameter (for example, a minimum threshold value of water mass flow required to be circulating through the system for its successful performance) and system state (i.e., the actual value of water mass flow circulating), to reflect the uncertainties in both the safety thresholds for failure and the actual conditions of the system state. Thus the mission of the passive system defines which parameter values are considered a failure by comparing the corresponding pdfs according to defined safety criteria. The main drawback in the last method devised by ENEA lies in the selection and definition of the probability distributions that describe the characteristic parameters, based mainly on subjective/engineering judgment.

Every one of three methods devised by ENEA shares with the main RMPS approach the issue related to the uncertainties affecting the system performance assessment process. With respect to the RMPS a greater simplicity is introduced, although detrimental to the relevance of the approaches themselves: this is particularly relevant as far as the approach based on hardware components failure is concerned.

Finally a different approach is followed in the APSRA (Assessment of Passive System ReliAbility) methodology developed by BARC (Bhabha Atomic Research Centre, India), see [15]. In this approach, a failure surface is generated by considering the deviation of all those critical parameters, which influence the system performance. Then, the causes of deviation of these parameters are found through root diagnosis. It is attributed that the deviation of such physical parameters occurs only due to a failure of mechanical components such as valves, control systems, etc. Then, the probability of failure of a system is evaluated from the failure probability of these mechanical components through classical PSA treatment. Moreover, to reduce the uncertainty in code predictions, BARC foresee to use in-house experimental data from integral facilities as well as separate.

With reference to the two most relevant methodologies (i.e. RMPS and APSRA), the RMPS consists mainly in the identification and quantification of parameter uncertainties in the form of probability distributions, to be propagated directly into a T-H code or indirectly in using a response surface; the APSRA methodology strives to assess not the uncertainty of parameters but the causes of deviation from nominal conditions, which can be in the failure of active or passive components or systems.

As a result, different approaches are used in the RMPS and APSRA methodologies. RMPS proposes to take into account, in the PSA model, the failure of a physical process. This problem is treated in using a best estimate T-H code plus uncertainty approach. APSRA includes in the PSA model the failure of those components which cause a deviation of the key parameters resulting in a system failure, but does not take into account possible uncertainties on these key parameters. As the consequence, the T-H code is used in RMPS to propagate the uncertainties and in APSRA to build a failure surface. APSRA incorporates an important effort on qualification of the model and use of the available experimental data. These aspects have not been studied in the RMPS, given the context of the RMPS project.

The following Table attempts to identify the main characteristics of the methodologies proposed so far, with respect to some aspects, such as the development of deterministic and probabilistic approaches, the use of deterministic models to evaluate the system performance, the identification of the sources of uncertainties and the application of expert judgment.

Table 3. Main features of the various approaches

Methodology	Probabilistic vs. deterministic	Deterministic Analysis	Uncertainties	Expert Judgment/Experimental data
REPAS/RMPS	Merge of probabilistic and thermal hydraulic aspects	T-H code adopted for uncertainty propagation	Uncertainties in parameters modelled by probability density functions	EJ adopted to a large extent; Statistical analysis when experimental data exist
APSRA	Merge of probabilistic and thermal hydraulic aspects	T-H code adopted to build the failure surface	parameters' deviations from nominal conditions caused by failure of active or passive components (root diagnosis)	Experimental data usage; EJ for root diagnosis
ENEA approaches	Only probabilistic aspects		Uncertainties in parameters	EJ adopted to a large extent (except the approach based on hardware failure)

6. OPEN ISSUES

From the exam of the various methodologies, which have been developed over these most recent years within the community of the safety research, and are currently available in the open literature, the following open questions are highlighted and consequently needs for research in all related areas are pointed out :

- The aspects relative to the assessment of the uncertainties related to passive system performance: they regard both the best estimate T-H codes used for their evaluation and system reliability assessment itself;
- The dependencies among the parameters, mostly T-H parameters, playing a key role in the whole process assessment.
- The integration of the passive systems within an accident sequence in combination with active systems and human actions.
- The consideration for the physical process and involved physical quantities dependence upon time, implying, for instance, the development of dynamic event tree to incorporate the interactions between the physical parameter evolution and the state of the system and/or the transition of the system from one state to another.

It's worth noticing that these two last aspects are correlated, but hey will be treated separately.

The comparison between active and passive systems, mainly on a functional viewpoint.

All of these points are elaborated in the following, in an attempt to cover the entire spectrum of issues related to the topic, and capture all the relevant aspects to concentrate on and devote resources towards for fulfilling a significant advance.

6.1. Uncertainties

The quantity of uncertainties affecting the operation of the T-H passive systems affects considerably the relative process devoted to reliability evaluation, within a probabilistic safety analysis framework, as recognized in [7].

These uncertainties stem mainly from the deviations of the natural forces or physical principles, upon which they rely (e.g., gravity and density difference), from the expected conditions due to the inception of T-H factors impairing the system performance or to changes of the initial and boundary conditions, so that the passive system may fail to meet the required function. Indeed a lot of uncertainties arise, when addressing these phenomena, most of them being almost unknown due mainly to the scarcity of operational and experimental data and, consequently, difficulties arise in performing meaningful reliability analysis and deriving credible reliability figures. This is usually designated as phenomenological uncertainty, which becomes particularly relevant when innovative or untested technologies are applied, eventually contributing significantly to the overall uncertainty related to the reliability assessment.

Actually there are two facets to this uncertainty, i.e., "aleatory" and "epistemic" that, because of their natures, must be treated differently. The aleatory uncertainty is that addressed when the phenomena or events being modelled are characterized as occurring in a "random" or "stochastic" manner and probabilistic models are adopted to describe their occurrences. The epistemic uncertainty is that associated with the analyst's confidence in the prediction of the PSA model itself, and it reflects the analyst's assessment of how well the PSA model represents the actual system to be modelled. This has also been referred to as state-of-knowledge uncertainty, which is suitable to reduction as opposed to the aleatory which is, by its nature, irreducible. The uncertainties concerned with the reliability of passive system are both stochastic, because of the randomness of phenomena occurrence, and of epistemic nature, i.e. related to the state of knowledge about the phenomena, because of the lack of significant operational and experimental data.

For instance, as initial step, the approach described in [16]. allows identifying the uncertainties pertaining to passive system operation in terms of critical parameters driving the modes of failure, as, for instance, the presence of non-condensable gas, thermal stratification and so on. In this context the critical parameters are recognized as epistemic uncertainties.

The same reference points out, as well, the difference between the uncertainties related to passive system reliability and the uncertainties related to the T-H codes (e.g. RELAP), utilized to evaluate the performance itself, as the ones related to the coefficients, correlations, nodalization, etc.: these specific uncertainties, of epistemic nature, in turn affect the overall uncertainty in T-H passive system performance and impinge on the final sought reliability figure.

A further step of the matter can be found in[11], which attempts to assign sound distributions to the critical parameters, to further develop a probabilistic model. As is of common use when the availability of data is limited, subjective probability distributions are elicited from expert/engineering judgment procedure, to characterize the critical parameters.

Three following classes of uncertainties to be addressed are identified:

- Geometrical properties: this category of uncertainty is generally concerned with the variations between the as-built system layout and the design utilized in the analysis: this is very relevant for the piping layout (e.g. suction pipe inclination at the inlet of the heat exchanger, in the isolation condenser reference configuration) and heat loss modes of failure.

- Material properties: material properties are very important in estimating the failure modes concerning for instance the undetected leakages and the heat loss.
- Design parameters, corresponding to the initial/boundary conditions (for instance, the actual values taken by design parameters, like the pressure in the reactor pressure vessel).
- Phenomenological analysis: the natural circulation failure assessment is very sensitive to uncertainties in parameters and models used in the thermal hydraulic analysis of the system. Some of the sources of uncertainties include but are not limited to: the definition of failure of the system used in the analysis, the simplified model used in the analysis, the analysis method and the analysis focus on failure locations and modes and finally the selection of the parameters affecting the system performance.

The first, second and third groups are part of the category of aleatory uncertainties because they represent the stochastic variability of the analysis inputs and they are not reducible.

The fourth category is referred to the epistemic uncertainties, due to the lack of knowledge about the observed phenomenon and thus suitable for reduction by gathering a relevant amount of information and data. This class of uncertainties must be subjectively evaluated, since no complete investigation of these uncertainties is available.

A clear prospect of the uncertainties as shown in Table 4 [5].

As emphasized above, clearly the epistemic uncertainties address mostly the phenomena underlying the passive operation and the parameters and models used in the T-H analysis of the system (including the ones related to the best estimate code) and the system failure analysis itself. Some of the sources of uncertainties include but are not limited to the definition of failure of the system used in the analysis, the simplified model used in the analysis, the analysis method and the analysis focus of failure locations and modes and finally the selection of the parameters affecting the system performance. With this respect, it is important to underline, again, that the lack of relevant reliability and operational data imposes the reliance on the underlying expert judgment for an adequate treatment of the uncertainties, thus making the results conditional upon the expert judgment elicitation process. This can range from the simple engineering/subjective assessment to a well structured procedure based on expert judgment elicitation, as reported in [17], which outlines the main aspects of the REPAS procedure.

Table 4. Categories of uncertainties associated with T-H passive systems reliability assessment

Aleatory
Geometrical properties
Material properties
Initial/boundary conditions (design parameters)
Epistemic
T-H analysis
Model (correlations)
Parameters
System failure analysis
Failure criteria
Failure modes (critical parameters)

In ref. [17], in order to simplify both the identification of the ranges and their corresponding probabilities, initially discrete values have been selected. As a general rule, a central pivot has been identified, and then the range has been extended to higher and lower values, if applicable. The pivot value represents the nominal condition for the parameter. The limits have been chosen in order to exclude unrealistic values or those values representing a limit zone for the operation demand of the passive system. Once the discrete ranges have been set up, discrete probability distributions have been associated, to represent the probabilities of occurrence of the values. As in the previous step, the general rule adopted is that the higher probability of occurrence corresponds to the nominal value for the parameter. Then lower probabilities have been assigned to the other values, as much low the probability as much wide the distance from the nominal value, as in a sort of Gaussian distribution.

Ultimately, as underlined in the previous section, the methodologies proposed in RMPS and within the studies conducted by MIT address the question by propagating the parameter and model uncertainties, by performing Monte Carlo simulations on the detailed T-H model based on a mechanistic code, and calculating the distribution of the safety variable and thus the probability of observing a value above the defined limit, according to the safety criterion.

6.2. Dependencies

Alike some other types of analyses for nuclear power plants, the documented experience with PSS reliability seems to focus on the analysis of one passive attribute at a time. In many cases, this may be sufficient, but for some advanced designs with multiple passive features, modelling of the synergistic effects among them is important. For example, modelling of a passive core cooling system may require simultaneous modelling of the amount of non condensable gases which build up along the circuit during extended periods of operation, the potential for stratification in the cooling pool, and interactions between the passive core cooling system and the core. Analysis of each of these aspects independently may not fully capture the important boundary conditions of each system. For instance, with regard to the aforementioned methodologies, the basic simplifying assumption of independence among system performance relevant parameters, as the degradation measures, means that the correlation among the critical parameter distributions is zero or is very low to be judged significant, so that the assessment of the failure probability is quite straightforward. If parameters have contributors to their uncertainty in common, the respective states of knowledge are dependent. As a consequence of this dependence, parameter values cannot be combined freely and independently. Instances of such limitations need to be identified and the dependencies need to be quantified. If the analyst knows of dependencies between parameters explicitly, multivariate distributions or conditional subjective pdfs (probability density functions) may be used. The dependence between the parameters can be also introduced by covariance matrices or by functional relations between the parameters.

As observed in [15], both REPAS and RMPS approaches adopt a probability density function (pdf) to treat variations of the critical parameters considered in the predictions of codes. To apply the methodology, one needs to have the pdf values of these parameters. However, it is difficult to assign accurate pdf treatment of these parameters, which ultimately define the functional failure, due to the scarcity of available data, both on an experimental and operational ground. Moreover, these parameters are not really independent ones to have deviation of their own. Rather deviations of them from their nominal conditions occur due to failure/malfunctioning of other components or as a result of the combination with different concomitant mechanisms. Thus the hypothesis of independence among the failure driving parameters appears non proper.

With reference to the functional reliability approach set forth in [13], the selected representative parameters defining the system performance, for instance coolant flow or exchanged thermal power, are properly modelled through the construction of joint probability functions in order to assess the correspondent functional reliability. A recent study shows how the assumption of independence between the marginal distributions to construct the joint probability distributions to evaluate system reliability adds conservatism to the analysis, [18]: for this reason the model is implemented to incorporate the correlations between the parameters, in the form of bivariate normal probability distributions. That study has the merit to highlight the dependence among the parameters underlying the system performance: further studies are underway, with regard, for instance to the approach based on independent failure modes. As described in the previous section 2, this approach begins by identifying critical parameters, properly modelled through probability functions, as input to basic events, corresponding to the failure modes, arranged in a series system configuration, assuming non-mutually exclusive independent events. It introduces a high level of conservatism as it appears that the probability of failure of the system is relevantly high to be considered acceptable, because of the combination of various modes of failure, where a single fault is sufficient to challenge the system performance. Initial evaluations, [19], reveal that the critical parameters are not suitable to be chosen independently of each other, mainly because of the expected synergism between the different phenomena under investigation, with the potential to jeopardize the system performance. This conclusion allows the implementation of the proposed methodology, by properly capturing the interaction between various failure modes, through modelling system performance under multiple degradation measures. It was verified that when the multiple degradation measures in a system are correlated, an incorrect independence assumption may overestimate the system reliability, according to a recent study, [20].

6.3. Incorporation of passive system within probabilistic safety assessment

PSA has been introduced for the evaluation of design and safety in the development of those reactors. A technology-neutral framework, that adopts PSA information as a major evaluation tool, has been proposed as the framework for the evaluation of safety or regulation for those reactors [21,22]. To utilize this framework, the evaluation of the reliability of Passive Systems has been recognized as an essential part of PSA.

In PSA, the status of individual systems such as a passive system is assessed by an accident sequence analysis to identify the integrated behaviour of a nuclear system and to assign its integrated system status, i.e. the end states of accident sequences. Because of the features specific of a passive system, it is difficult to define the status of a passive system in the accident sequence analysis. In other words, the status of a passive system does not become a robust form such as success or failure, since "intermediate" modes of operation of the system or equivalently the degraded performance of the system (up to the failure point) is possible. This gives credit for a passive system that "partially works" and has failed for its intended function but provides some operation: this operation could be sufficient to prolong the window for opportunity to recover a failed system, for instance through redundancy configuration, and ultimately prevent or arrest core degradation [19]. This means that the status of a passive system can be divided into several states, and each status is affected by the integrated behaviour of the reactor, because its individual performance is closely related with the accident evolution and whole plant behaviour.

Ref. [23] lays the foundations to outline a general approach for the integration of a passive system, in the form of a front line system and in combination with active ones and/or human actions, within a PSA framework.

In [7] a consistent approach, based on an event tree representation, has been developed to incorporate in a PSA study the results of reliability analyses of passive systems obtained on specific accident sequences. In this approach, the accident sequences are analyzed by taking into account the success or the failure of the components and of the physical process involved in the passive systems. This methodology allows the probabilistic evaluation of the influence of a passive system on a definite accident scenario and could be used to test the advantage of replacing an active system by a passive system in specific situations.

However in order to generalize the methodology, it is important to take into account the dynamic aspects differently than by their alone modelling into the T-H code. Indeed in complex situations where several safety systems are competing and where the human operation cannot be completely eliminated, this modelling should prove to be impossible or too expensive in computing times. It is thus interesting to explore other solutions already used in the dynamic PSA, like the method of the dynamic event trees, in order to capture the interaction between the process parameters and the system state within the dynamical evolution of the accident.

In the PSA of nuclear power plants (NPPs), accident scenarios, which are dynamic in nature, are usually analyzed with event trees and fault trees.

The current PSA framework has some limitations in handling the actual timing of events, whose variability may influence the successive evolution of the scenarios, and in modelling the interactions between the physical evolution of the process variables (temperatures, pressures, mass flows, etc.,) and the behaviour of the hardware components. Thus, differences in the sequential order of the same success and failure events and the timing of event occurrence along an accident scenario may affect its evolution and outcome; also, the evolution of the process variables (temperatures, pressures, mass flows, etc.,) may affect the event occurrence probabilities and thus the developing scenario. Another limitation lies in the binary representations of system states (i.e., success or failure), disregarding the intermediate states, which conversely concern the passive system operation, as illustrated above.

To overcome the above-mentioned limitations, dynamic methodologies have been investigated which attempt to capture the integrated response of the systems/components during an accident scenario [24].

The most evident difference between dynamic event trees (DETs) and the event trees (ETs) is as follows. ETs, which are typically used in the industrial PSA, are constructed by an analyst, and their branches are based on success/ failure criteria set by the analyst. These criteria are based on simulations of the plant dynamics. On the contrary, DETs are produced by a software that embeds the models that simulates the plant dynamics into stochastic models of components failure. A challenge arising from the dynamic approach to PSA is that the number of scenarios to be analyzed is much larger than that of the classical fault/event tree approaches, so that the a posteriori information retrieval can become quite onerous and complex.

This is even more relevant as far as thermal hydraulic natural circulation passive systems are concerned since their operation is strongly dependent, more than other safety systems, upon time and the state/parameter evolution of the system during the accident progression.

Merging probabilistic models with T-H models, i.e. dynamic reliability, is required to accomplish the evaluation process of T-H passive systems in a consistent manner: this is particularly relevant with regard to the introduction of a passive system in an accident sequence, since the required mission could be longer than 24 h as usual level 1 PSA mission time. In fact for design basis accidents, the passive systems are required to establish and maintain core cooling and containment integrity, with no operator intervention or requirement for a.c. power for 72 h, as a grace time [25].

The goal of dynamic PRA is to account for the interaction of the process dynamics and the stochastic nature/behavior of the system at various stages: it associates the state/parameter evaluation capability of the thermal hydraulic analysis to the dynamic event tree generation capability approach. The methodology should estimate the physical variation of all technical parameters and the frequency of the accident sequences when the dynamic effects are considered. If the component failure probabilities (e.g. valve per-demand probability) are known, then these probabilities can be combined with the probability distributions of estimated parameters in order to predict the probabilistic evolution of each scenario outcome.

A preliminary attempt in addressing the dynamic aspect of the system performance in the frame of passive system reliability is shown in [26], which introduces the T-H passive system as a non-stationary stochastic process, where the natural circulation is modeled in terms of time-variant performance parameters, (as for instance mass flow-rate and thermal power, to cite any) assumed as stochastic variables. In that work, the statistics associated with the stochastic variables change in time (in terms of associated mean values and standard deviations increase or decrease, for instance), so that the random variables have different values in every realization, and hence every realization is different.

6.4. Comparative assessment between active and passive systems

The design and development of future water-cooled reactors address the use of passive safety systems, i.e. those characterized by no or very limited reliance on external input (forces, power or signal, or human action) and whose operation takes advantage of natural forces, such as free convection and gravity, to fulfil the required safety function and to provide confidence in the plant's ability to handle transients and accidents. Therefore, they are required to accomplish their mission with a sufficient reliability margin that makes them attractive as an important means of achieving both simplification and cost reduction for future plants while assuring safety requirements with lesser dependence of the safety function on active components like pumps and diesel generators.

On the other hand, since the magnitude of the natural forces, which drive the operation of passive systems, is relatively small, counter-forces (e.g. friction) can be of comparable magnitude and cannot be ignored as is generally the case with pumped systems. This concern leads to the consideration that, despite the fact that passive systems "should be" or, at least, are considered, more reliable than active ones - because of the smaller unavailability due to hardware failure and human error - there is always a nonzero likelihood of the occurrence of physical phenomena leading to pertinent failure modes, once the system enters into operation.

These characteristics of a high level of uncertainty and low driving forces for heat removal purposes justify the comparative evaluation between passive and active options, with respect to the accomplishment of a defined safety function (e.g. decay heat removal) and the generally accepted viewpoint that passive system design is more reliable and more economical than active system design has to be discussed [27].

Here are some of the benefits and disadvantages of the passive systems that should be evaluated vs. the correspondent active system.

Advantages

- No external power supply: no loss of power accident has to be considered.
- No human factor, implying no inclusion of the operator error in the analysis.
- Better impact on public acceptance, due to the presence of "natural forces".
- Less complex system than active and therefore economic competitiveness.
- Passive systems must be designed with consideration for ease of ISI, testing and maintenance so that the dose to the worker is much less.
- Drawbacks
- Reliance on "low driving forces", as a source of uncertainty, and therefore need for T-H uncertainties modeling.
- Licensing requirement (open issue), since the reliability has to be incorporated within the licensing process of the reactor. For instance the PRA's should be reviewed to determine the level of uncertainty included in the models.
- Need for operational tests, so that dependence upon human factor can not be neglected.
- Time response: the promptness of the system intervention is relevant to the safety function accomplishment. It appears that the inception of the passive system operation, as the natural circulation, is conditional upon the actuation of some active components (as the return valve opening) and the onset of the conditions/mechanisms for natural circulation start-up
- Reliability and performance assessment in any case. Quantification of their functional reliability from normal power operation to transients including accidental conditions needs to be evaluated. Functional failure can happen if the boundary conditions deviate from the specified value on which the performance of the system depends.
- Ageing of passive systems must be considered for longer plant life; for example corrosion and deposits on heat exchanger surfaces could impair their function.
- Economics of advanced reactors with passive systems, although claimed to be cheaper, must be estimated especially for construction and decommissioning.

The question whether it is favourable to adopt passive systems in the design of a new reactor to accomplish safety functions is still to be debated and a common consensus has not yet been reached, about the quantification of safety and cost benefits which make nuclear power more competitive, from potential annual maintenance cost reductions to safety system response.

7. FINAL REMARKS

Based on the analysis of the critical aspects related to the open points discussed in the previous section a qualitative analysis, on the basis of the author's opinion, reported in Table 5 below aims at identifying for each of the above items both the criticality with respect to the passive system reliability assessment process, in terms of the relative importance and the existing advancement, according to Table 6 which ranks the relative level of both the importance and progress.

Table 5. Importance analysis

Item	Importance	Advance
Uncertainties	H	L
Dependencies	M	L
Integration within PSA	M	L
Passive vs. Active	H	L

Table 6. Grade rank for importance and advancement analysis

	Grade	Definition
Importance	H	The item is expected to have a significant impact on the system failure
	M	The item is expected to have a moderate impact on the system failure
	L	The item is expected to have only a small impact on the system failure
Advance	H	The issue is modelled in a detailed way with adequate validation
	M	The issue is represented by simple modelling based on experimental observations or results.
	L	The issue is not represented in the analysis or the models are too complex or inappropriate which indicates that the calculation results will have a high degree of ambiguity

It is clear that he worst case is characterized by "high "and "low" rankings relative respectively to the importance and the advancement aspects, thus making the correspondent item development a critical challenge.

Based on this, the results of this qualitative analysis show the relevance relative to the uncertainties and the comparison between active and passive, as most critical points to be addressed in the application of the PRA to the evaluation of the passive system performance assessment. This allows the analyst to track a viable R&D program to deal with these issues and limitations and to steer the relative efforts towards their implementation.

8. CONCLUSIONS

Due to the specificities of passive systems that utilize natural circulation (small driving force, large uncertainties in their performance, lack of data...), there is a strong need for the development and demonstration of consistent methodologies and approaches for evaluating their reliability. This is a crucial issue to be resolved for their extensive use in future nuclear power plants. Recently, the development of procedures suitable for establishing the performance of a passive system has been proposed: the unavailability of reference data makes troublesome the qualification of the achieved results. These procedures can be applied for evaluating the acceptability of a passive system, specifically when nuclear reactor safety considerations are important for comparing two different systems having the same mission and, with additional investigation, for evaluating the performance of an active and passive system on a common basis. The study while identifying limitations of the achieved results or specific significant aspects that have been overlooked has suggested areas for further development or improvements of the procedures:

- In order to get confidence in the achieved results, the reduction of the so identified level of uncertainty pertaining to the passive system behaviour, and regarding in particular the phenomenological uncertainty. In fact, it's worth noting that these uncertainties are mainly related to the state of knowledge about the studied object/phenomenon, i.e., they fall within the class of epistemic uncertainties, thus suitable for reduction by gathering and analyzing a relevant quantity of information and data.

- The determination of the dependencies among the relevant parameters adopted to analyze the system reliability.

- The study of the dynamical aspects of the system performance, because the inherent dynamic behaviour of the system to be characterized: this translates into the development of the dynamic event tree.
- The comparison against the active system, also to evaluate the economical competitiveness, while assuring the same level of safety.

Future research in nuclear safety addressing this specific topic relevant to advanced reactors should be steered towards all these points in order to foster and add credit to any proposed approach to address the issue and to facilitate the proposed methods endorsement by the scientific and technical community.

REFERENCES

1. IAEA TEC-DOC-626, 1991 Safety Related Terms for Advanced Nuclear Power Plants. September 1991.
2. IAEA TEC DOC-1474, 2005 Natural Circulation in Water Cooled Nuclear power Plants. *Phenomena, models, and methodology for system reliability assessments*, November 2005.
3. IAEA TECDOC-1624, 2009 Passive Safety Systems and Natural Circulation in Water Cooled Nuclear Power Plants. November 2009
4. United States Nuclear Regulatory Commission's (U.S. NRC) Reactor Safety Study (WASH-1400, 1975
5. Zio, E. , N. Pedroni, 2009 Building Confidence in the Reliability Assessment of Thermal hydraulic Passive Systems. Reliability Engineering and System Safety, 94 268-281.
6. Jafari, J. , F. D'Auria, et al. 2003 Reliability Evaluation of a Natural Circulation System. Nuclear Engineering and Design 224 79-104.
7. Marques, M. , L. Burgazzi, et al. 2005 Methodology for the Reliability Evaluation of a Passive System and its Integration into a Probabilistic Safety Assessment. Nuclear Engineering and Design 235 2612-2631.
8. Lorenzo G., et al., Assessment of an Isolation Condenser of an Integral Reactor in View of Uncertainties in Engineering Parameters, Science and technology of Nuclear Installations, Volume 2011 Article ID 827354, 9 pages
9. Apostolakis, G. , L. Pagani, P. Hejzlar, 2005 The Impact of Uncertainties on the Performance of Passive Systems. Nuclear Technology 149 129-140
10. Apostolakis, G. , F. Mackay, Hejzlar. and, 2008 Incorporating Reliability Analysis into the Design of Passive Cooling System with an Application to a Gas-Cooled Reactor. Nuclear Engineering & Design 238, 217-228
11. Burgazzi, L. , 2007a Addressing the Uncertainties related to Passive System Reliability. Progress in Nuclear Energy 49 93-102.
12. Burgazzi, L. , 2002 Passive System Reliability Analysis: a Study on the Isolation Condenser, Nuclear Technology 139 3-9.
13. Burgazzi, L. , 2003 Reliability Evaluation of Passive Systems through Functional Reliability Assessment, Nuclear Technology 144 145-151.
14. Burgazzi, L. , 2007b Thermal-hydraulic Passive System reliability-based design approach, Reliability Engineering and System Safety 92 (9), 1250-1257.
15. Nayak, A. K. , et al. 2008 Passive System Reliability Analysis using the APSRA Methodology. Nuclear Engineering and Design 238 1430-1440.
16. Burgazzi, L. , 2004 Evaluation of Uncertainties related to Passive Systems Performance. Nuclear Engineering and Design 230 93-106.

17. Ricotti, M. E. , E. Zio, F. D'Auria, G. Caruso, 2002 Reliability Methods for Passive Systems (RMPS) Study- Strategy and Results, in proceedings of the NEA CSNI/WGRISK Workshop on Passive System Reliability. A Challenge to Reliability Engineering and Licensing of Advanced Nuclear Power Plants, 146-163
18. Burgazzi, L. , 2008a Reliability Prediction of Passive Systems based on Bivariate Probability Distributions, Nuclear Technology 161 1-7.
19. Burgazzi, L. , 2009 Evaluation of the Dependencies related to Passive System Failure. Nuclear Engineering and Design 239 3048-3053
20. Burgazzi, L. , 2011 Reliability Prediction of Passive Systems with Multiple Degradation Measures, Nuclear Technology 173 153-161.
21. USNRC, 2007 Feasibility study for a risk-informed and performance-based regulatory structure for future plant licensing. US Nuclear Regulatory Commission, NUREG-1860.
22. IAEA, 2007 Proposal for a technology-neutral safety approach for new designs. International Atomic Energy Agency, TECDOC-1570, Vienna.
23. Burgazzi, L. , 2008b Incorporation of Passive Systems within a PRA Framework. Proceedings of PSAM9, 9th International Probabilistic, Safety Assessment and Management Conference, Hong Kong, 18-23 May 2008.
24. Mercurio, D., Podofillini, L., Zio, E., Identification and Classification of Dynamic Event Tree Scenarios via Possibilistic Clustering: Application to a Steam Generator Tube Rupture Event. Accident Analysis and Prevention 41 2009 1180-1191
25. Matzie, R. A. , A. Worrally, A. The, 1000 reactor-the Nuclear Renaissance Option. Nuclear Energy, 2004, 43, 1 Feb., 33-45
26. Burgazzi, L. , 2008c About Time-variant Reliability Analysis with Reference to Passive Systems Assessment. Reliability Engineering and System Safety 93 1682-1688.
27. JiYong Oh and Golay, M., 2008 Methods for Comparative Assessment of Active and Passive Safety Systems with respect to Reliability, Uncertainty, Economy and Flexibility. Proceedings of PSAM9, 9th International Probabilistic, Safety Assessment and Management Conference Hong Kong, 18-23 May 2008.

CHAPTER 5

The effect of organic retarders on grout thickening and setting during deep borehole disposal of high-level radioactive waste

Nicholas C. Collier [a], *, Neil B. Milestone [b], Karl P. Travis [a], Fergus.G.F. Gibb [a]

[a] *Deep Borehole Disposal Research Group, Immobilisation Science Laboratory, Department of Materials Science & Engineering, The University of Sheffield, Sheffield S1 3JD, United Kingdom*

[b] *Callaghan Innovation, 69 Gracefield Road, PO Box 31310, Lower Hutt 5040, New Zealand*

ABSTRACT

Deep borehole disposal (DBD) is being increasingly seen as a viable and potentially superior alternative to comparatively shallow mined repository concepts for disposal of some high-level radioactive wastes. We report here details of proof-of-concept investigations into the use of cementitious grouts as sealing/support matrices for use in low temperature DBD scenarios. Using the cementitious grout to fill annular space within the disposal zone will not only support waste packages during placement, but will also provide a low permeability layer around them which will ultimately enhance the safety case for DBD. Grouts based on Class G oil well cement are being developed. The use of retarders to delay the accelerated onset of thickening and setting (caused by the high temperature and pressure in the borehole) is being investigated experimentally. Sodium gluconate and a polycarboxylate additive each provide sufficient retardation over the range 90–140 °C in order to be considered for this application. Phosphonate and sulphonate additives provide desirable retardation at 90 °C. The additives did not affect grout composition at 14 days curing and the phases formed are durable at elevated temperature and pressure.

Keywords: Radioactive waste; Deep borehole disposal; Cement; Retarder; Organic

1. INTRODUCTION

Advantages associated with safety, cost, and ease of implementation, and the ability to drill deeper larger diameter holes (Juhlin and Sandstedt, 1989; Beswick, 2008; Beswick and Forrest, 1982 ; Exxon Neftegas,), means that the use of deep boreholes to dispose of high level radioactive wastes (HLW, including spent nuclear fuel (SF)) is now being increasingly seen as a viable alternative to emplacement in geologically shallow, mined repositories (Chapman and Gibb, 2003 ; Beswick et al., 2014). The disposal of wastes generated during the production of nuclear energy is of significant importance to the overall nuclear fuel cycle and is currently receiving particular attention around the world. Even though considerable research has been performed in developing waste repositories several hundreds of meters below ground, there is currently no operational facility to provide ultimate waste disposal. Therefore, the development of an alternative more advantageous concept for the disposal of HLW is of particular interest to those involved in the nuclear fuel cycle.

The concept of deep borehole disposal (DBD) of radioactive waste is a multi-barrier approach that places greater emphasis on the performance of geological barriers rather than engineered systems. A large diameter borehole (up to ~0.65 m) is drilled up to 6 km deep into the granitic basement of the continental crust and is subsequently cased (Beswick et al., 2014). Packages of radioactive waste are then emplaced into the bottom 1–2 km of the borehole (the disposal zone) within which they are sealed using materials known as sealing and support matrices (SSMs). These SSMs fill the annular space between the waste packages and the casing, and between the casing and the borehole wall. SSMs provide mechanical support against buckling and damage caused by the load stresses from overlying packages, act as a seal/barrier to the ingress of saline groundwater to the waste container, thus prolonging container life, and provide a barrier to the migration path for any gaseous corrosion products. The borehole itself is then permanently sealed above the disposal zone to the surface.

DBD has significant advantages over disposal in repositories only a few hundred meters deep (such as geological disposal facilities, GDFs) (Gibb, 2010): (i) the safety of the waste is ensured for millions of years due to the isolation provided by the much greater geological barrier, (ii) the boreholes are very deep and waste packages are sealed within a stable geological and hydrogeological barrier, (iii) costs for DBD are estimated to be 20% of those for a GDF per tonne of SF, and (iv) DBD could be implemented approximately 50 years earlier than a GDF for disposal of HLW. The strength of this geological barrier in providing enhanced safety is multi-fold and is based on:

a) The great depth means that the return path to surface for water-borne wastes is an order of magnitude longer.

b) Hydraulic conductivity of the rock at depth is very low.

c) Groundwater density stratification will prevent vertical movement of waste ions and has long-term stability.

d) Isolation from near-surface processes like glaciations is provided.

Building on pioneering work on the disposal of radioactive waste in deep boreholes over the past 25 years ((Gibb et al., 2008), and references therein) the DBD Research Group at The University of Sheffield in the UK is now investigating the use of cementitious grouts as SSMs for emplaced waste packages. These grouts are for use in disposals where temperatures at the surfaces of the packages are below ~190 °C (Beswick et al., 2014). Above this temperature other types of SSMs can be employed (Gibb et al., 2008). The presence of a cement grout around the packages could also help to retard radionuclide migration when the container does eventually fail.

Cementation using specially designed Portland cement grouts is most commonly used within a hydrocarbon or geothermal well to secure the casing after drilling, and to provide a degree of separation between the different fluid chemistries and rock formations through which the borehole passes (Taylor, 1997 ; Nelson and Guillot, 2006). Because of both the depth and the decay heat from the radioactive waste packages, the temperature reached in the DBD application is similar to the cementing jobs in some hydrocarbon and many geothermal well applications, so the experience of cementing oil and geothermal wells is of benefit to the application to dispose of specific radioactive wasteforms.

Challenges are associated with down-hole cementing operations created by the elevated temperatures and pressures at the bottom of the boreholes.

These high temperature and pressure conditions are due to (i) the local geological environment where ambient temperatures in the continental crust would be in the range of 80–130° C for the depths being considered for DBD (Best, 2003), and (ii) the hydrostatic pressure caused by the head of borehole fluid present. Grout deployment should be within a few hours of package placement so the radioactive decay heat from that package will not have built up significantly during placement and setting (Gibb et al., 2012). Exposure to these adverse conditions affects the properties of a cementitious grout, giving rise to a reduction in thickening and setting time (Scherer et al., 2010;Jupe et al., 2008) in the fresh state before setting, and after setting and hardening when different crystalline products are formed. Elevated temperature and pressure accelerate the hydration reactions in a cement grout reducing the time to thicken and set (Taylor, 1997; Nelson and Guillot, 2006; Bensted et al., 2008; Shariar and Nehdi, 2012; Zhang et al., 2010). Elevated temperature has a greater influence than elevated pressure over how quickly cement hydration reactions occur (Nelson and Guillot, 2006; Scherer et al., 2010; Jupe et al., 2008). It will also have a greater effect on the composition of the main cement hydrate phases formed (Taylor, 1997 ; Nelson and Guillot, 2006). Phase composition of the hardened cement grout is important in terms of the lifetime of the SSM, and the most durable hydrate phases need to be formed.

The work presented here reports on Portland cement-based grouting systems that are being developed for DBD cementing operations. A proof-of-concept study has been undertaken, and the influence of a range of additives on grout performance has been assessed. Rheological properties and setting characteristics of fresh grouts have been studied at elevated temperature and pressure, and the early age phase composition has been investigated to confirm the formation of desirable hydrated phases. Four different organic additives were chosen to study the influence on grout thickening/setting properties under conditions representative of those found in deep boreholes. Two of these additives are marketed as retarders and two as superplasticisers or dispersants. The latter have been chosen because superplasticisers restrict chemical reaction between cement particles and mix water, and in doing so may also cause retardation of thickening and setting. Even though the presence of organic compounds will complex any radionuclides and increase their solubility, the release of any waste ions as a result of container corrosion will only occur many years after the borehole has been sealed; this creates a geological barrier where any release of radioactive material will take millions of years to return to the human environment making it radiologically harmless. The results obtained in this study have been used to assess the applicability of using cementitious grouts in this DBD application.

2. MATERIALS AND METHODS

A Class G oil well cement (BS EN ISO, 2009) partially replaced with silica flour was used to make the grout. The cement was supplied by Holcim and was manufactured to BS EN ISO 10426-1/API Spec 10A, and the silica flour was supplied by Sibelco Ltd (Unimin Silica Flour 350G). Detailed information concerning the oxide composition of the cement and silica flour, the phase composition of the cement, and the particle size characteristics of both powders is given in Table 1, Table 2; Table 3 respectively. To enable the grout to flow through water without dispersion, an underwater additive (UCS Pak) supplied by Sika Ltd was used which is reported to contain silica and organic compounds. The products assessed that were marketed as retarders were sodium gluconate and Sika Retarder (a proprietary phosphonate product), whereas those marketed as superplasticisers were Viscocrete 3110 (an aqueous solution of polycarboxylate co-polymers supplied by Sika Ltd) and CD-33L (a sulphonated organic polymer supplied by Baker Hughes). Hereafter, all four additives are referred to as retarders. Each retarder was added to the grout mix water (tap water) by weight of cement (BWOC) prior to the addition of the pre-blended powders. The density of the grout (excluding any retarder) was 1.892 kg/m3.

Table 1 Oxide composition of Class G cement and silica flour.

Oxide	Compositional content (wt%)	
	Class G cement	Silica flour
SiO_2	21.19	98.56
Al_2O_3	3.80	0.27
Fe_2O_3	5.34	0.01
MnO	0.05	0.01
MgO	1.82	0.03
CaO	63.18	0.02
Na_2O	0.07	0.07
K_2O	0.47	0.03
TiO_2	0.95	0.11
P_2O_5	0.07	<0.01
SO_3	2.08	<0.01
LoI*	0.89	0.16
Total	99.91	99.27

Notes. * – Loss on Ignition (ASTMD7348-13).

Table 2 Cement phase composition of Class G cement

Cement phase*	Phase content (wt%)
C_3S	55.11
β-C_2S	19.18
C_3A	1.03
C_4AF	16.25

Notes. * – calculated from ASTM C150 (ASTM C150/C150M), and using cement chemist nomenclature where C = CaO, S = SiO_2, A = Al_2O_3 and F = Fe_2O_3.

Grout consistency was determined using a high pressure, high temperature, consistometer (Cement Testing Equipment 40–600 HTHP Consistometer) operating at 90, 120 and 140 °C and 50 MPa with a linear 4 h heating/pressurizing regime, after which time the testing temperature and pressure were maintained until the sample consistency exceeded the upper limit of measurement. The times at which changes in consistency occurred were recorded; t_1 was the time at which minimum consistency occurred and t_2 was the time for consistency to reach 70 Bearden units (Bc), the limit of pumpability (LoP) accepted in well cementing applications (Nelson and Guillot, 2006). This value has been taken as the maximum consistency at which a grout will flow around a waste package located at the bottom of a borehole. Drilling engineers are confident that grout could be delivered to the bottom of a borehole approximately 5 km deep in under 4 h, so this was taken as the minimum delay required to reach a consistency of 70 Bc (t_2), although options for faster delivery times are being explored. For each grout where t_2 was greater than 4 h, initial and final set were checked 24 h after mixing using manual Vicat testing equipment (ASTMC191-13). Cement hydration reactions were investigated using an isothermal calorimeter set at 120 °C and flow characteristics were investigated at room temperature (22 °C) and pressure (1 atm) using a modified ASTM flow cone test (ASTMC939-10) recording the time to flow through the testing cone. Once the amounts of retarders that gave acceptable levels of grout consistency had been determined, viscosity was obtained using an Ofite 900 Viscometer operating at temperatures of 22, 45 and 70 °C and at a range of different shear rates (~5–1000 s^{-1}). A qualitative assessment of particle dispersion for each grout mix in water was investigated by pouring a stream of freshly mixed grout into a beaker of water held at approximately 63 °C, and observing any particle dispersion.

Table 3 Physical characteristics of powder particles.

Powder	Particle diameter (µm)				
	Median	Mode	10th Percentile	50th Percentile	90th Percentile
Class G Cement	16.569	25.332	2.726	16.569	43.554
Silica Flour	16.951	25.332	2.018	16.951	55.958

Early age composition was investigated by casting samples of each grout and hydrothermally curing them under water in sealed containers at 120 °C for 3, 7 and 14 days. After each time interval, hydration was arrested using a solvent replacement technique (Collier et al., 2008) and the composition determined by X-ray diffraction (XRD) and thermogravimetric analysis/derivative thermogravimetric analysis (TGA/DTG). The Philips PW1729 X-Ray diffractometer used a Cu Kα radiation source and was operated between 4 and 80° 2θ at 3°/min and the Alphatech SDT Q600 combined TGA/Differential Scanning Calorimeter (DSC) analyzer was operated between 0 and 1000 °C at 10 °C/min with a nitrogen atmosphere.

3. RESULTS

3.1. Grout thickening and setting

The objective of this part of the study was to determine the retarder addition levels required to give t_2 greater than 4 h. Grout consistency is a function of viscosity (BS EN ISO, 2009), but in this investigation it was used simply to determine the retardation effect of the additives. Consistency plots for grouts containing either sodium gluconate or CD-33L are shown in Fig. 1, with the data from all tests summarized graphically in Fig. 2. Upper addition levels for all the proprietary products were based on the manufacturer's recommendations. While carrying out tests at lower temperatures if it became apparent that successful retardation would not be achieved at higher temperatures, testing at the higher temperatures was not undertaken.

As shown in Fig. 1, the progression of consistency for each grout was similar, increasing over the first 10 min of testing (when the temperature was between 20 and 25 °C); this increase is most likely due to the gelling effect of the organic retarders. Following this early thickening, the consistency of all grouts, apart from those without retarders or those made with CD-33L, decreased with increasing time and temperature towards zero Bc (which equates to a torque of 0.00782 Nm, and indicates a very low viscosity, virtually at the limit of measurable consistency for the equipment, hence the flat-lining of the trace for the sample containing 0.75% sodium gluconate). Towards the end of the testing period, the consistency of each grout rapidly increased towards 100 Bc at which point the test was halted.

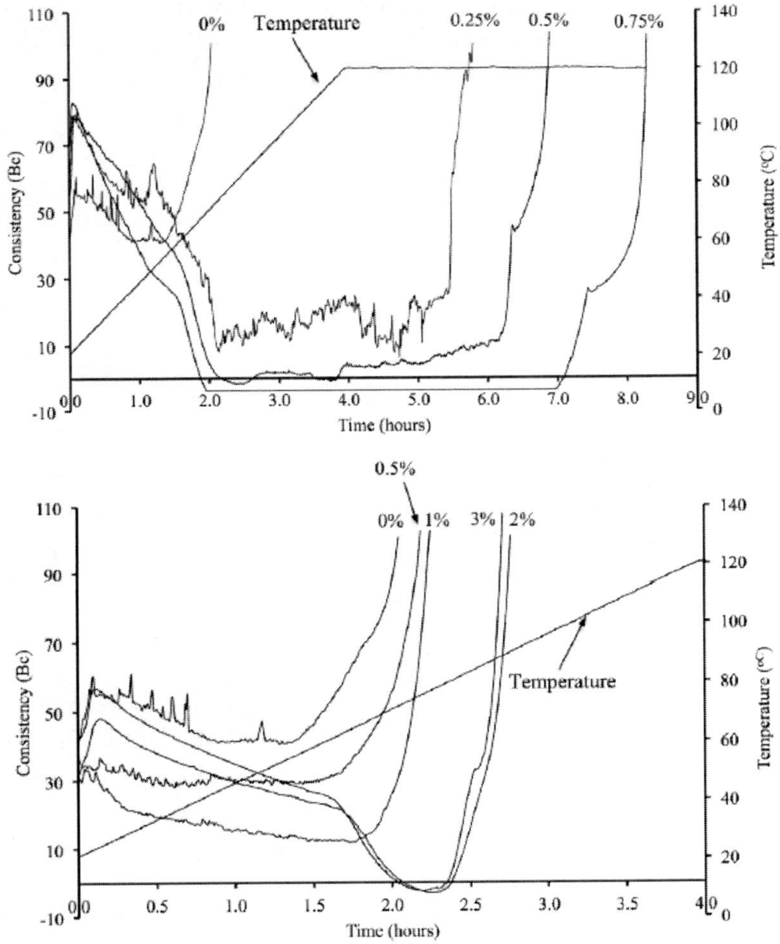

Figure. 1. Comparison of consistency results for grouts containing sodium gluconate (top) (Collier et al., 2015a) and CD-33L (bottom) at 120 °C.

As expected, t_2 for the control grout (without retarder) was reached in less than 4 h, confirming the need for retardation. Fig. 2 shows that, to achieve a t_2 greater than 4 h at 90, 120 and 140 °C, sodium gluconate addition levels of at least 0.05, 0.25 and 0.25% BWOC respectively were required. Considerably more Viscocrete 3110 was required to achieve a t_2 greater than 4 h, with at least 0.75, 1.0 and 2.0% BWOC additions needed at the same testing temperatures respectively. With Sika Retarder, only the grout tested at 90 °C with an addition of ≥0.5% BWOC gave a t_2 greater than 4 h and it was not possible to produce sufficient retardation at 120 °C with the maximum recommended addition. It was also not possible to produce a grout containing CD-33L where t_2 was greater than 4 h at 90 or 120 °C.

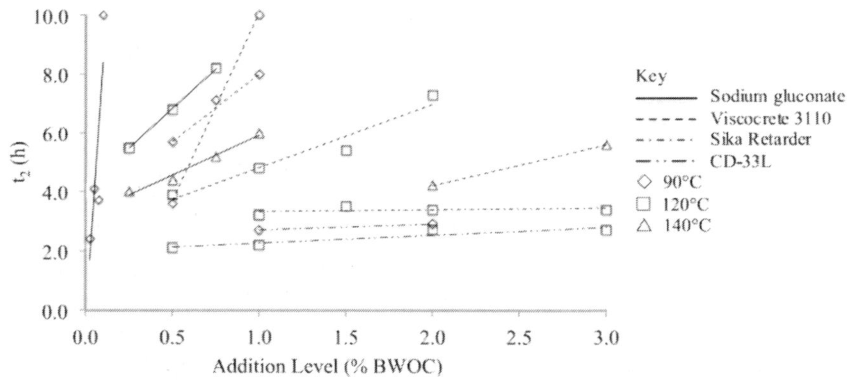

Figure. 2. Changes in t_2 times plotted for all grouts. Notes. t_2 was the time for consistency to reach 70 Bc. Linear trend lines have been added as a visual guide only.

In all grouts where t_2 was greater than 4 h, final set had occurred within 24 h, providing confidence that waste package deployment rates of the order of one a day could be achievable.

3.2. Heat flow during grout thickening and setting

Calorimetry curves for grouts containing sodium gluconate and CD-33L are shown in Fig. 3 with all heat flow events summarized in Table 4. Because the samples were mixed externally to the calorimeter, any heat flow events that occurred before the samples reached thermal equilibrium within the calorimeter were not recorded (time zero was when the grouts were placed in the calorimeter immediately after mixing). The quantity of each retarder used in these tests was based on the consistency results previously obtained.

Increasing the amounts of retarder added changed the heat generation characteristics of the grout, especially for the samples containing sodium gluconate. It is difficult to identify any consistent trend in the magnitude of the specific heat flow peaks for each grout in relation to retarder addition level, but as this amount increased, the times at which the heat flow events occurred also increased, particularly when using sodium gluconate.

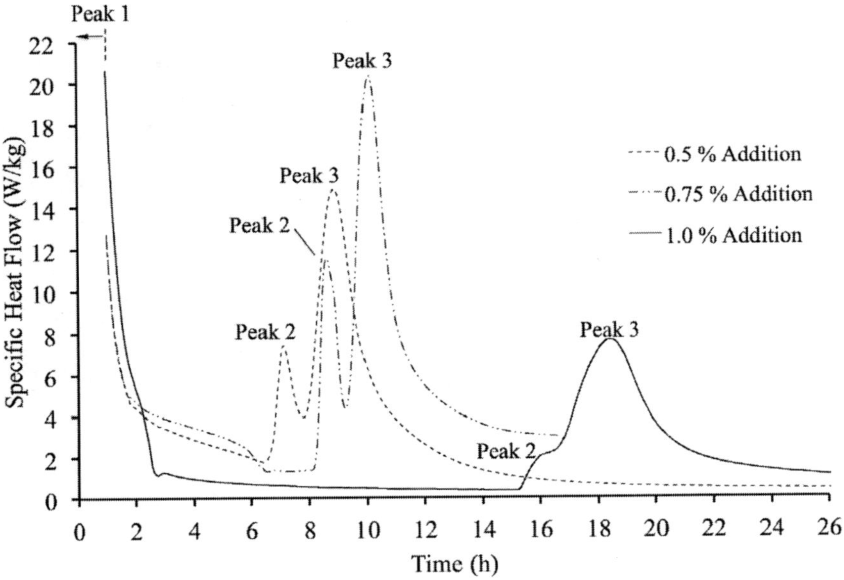

Figure. 3. Calorimetry results for grouts containing sodium gluconate (top) and CD-33L (bottom) at 120 °C.

Three specific heat flow maxima (labeled peaks 1, 2 and 3) were identified for all grouts although peak 3 often merged with peak 2. Peak 1 always occurred at around 1 h, before the sample had reached thermal equilibrium. This peak has been ascribed to the very early hydration reactions between the cement and water or wetting of the cement powder (Taylor, 1997). Sodium gluconate was the most effective retarder for both peaks 2 and 3 with retardation times from 7.0 to 15.6 h for peak 2 and 9.0–18.3 h for peak 3. This compares with 1.6–3.2 h and 3.1–7.0 h for peaks 2 and 3 respectively for the other retarders. Also significantly less sodium gluconate was needed than any of the other retarders.

Several of the calorimeter traces showed marked differences when addition levels were increased. For example, in the trace for the grout containing 1.0% sodium gluconate, peaks 2 and 3 were significantly smaller than at lower addition levels, and the times at which they occurred were much later than at lower levels. Additionally, no peak 3 heat events could be identified for either of the 1.0 and 2.0% CD-33L samples, and only a hint of a peak 3 heat event could be identified in the sample containing 0.5% CD-33L.

Table 4. Summary of calorimetry data.

Retarder	Heat flow peaks	Magnitude and time of heat flow events (W/kg) (h)					
		Addition level (% BWOC*)					
		0.5%		0.75%		1.0%	
Sodium Gluconate	Peak 1	>12.5	<1.0	>12.5	<1.0	>20.6	<1.0
	Peak 2	7.0	7.0	11.2	8.6	1.4	15.6
	Peak 3	14.6	9.0	19.9	10.0	7.6	8.3
		Addition Level (% BWOC*)					
		0.5%		1.0%		2.0%	
Sika Retarder	Peak 1	>9.8	<1.0	17.9	1.1	>9.8	<1.0
	Peak 2	13.4	2.9	16.0	3.1	7.2	2.6
	Peak 3	4.4	5.6	3.8	7.0	16.2	4.9
Viscocrete 3110	Peak 1	>20.1	<1.0	>15.5	<1.0	>18.1	<1.0
	Peak 2	13.9	1.6	18.4	2.1	14.8	2.6
	Peak 3	3.9	3.4	5.1	3.9	5.1	4.1
CD-33L	Peak 1	>14.4	<1.0	>18.1	<1.0	>16.5	<1.0
	Peak 2	16.0	1.8	13.5	1.6	15.7	3.2
	Peak 3	6.7	3.1	**	**		

Notes. * – By Weight of Cement. ** – it was not possible to identify this heat event.

3.3. Grout rheology

Fig. 4 shows the results of the flow cone tests. The flow time of the grout containing Sika Retarder increases with increasing addition level up to ~0.75% after which is decreases. For sodium gluconate, Viscocrete 3110 and CD-33L additions, the results show a decrease in flow time with increasing addition levels with the exception that for sodium gluconate additions beyond 0.6% BWOC there is an increase in flow time.

These results demonstrate the potential for reducing the water content in all grouts except for those containing Sika Retarder. Reducing the quantity of mix water in any grout will produce a less permeable matrix when hardened, which would provide better sealing of waste packages in a borehole against the ingress of saline groundwater.

A summary of the viscosity results for all grouts is shown in Fig. 5, which displays viscosity at each testing temperature when measured using a shear rate of 5.1 s^{-1}. This low shear rate was used to replicate the low shear experienced by the grouts after deployment in the borehole without any additional pressure applied by pumping.

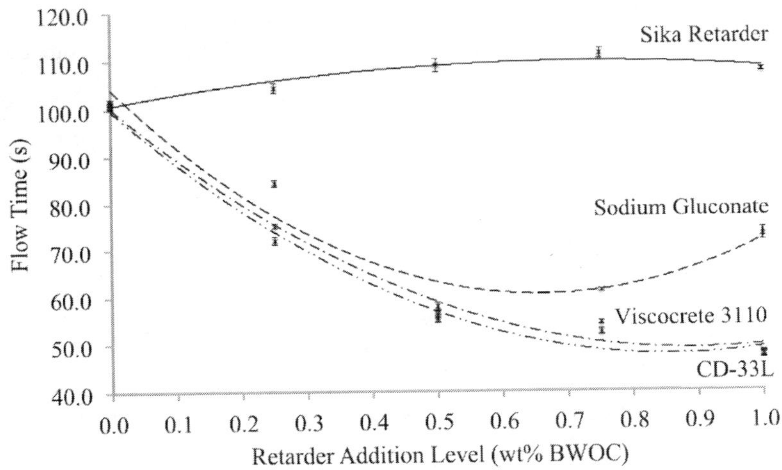

Figure. 4. Grout flow test results. Notes. The curves represent polynomial fits to the data, added as a visual guide only. Previously published data for Viscocrete 3110 and CD-33L (Collier et al., 2015b) are displayed here for comparative purposes.

The results show that at 22 °C, the viscosity of the grout containing either Sika Retarder or sodium gluconate is significantly higher than that of the control. The viscosity of the Viscocrete 3110 grout was similar to that of the control, while the viscosity of the CD-33L grout is lower. Increasing the testing temperature to 45 °C reduced the viscosities of all the retarded grouts to less than that of the control, particularly for the mixes containing Viscocrete 3110 or CD-33L. At the highest testing temperature (70 °C) the viscosities of all retarded grouts were very similar and were all significantly lower than the control, approximately 30 cP compared to approximately 70 cP respectively. The effect of temperature on the control grout was different to those containing Sika Retarder, sodium gluconate and Viscocrete 3110, but was similar to that containing CD-33L grout although no meaningful trend could be identified in terms of the effect of temperature on the viscosities of the control and CD-33L grouts. It should be noted that the flow results obtained when testing at room temperature are not always consistent with the viscosity results. A reduction in flow time, which should correlate with a reduction in viscosity, was not always observed when using a retarder, as with the flow and viscosity results measured at room temperature for the samples containing Sika Retarder, sodium gluconate and to a lesser degree Viscocrete 3110 (this anomaly is not observed at elevated temperature). This may be due to the differences in shear stresses applied to the grouts during the testing processes; the level of shear applied in the viscosity testing may be greater than that in the flow testing where the shear applied in the latter test only results from the gravitational flow of the grout itself. This suggests that the grouts exhibit shear thickening (rheopectic) characteristics.

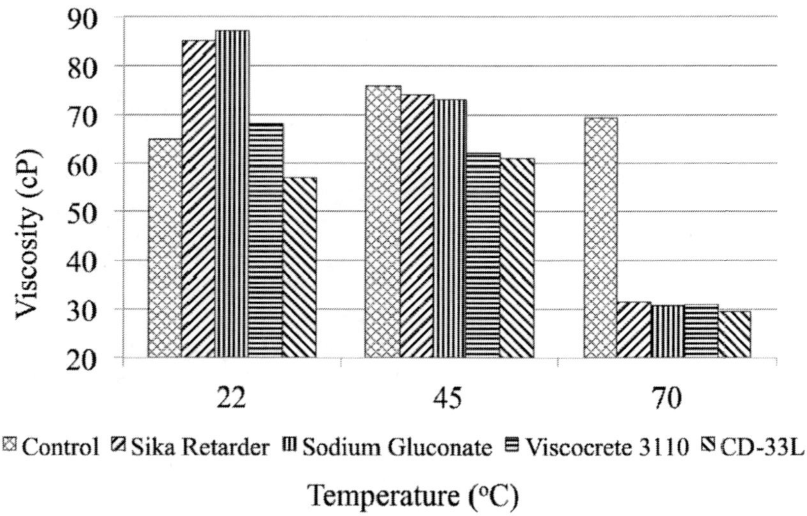

Figure. 5. Grout viscosity results measured using a shear rate of 5.1 s−1. Notes. Addition levels used were 1.0% Sika Retarder, 0.5% sodium gluconate, 1.0% Viscocrete 3110 and 1.0% CD-33L, all BWOC.

Any grout with t_2 consistency time greater than 4 h was poured through water held at 63 °C and its dispersion observed. Generally all the grouts flowed freely through the water with only a small amount of particle dispersion. Any particles that did disperse soon came to rest on the top surface of the settled grout.

3.4. Phase composition of hardened cement paste

The waste packages will be located in the lower 1–2 km of the boreholes (the disposal zone) and, having sealed the borehole above the disposal zone, any radionuclides released will take a very long time, depending on the route of migration, to return to the biosphere (of the order of a million years or more (Chapman and Gibb, 2003 ; Brady et al., 2009)). Consequently, the use of SSMs to seal individual waste packages within the disposal zone is not absolutely essential. However, any sealing of the packages which prolongs the life of the containers, and hence the primary containment, as well as restricting the migration path of any gaseous corrosion products, will be advantageous when making the safety case for the overall DBD process. This can be achieved by designing the grouting system to form stable and durable cement hydrate phases at the elevated temperatures.

Example XRD diffractograms and TGA/DTG traces for grout samples containing sodium gluconate are shown in Fig. 6 ; Fig. 7 respectively.

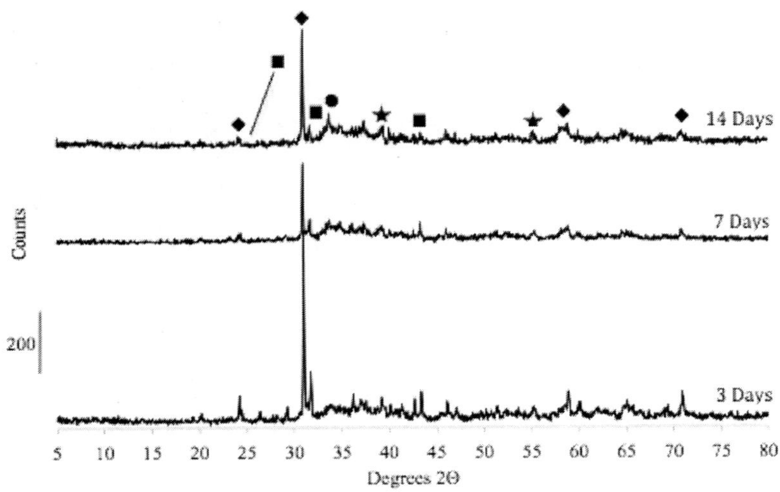

Figure. 6. XRD diffractograms for samples containing sodium gluconate. Notes. ◆ – quartz (SiO_2); ■ – α-C_2SH ($Ca_2(SiO_4)H_2O$); ● – tobermorite-11 Å ($Ca_5Si_6(OH)18.5(H_2O)$); ★ – brownmillerite ($Ca_2FeA_lO_5$).

The main crystalline phase identified in all samples was quartz (SiO2) from the unreacted silica flour. Two crystalline calcium silicate hydrates were identified by XRD, α-C_2SH ($Ca_2(SiO_4)H_2O$) and Tobermorite-11 Å ($Ca_5Si_6(OH)_{18}.5(H_2O)$), along with an amorphous material, usually referred to as C–S–H (Taylor, 1997), which was responsible for a "hump" centered at approximately 37° 2θ in all XRD traces. While crystalline portlandite ($Ca(OH)_2$) was not detected by XRD, the TGA/DTG results identified a loss centered at approximately 450 °C in all samples suggesting it is amorphous (Midgely, 1979), and this phase will carbonate as evidenced by the detection of calcite by TGA/DTG. A small quantity of unreacted brownmillerite (Ca_2FeAlO_5) was detected in each sample. The same phases were identified in each sample type, which suggests that none of the retarders influenced composition when added at the addition levels used here. Although not displayed here, there was no discernable difference in the compositions of the hardened cement pastes due to the presence of any quantity of each of the retarders. For each grout, the intensity of the quartz XRD reflections reduced slightly between 3 and 7 days suggesting the quartz had reacted slowly with time. α-C_2SH was identified in all samples cured for 3 and 7 days with the intensities of its main reflections reducing with time making detection after 14 days difficult. Tobermorite-11 Å was not detected in any sample after 3 days curing, but small XRD reflections were identified in all samples cured for 7 and 14 days (one of the strongest reflections for this phase at 9.2° 2θ was difficult to identify, but the other main reflections were identified). The quantities of the crystalline calcium silicate hydrate phases increased with time whilst the amount of portlandite decreased. These progressive phase changes follow those suggested in the literature with (1) quartz reacting slowly with calcium hydroxide to form calcium silicate hydrate phases, (2) the early formation of α-C_2SH, the quantity of which gradually reduces with time, and (3) the formation of tobermorite-11 Å following the decline of α-C2SH (Nelson and Guillot, 2006 ; Bensted et al., 2008). Although not presented here, TGA/DTG data showed that the quantities of the calcium silicate hydrate phases in samples containing any of the retarders were less than those formed in the control samples.

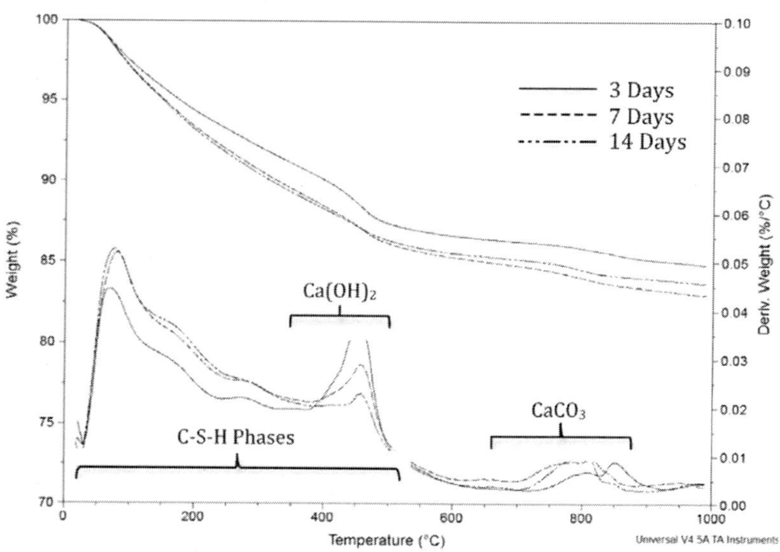

Figure. 7. TGA/DTG traces for samples containing sodium gluconate.

The phases formed in the hardened grouts after 14 days hydrothermal curing show no effect due to the presence of any retarder, and are typical of those found in oil and geothermal well cementing applications, which suggests a high level of durability for the grouting system.

4. DISCUSSION

4.1. Influence on paste properties

Temperature has a significant influence on the progression of consistency with time, with increasing temperature accelerating the onset of grout thickening in line with published results (Shariar and Nehdi, 2012 ; Zhang et al., 2010). Elevated pressure also accelerates grouts thickening, but not as significantly as elevated temperature (Scherer et al., 2010). How the grout will be placed in the borehole will require investigation, and should include assessment of whether cooling or agitation during deployment may be required.

The consistency and setting data obtained from this work demonstrates that cementitious grout formulations not too dissimilar to those used in oil well and geothermal well applications could be suitable for DBD. All of the retarders studied were able to delay the onset of thickening and setting, with sodium gluconate having the greatest influence. However, only sodium gluconate and Viscocrete 3110 appear to provide the level of retardation required for the DBD process up to 140 °C when addition levels similar to those recommended by the manufacturers were used. While Sika Retarder and CD-33L did not delay the onset of thickening enough for the DBD application, their use as plasticizers should not be discounted if they were used in combination with another retarder.

The torque required to stir the wet cementitious paste (which is measured during the consistency testing process) is directly related to consistency by a mathematical relationship described in BS EN ISO 10426 (BS EN ISO, 2009). We intend to exploit this relationship and extract other paste properties such as viscosity and yield stress in further work.

Although both the consistency and calorimetry data indicate that all four additives retard grout thickening time, it is difficult to correlate the two results because the experimental operating conditions are different; similar difficulties have been reported by other workers (Jupe et al., 2008). Consistency testing best replicates the DBD application of the fresh grout because the heating rate from room to test temperature, and the resultant increase in testing pressure, can be used to represent the change in temperature and pressure on the grout as it travels down the borehole before encapsulating the waste package.

The flow data for the grouts obtained at room temperature are useful in comparing the influence of retarder type, but ideally these data need to be obtained and compared at temperatures and pressures representative of those down the borehole. Additionally, the overall sealing of waste containers should be confirmed at temperatures and pressures representative of those in an actual disposal. The applicability of setting the LoP (70 Bc) as the limit in DBD for the ability of the grout to flow around waste containers can also only be assessed by carrying out waste encapsulation tests in conditions representative of those down a borehole.

The method of grout mixing, which is likely to take place at the wellhead, also requires investigation. Although the major oil well cement testing standard (BS EN ISO, 2009) stipulates that any grout testing should be performed using vertical mixers similar to that used here, the most common mixing systems used in the field for oil and geothermal well cementing applications are based on a jet mixing process where pressurized water is combined with cement powder to form a slurry (Nelson and Guillot, 2006). Assessment of the potential for using this type of jet mixing process, or any other type of process, in DBD and its subsequent effect on grout rheology, will be required. This is likely to be complex because of the differences in grout rheology during mixing (at atmospheric temperature and pressure) and during deployment in the borehole, and will need to be conducted at high temperature and pressure.

To provide a good seal for the waste packages against groundwater ingress, low permeability of the hardened grout will be required which means that a low grout water content and use of a superplasticiser may be necessary. However, this will also be governed by the ability to mix the grout under atmospheric conditions at the wellhead.

Further work is required to evaluate how the grout should be mixed and deployed, how readily it will encapsulate and seal waste packages down the borehole, and what its long-term durability will be in the DBD process. Because of issues associated with the difficulty of testing at high temperature and pressure, it is likely that small-scale trials of grout deployment and waste package encapsulation will be needed to fully investigate applicability.

4.2. Applicability to DBD

The results generated in this study demonstrate that the cementitious grout being developed can be used to encapsulate containers of HLW when located in the disposal zone. The grout will remain sufficiently fluid to flow around emplaced waste packages, will set within the desired time, and the cement hydrate phases formed will have high durability similar to those formed in hydrocarbon and geothermal well cementing applications.

Various DBD concepts are being developed internationally (Beswick et al., 2014 ; Brady et al., 2009), and some of these utilize materials based on well drilling fluids (mixtures of water and bentonite) to fill the annuli in the waste package disposal zone, but this use of such fluids provides very little package support and provides no barrier to corrosion or gas migration (Brady et al., 2009). By contrast, using a cementitious grout to fill these annuli has several advantages: 1) the waste packages will be physically supported during the disposal zone loading process which will prevent/reduce container buckling and reduce the subsequent risk of container breaching, 2) the hardened grout will provide a low permeability layer around individual waste packages restricting/limiting ingress by groundwater and delaying corrosion of the container, 3) the hardened grout will restrict the movement of any gases produced as a consequence of container corrosion when it does occur, and 4) cement hydrates have the potential to immobilize many waste ions (Gougar et al., 1996) by surface adsorption or lattice incorporation that may eventually be released from the waste packages.

The strength of the geological barrier in the DBD concept is multi-fold and is based on the provision of a longer waste ion return path due to the great depth of the borehole, the low hydraulic conductivity of the rock at depth, the groundwater density stratification and its long-term stability, and the isolation of the waste packages from near-surface processes like glaciations. The work presented here demonstrates that the cement grouts being developed can be used as SSMs to fill annuli in the disposal zone, which will add to near-field safety, augment the overall far-field geological barrier, and thereby enhance the overall safety provided by DBD.

5. CONCLUSIONS

The following specific conclusions can be drawn from this work:

- Sodium gluconate and Viscocrete 3110 offer the potential to provide the amount of retardation required for use in DBD and also increase fluidity. The quantity of sodium gluconate required to be added to the grout to ensure it retains the desired fluidity for 4 h at 90, 120 and 140 °C are 0.05, 0.25 and 0.25% BWOC respectively. Similarly, 0.75, 1.0 and 2.0% additions BWOC of Viscocrete 3110 are required to ensure sufficient fluidity for 4 h at 90, 120 and 140 °C respectively. At the addition levels recommended by the manufacturer, Sika Retarder only provides the desirable retardation properties at 90 °C and has little influence on fluidity. CD-33L increases grout fluidity and retards consistency, but the amount of retardation is insufficient for DBD applications at the addition levels recommended by the manufacturer.

- The phase composition of the hardened cement paste after 14 days hydrothermal curing is not affected when compared to the retarder-free control grout. All phases present in the hardened cement pastes are in line with those reported in the literature and have demonstrated durability at temperatures and pressures similar to those in the DBD application.
- In all grouts where retardation was greater than 4 h, final set is achieved within 24 h.

This work demonstrates that cementitious grouts similar to those used in oil well and geothermal well applications, where the onset of grout thickening and setting can be delayed for the required time period, would be suitable for application as SSMs in DBD up to temperatures of 140 °C. Using the grouts to fill annular space within the disposal zone will not only support waste packages during placement, but will also provide a low permeability layer around the packages which will ultimately enhance the safety case for disposing of high-level radioactive waste using this DBD concept.

ACKNOWLEDGMENTS

This work was financially supported by the UK Engineering and Physical Science Research Council (Grant number EP/K039350/1. The authors acknowledge Holcim, Sibelco and Sika for the supply of cement, silica flour and additives (UCS Pak and Viscocrete 3110) respectively, and Baker Hughes for the supply of CD-33L.

REFERENCES

1. ASTM C150/C150M, Standard Specification for Portland Cement, American Society for Testing and Materials.
2. ASTM C191-13, Standard Test Methods for Time of Setting of Hydraulic Cement by Vicat Needle, American Society for Testing and Materials.
3. ASTM C939-10, Standard Test Method for Flow of Grout for Preplaced-Aggregate Concrete (Flow Cone Method), American Society for Testing and Materials.
4. ASTM D7348-13, Standard Test Methods for Loss on Ignition (LOI) of Solid Combustion Residues, American Society for Testing and Materials.

5. Bensted, J., 2008. Development with oilwell cements. In: Bensted, J., Barnes, P. (Eds.), Structure and Performance of Cements, second ed. Spon Press, London.
6. Best, M.G., 2003. Igneous and Metamorphic Petrology. Blackwell Science, Malden, Massachusetts, USA.
7. Beswick, J., 2008. Status of Technology for Deep Borehole Disposal, NP 01185. Report by EPS International for the Nuclear Decommissioning Authority, UK.
8. Beswick, J., Forrest, J., 1982. New low-speed high-torque motor experience in Europe. In: Proceedings of the SPE Annual Technical Conference, New Orleans, LA, USA.
9. Beswick, J., Gibb, F., Travis, K., 2014. Deep borehole disposal of nuclear waste: engineering challenges. Proc. ICE Energy 167, 47-66.
10. Brady, P.V., Arnold, B.W., Freeze, G.A., Swift, P.N., Bauer, S.J., Kanney, J.L., Rechard, R.P., Stein, J.S., 2009. Deep Borehole Disposal of High Level Radioactive Waste. Sandia Report SAND2009-4401. Sandia National Laboratories, Albuquerque, New Mexico, USA.
11. BS EN ISO 10426-10431:2009, 2009. Petroleum and Natural Gas Industries, Cements and Materials for Well Cementing, Specification. British Standard Institute.
12. Chapman, N., Gibb, F., 2003. A ruly final waste management solution e is very deep borehole disposal a realistic option for high-level waste or fissile materials? Radwaste Solut. 10, 26-37.
13. Collier, N.C., Sharp, J.H., Milestone, N.B., Hill, J., Godfrey, I.H., 2008. The influence of water removal techniques on the composition and microstructure of hardened cement pastes. Cem. Concr. Res. 38, 737-744.
14. Collier, N.C., Travis, K.P., Gibb, F.G.F., Milestone, N.B., 2015a. Characteristics of cementitious paste for use in deep borehole disposal of spent fuel and high level wasteforms. In: Scientific Basis for Nuclear Waste Management XXXVIII. Materials Research Society Symposium Proceedings, vol. 1744. http://dx.doi.org/10.1557/opl.2015.314.
15. Collier, N.C., Travis, K.P., Gibb, F.G.F., Milestone, N.B., 2015b. Cementitious grouts for disposal of nuclear wasteforms in deep boreholes. In: Proceedings of American Nuclear Society International High-level Radioactive Waste Management, Charleston, SC, USA, April 12-16.
16. Exxon Neftegas; Sakhalin-1 Drilling Programme Marks 10th Anniversary with World Records, October, 2013; Upstreamonline Website; http://www.upstreamonline.com.
17. Gibb, F., 2010. Looking Down the Bore. Nuclear Engineering International, pp. 21-22. February.
18. Gibb, F., McTaggart, N., Travis, K., Burley, D., Hesketh, K., 2008. High-density support matrices: key to the deep borehole disposal of spent nuclear fuel. J. Nucl. Mater. 374, 370-377.
19. Gibb, F.G.F., Travis, K.P., Hesketh, K.W., 2012. Deep borehole disposal of higher burn up spent nuclear fuels. Min. Magn. 76, 3003-3017.
20. Gougar, M.L.D., Scheetz, B.E., Roy, D.M., 1996. Ettringite and CeSeH Portland cement phases for waste ion immobilization: a review. Waste Manag. 16, 295-303.
21. Juhlin, C., Sandstedt, H., 1989. Storage of Nuclear Waste in Very Deep Boreholes, Feasibility Study and Assessment of Economic Potential. SKB Technical Report 89–39, Stockholm, Sweden.
22. Jupe, A., Wilkinson, A., Luke, K., Funkhouser, G., 2008. Class H cement hydration at 180 C and high pressure in the presence of added silica. Cem. Concr. Res. 38, 660-666.
23. Midgely, H., 1979. The determination of calcium hydroxide in set portland cements. Cem. Concr. Res. 9, 77-82.
24. Nelson, E., Guillot, G. (Eds.), 2006. Well Cementing, second ed. Schlumberger, USA.
25. Scherer, G., Funkhouser, G., Peethamparan, S., 2010. Effect of pressure on early hydration of Class H and white cement. Cem. Concr. Res. 40, 845-850.
26. Shariar, A., Nehdi, M., 2012. Rheological properties of oil well cement slurries. Proc. ICE Constr. Mater. 165, 25-44.
27. Taylor, H.F.W., 1997. Cement Chemistry, second ed. Thomas Telford, London, UK.
28. Zhang, J., Weissinger, E., Peethamparan, S., Scherer, G., 2010. Early hydration and setting of oil well cement. Cem. Concr. Res. 40, 1023-1033.

CHAPTER 6

The Future Role of the Nuclear Energy in Brazil in a Transition Energy Scenario

Fábio Branco Vaz de Oliveira, Kengo Imakuma, Delvonei Alves de Andrade

Nuclear and Energy Research Institute, Cidade Universitária, São Paulo, Brazil

ABSTRACT

This paper discusses and presents figures about the future power consumption in the world and, especially in Brazil, based on the current world and Brazilian's energy scenarios. Emphasis is given to the scenarios of nuclear power and uranium resources demand. A discussion on the future roles of thorium and uranium fuels in the replacement of the traditional resources like oil and gas is also presented, as it is the role of the new nuclear power plants, planned to be built in a short term time horizon. This paper considers two different indexes for future projections, and the results obtained indicated a strong dependence on them. The time horizon for the analysis was fixed on the time estimated for Brazil to reach its maximum in population, and parameters evaluated were taken from the Brazilian's governmental and world data on the population growth, energy consumption and energy consumption per capita. Calculations show that the power consumption projections for Brazil, for the adopted time horizon and working with global indexes, become overestimated, when compared with the results considering the national indexes. According to our approach, power consumption estimates using global indexes becomes approximately 4.5 times higher than the estimates presented by the Brazilian indexes. This was the motivation to the discussion between the

Brazilian and world energy demand scenarios, and also the roles of nuclear energy in the future transition from the current conventional to alternative sources.

Keywords: Nuclear Energy, Renewables, Future Energetic Resources, Uranium, Projections

1. INTRODUCTION

The shortage of the most common used energy resources, such as oil, coal and water, and their growing consumption, together with the population growth and global warming, are phenomena clearly affecting the planet. The recent formation of tornadoes in the coast of Santa Catarina, which never happened or had been reported in all Brazilian history, the excessive melting of the polar ice caps, for example, can be foreshadows of a large scale global manifestations of uncontrolled and unpredictable climate changes, due to the exaggeration of the consumption of natural resources. Also, the increasing gap between economical and social classes and the pressure it exerts in the production of more expensive products, which also generates an enhancement in the energy consumption, clearly become an important issue on the countries future energetic planning. Besides the social and economical needs, planning must take into consideration the structure of the national energy matrix, and has to be based on what it can provide to support the country's growing perspectives. In this scenario, studies concerning the use and improvement in the technologies related to renewable or clean energies are growing and nuclear energy, among them, plays an important role.

Energy from renewable and nuclear resources, are considered candidates to reduce an incoming world energy crisis. Nuclear energy is one of the most promising candidates, despite the recent accident in Fukushima, since the technology is well developed and has a strong knowledge basis. Some countries are declining to build new nuclear plants. Nonetheless, it is interesting to note, for example, the acceleration of the construction of more secure power plants and the adequacy of the older ones to the new and more stringent security criteria in the world and in particular in United States. It shows a pressure in the production of energy, at least for this country. This can be an image of what might occur in the world in terms of energy consumption, but in a lesser scale, since United States, together which China, are the countries exerting the highest pressure in the consumption of the global energy resources [1] making their demands for energy more urgent. An argument which favors the utilization of nuclear energy is given, for example, in the work of Haratyk and Forsberg [2]. They devised a coupling between nuclear energy and renewable facilities. According to them, renewables cannot supply the increasing demand of energy in the short term, given the current countries technological status.

Thus, it is expected that nuclear energy will replace in a large scale and in a short term, the conventional resources, increasing its participation in international and national scenarios of energy demand. But it cannot be accomplished without a deep study about the future supply of nuclear fuels, mainly uranium and, for a more advanced perspective, thorium.

Both minerals are abundant in Brazil, and for the national projections, uranium is available in sufficient amounts to supply the four planned PWR reactors, to be constructed up to 2050 [3]. Even having the technology knowledge of all the fuel cycle, the energy matrix in Brazil is mainly hydroelectric and nuclear represents 3%. In the world, nuclear energy is about 17% of the total generated electricity, and about 5.4% of the total world energy consumption. However, in a transition energy scenario, even with the recent discoveries of oil reserves in pre-salt formations in the Brazilian coast, having a potential to increase power in our energy matrix, with the construction of the Belo Monte hydroelectric power facility, abundance of solar radiation over the entire year covering all our territory, etc., nuclear power could be a short term solution of a possible energy supply delay, expected for the years to come, up to the development of the renewables technology to reach efficiency levels, suitable for commercial exploration. Besides, the development of the technology of the extraction and purification of the small amounts of uranium from the recent discovered phosphatic deposits [4], which covers almost one third in area of the Brazilian territory, will improve the reserves and also the capabilities of the expansion of its nuclear and, thus, energetic matrix.

The construction of nuclear power facilities is also a matter of public discussions, mainly concerned about the safety requirements and accident risks, which rises again after the recent Fukushima's accident. From the arguments above and the discussions to be presented in the next items, it is easy to see and, according to our opinion, that Brazil has a privileged position in terms of the availability of energy resources, including uranium. This privileged position could be used to overcome public rejection and to support the acceptance, even in the long term, to the Brazilian's initiative in the direction of the construction of new nuclear power facilities in Brazil, since massive investments in nuclear power, like those in United States and China, are not planned [3]. However, for a long term planning, energy consumption estimates must be calculated on reliable basis, emphasizing both the global and national scenarios in the current energy context.

This paper analyzes and estimates the future power consumption in the world and in Brazil, based on the current worlds and Brazilian's energy scenarios, with emphasis to the scenarios of nuclear power and uranium resources demand. A discussion about the future roles of thorium and uranium fuels in the replacement of the traditional resources like oil and gas, the advantages of thorium in terms of energetic efficiency, is also presented. As a consequence of the uses and availability of the nuclear fuels, the role of the new nuclear power plants, planned to be built in a short term time horizon, is also discussed in this same scenario. Emphasis was given to the called 2DS scenario [5], which means a scenario where it is predicted a minimum of 2°C for the increasing in the global mean temperature, up to 2025. Actions and plans are under discussion, each one of the possible new technologies having specific advantages and drawbacks [5].

The comparison presented in this paper pointed out a strong dependence on the two different indexes, used for future projections. Global indexes were extracted from the projections made in the Tomabechi's work [6] for the world consumption, which deals with the relation between energy consumption indexes and world's population growth as the parameter to be evaluated. These results were extended to make projections for the Brazilian case. Brazilian indexes were based on the national projections for the population growth, listed by IBGE [7] and were used for the future projections. To check for the importance of domestic scenarios in the projections, graphics are given to show the evolution and to compare global and domestic parameters such as population growth, power consumption per capita and total power consumption in some countries. The time horizon for the analysis was fixed on the time estimated for Brazil to reach its maximum population, and parameters evaluated were taken from the Brazilian's governmental and world data on the population growth, energy consumption and energy consumption per capita.

This was the motivation to both discussions carried out here, one associating Brazil within a global transition scenario, and the other which dissociates our country from the global considerations, discussing our future projection results in a domestic perspective. Since nuclear is the prompt alternative instead of the undeveloped renewable technology, it is also discussed its role as the main alternative to the replacement of the current conventional ones.

2. NUCLEAR ENERGY IN A TRANSITION ENERGY SCENARIO

Concerning today's debates regarding energy demand and global warming, a transition energy scenario, according to reference [5], means that world needs to change from the prevalent high-carbon/high-pollutant technologies to the incoming low-carbon/low pollutant ones, from the exploitation of the energy resources through the generation of energy. However, considering the phenomena of the development of those new technologies as a dynamical process, demanding time and consuming resources (energetic and monetary), it is known that it usually does not evolves with time at the same rate as that necessary to overcome the problems for the today's climate changes and energy consumption.

Since earth is a dynamical system, it is continuously changing, through changes in its observable properties with time. However, since the time human beings are exploring its resources, they become one of the major causes of changes on earth's properties, the most sensitive of them and most easily visualized and studied are those related to weather. The problem is, at the today's rates of consumption, these changes are possibly leading humanity to an irreversible path and, the worst of all, in a short period of time. Technology and human intelligence solved several problems of our everyday life, enabling us to go to the moon, to explore planets, the universe and the subatomic world, to defeat diseases, communication problems, etc., but its side effects is presenting itself as a very harm one.

Based on the current status of the renewable technology development and the growth of the mean earth's temperature, scientists devised 3 possible states for the future global weather, based on 3 possible values for the increase in the mean earth's temperature. Based on that, governments can take action plans to deal with these three possible scenarios; changes in the energy policy are necessary to be planned, defined and studied to provide approaches for the solution of the incoming energy problems. The above scenarios go from the minimum to the maximum harm for the living beings in the planet. They are called XDS scenarios, where X = 2, 4 or 6 and stands for the expected increase in the mean earth's temperature, in degrees C, for a fixed time horizon of 2025.

It is pointed out by reference [5] that energy technologies like solar and photovoltaic are promising to reach the 2DS requirements, but the ones with greatest potential to reduce carbon emissions, like the nuclear, are making the slowest progress, since security is the main topic of interest for the public acceptance and for its full utilization. It is expected a growth of 600 GW of nuclear installed capacity to reach the 2DS requirements, but after the Fukushima's accident it is also expected a deployment in 100 GW, becoming this process a complicated one.

For the projections of the future of the energy supply by nuclear energy, the repercussion of the 011 accident was fundamental, leading United States to review the current security criteria for the construction of the 22 planned new nuclear power plants, which are based on the intrinsically safe AP1000PWR design, mostly from 100 MW to 300 MW power. Despite of its safety aspects, however, some authors still criticize its design, like Piore [8], who emphasizes that the security factors for the pressure vessel's project is undersized. For the breeder reactors, another technology being considered which enables the extension of the fuel's lifetime, Cochran [9] emphasizes the problem of the isolation of sodium and water lines, critical due to the risk of explosions. Those are issues that testify against the use of such technologies.

However, according to the World Nuclear Association data [10], in addition to the today's 441 nuclear reactors in the world, more 338 nuclear reactors are planned to be constructed, four in Brazil [10]. Based on the same report [10], in 2011 83 reactors were under construction, most of them are PWR, according to **Table 1**.

On June 1st, 2010, Brazil restarts the construction of the Angra III nuclear power reactor and, together with the planned four, as stated recently in 2012. Our supply of nuclear power will increase in approximately 5300 MW. More ambitious construction plans are presented by China, Russia, United States and Ukraine which are planning to build, respectively, 115, 40, 30, 28 and 20 more nuclear power reactors, a total of 223, mostly based on the LWR technologies. Thus, despite the accident and the rising of the unfavorable public opinion against the construction of new nuclear power plants, it is observed [5] that most of the countries kept their nuclear programs active. But the only factor that has changed is the speed on which the new facilities are planned to be constructed.

Another favorable point for the installation of new nuclear plants was devised by WEC [11], also having as basis that this technology is well developed. In this document it was suggested that, to accelerate the transition from Green House Gas, GHG, to non-GHG technologies, nuclear expertise could be shared from the nuclear developed countries to the non-developed ones. According to WEC [11], this could be another way of global scale cooperation for the reduction of GHG emissions and to develop the 3rd world technological infrastructure, aiming the minimization of the time for an undeveloped country to reach an acceptable level of technology to deploy clean energy technologies. Also, [11] states that new rules for patent interchange and utilization must be redefined, for the humankind to adapt to the new survival requirements. Sharing costs could be viewed by the leading technological countries in the world as a logical path. Since global warming and energy consumption is a world problem, large-scale technology as well as costs should be diffused at same basis in order to share a solution in the direction of the reduction of the global costs of the pollution through the deployment, for example, of the clean energy technologies. This is an important point of debate in the more recent seminars on greenhouse gas emissions and the future of the climate of the planet

We have to keep in mind that important differences arise if the scenario for the estimates is taken locally. For example, countries with high number of nuclear reactors, large populations and with large energy consumption, like United States, China, European Union, etc., global indexes obviously do not reflect the truth about the national estimates for these countries, since they pull the mean world values of consumption/demand of electricity upwards. For example, in terms of the greenhouse gas emissions, it is observed that 50% of the total is due to oil and coal usage, mainly from United States and China. Thus, there is a mistake of sharing the indexes/costs. It is suggested here a criterion for the evaluation of a real "contribution" of a country for the global energy consumption.

For Brazil, the technology for clean energy deployment is still in a growing phase for some of the renewables, in terms of the domestic energy supply index. According to Cerri [12], 41% of Brazil's total domestic energy supply comes from renewable resources, against 14% as a mean value in the world and 6% in industrialized countries.

Table 1. Nuclear reactors under construction or "almost so", by design [10].

Reactor Design	Quantity
Fast Breeder Reactor	02
Pressurized Water Reactor	61 (China = 26)
Pressurized Heavy Water Reactor	11
Advanced Boiling Water Reactor	08
High Temperature Gas Reactor	01
Total	83

This outcome implies a reduced dependence for imports of energy sources, and an advantage in terms of energy supply for the future, even its nuclear park representing only 3% of the total energetic production. Thus, the use of local indexes could be a better choice, for a better accuracy for projections of energy consumption. But it is important to observe that Brazil holds the position to be one of the top green-house gas emitters in the world, mainly due to the deforestation to livestock and agricultural uses [12].

3. THE URANIUM SCENARIO

To see how uranium resources can provide a short term solution to face a possible energy supply crisis, it is convenient to compare the world's uranium reserves capabilities with those of oil and coal, the most used natural resources in the world.

For generating 375 GWe, the corresponding electricity generated by nuclear fissions, it is required 68,000 t of uranium per year. According to **Table 2** [13], and ignoring the so called secondary sources of uranium (uranium from nuclear weapons, reprocessing, etc.), if the total estimated amount of uranium metal is 5.5 Mt, at the above rate of consumption the uranium reserves will last 80.8 years. This amount is enough to reach the so called 2DS objectives up to 2025 [5], where the required power related to nuclear energy is about 600 GWe. The difference of 225 GW corresponds to 40,000 t of uranium per year, an enough amount to cover 2DS requirements. As explained below, the Fukushima accident changed some policies related to the construction and operation of new nuclear power plants, the value of 600 GWe must be reduced for this reason.

In terms of oil, based on the current consumption rate, 8.2×10^7 barrels/day, and considering it as constant, a non-realistic hypothesis, its length can be calculated in almost 41 years. United States is the planet's largest consumer of this resource, with approximately 1.9×10^7 barrels/day, nearly 25% of the world's consumption, and Brazil appears in seventh place, 2.5×10^6 barrels/day. Regarding coal, China is in the first place, with 1.31×10^9 tons/year, followed by the United States, with 1.06×10^9 tons/year, and together account for 50% of the world's total consumption, and also for the corresponding greenhouse gas emissions. Brazil appears in 19th place, with 2.3×10^7 tons/year, twice orders of magnitude less than the previous two [14]. In terms of proved coal reserves, considering that there are around 522 billion tons, the current consumption rate of 4.59×10^9 tons/day [14], and supposing that rate remains constant for the following years, we can estimate that the reserves will last 113 years. In both cases, uranium can be seen as the readiest resource for their possible replacement. For example, as previously mentioned, primary resources of uranium will end, at the current consumption rate, in about 80 years, in the mean value, accounted for the world, not for a particular country figure.

Table 2. Known recoverable resources of uranium (Reasonably Assured Resources plus Inferred Resources, to US$ 130/kg U, 1/1/09, from OECD NEA & IAEA [10]).

Country	Tons of uranium	% of world's total
Australia	1,673,000	31.0
Kazakhstan	651,000	12.0
Canada	485,000	9.0
Russian Federation	480,000	9.0
South Africa	295,000	5.5
Namibia	284,000	5.0
Brazil	279,000	5.0
Niger	272,000	5.0
United States	207,000	4.0
China	171,000	3.0
Jordan	112,000	2.0
Uzbekistan	111,000	2.0
Ukraine	105,000	2.0
India	80,000	1.5
Mongolia	49,000	1.0
Other countries	150,000	3.0
World total	5,404,000	100%

For Brazil, with the estimated growth scenarios for our nuclear matrix [15], from the current 1.95 GW to 3.5 up to 7 GW, our reserves will last 120 to 240 years (reserves of uranium considered at a total cost of less than US$ 130/kg, the criteria for an economical exploration of an uranium mine) and from 200 to 400 years, taking into account also the inferred reserves. A safer estimate, which predicts a maximum percentage of the total electricity generated by the Brazilian nuclear matrix of 5.7%, excluding the estimated by INB [4] inferred reserves of 800,000 tons, the duration of our uranium resources is, in the worst scenario, about 90 years. Taking into account these reserves, together with the 279,000 ton of uranium metal, the estimates are for the operation of 10 nuclear reactors like Angra III, for 200 years [16]. The problem is that most of the uranium in the inferred reserves comes from phosphates, and the technique for uranium extraction from phosphates is still uneconomical. The same problem in terms of the costs/economy of extraction is presented with the uranium obtained from sea water, whose amount is estimated in 3.3 ppb [6].

Since most of the energy generated in Brazil is from hydropower, and since the water resources are now limited due to considerations of preservation and other environmental aspects [16], to cope with the hydropower restraint in terms of space, nuclear expansion in terms of the self-sustained scenario would be helpful, according to the data presented in **Table 3** presented in the next item.

Thorium is another source of fissionable isotope, U^{233}, its reserves are estimated to be 3 times higher than uranium's. In the work of Ashley [17], it was stated that the amount of thorium as a by-product of rare earth processing would be enough to feed 200 nuclear ADTR ("Advanced Thorium Reactor"), without the need to open new mines for exploration and extraction, thus with no initial investments. Also, other advantage of thorium compared to uranium is its total usage as nuclear fuel as extracted from the mines, instead of the 0.7% relating to the fissile isotope U^{235}, present in U^{238} matrix. It means, according to Ashley [17] that for each GW of electricity produced, only 1 ton of mined thorium is necessary, whereas 200 tons of uranium from the mines or 3.5 Mt of coal would be required to produce the same amount of electricity.

Although there are still doubts about the real need for the construction of new nuclear power plants in Brazil, the debate is still open, and the numbers shown in the tables and the scenarios presented above are favorable. Considering the capabilities of the Brazilian nuclear reserves, it will be described and simulated, in the nextitems, some possible scenarios for its utilization in relation to two parameters, among the huge numbers of variables that usually affect a more complete analysis, population growth and energy consumption.

The scenarios for the national analysis were taken from the Plano Nuclear Brasileiro 2030 ("Brazilian Nuclear Plan 2030") [18], from EPE and Eletronuclear. A comparison will be made with results found in recent literature.

4. PROJECTIONS: ENERGY CONSUMPTION AND DEMAND, AND POPULATION GROWTH

Since population and the energy consumption growths are some of the most significant parameters to pressure energy demand, the projections for the Brazilian population growth, presented by the data of IBGE [7] in **Figure 1**, are firstly analyzed before the estimates of energy needs. Other sources for population growth projections are available, like the CIA fact book [1]. According to IBGE, the maximum in Brazilian population will occur in 2039 (219,124,700 inhabitants). However, between 2020 and 2030, the limits of the proposed energetic planning horizons [3] [19], our population will have approximately 212 million inhabitants.

Table 3. Electricity consumption generated via nuclear power, for inhabitant year, according to the scenarios of the PNB 2030 and using Equation (2).

Scenarios of growth	Year 2030 [MJ/(inhab. year)]	Year 2040 [MJ/(inhab. year)]
Minimum (2.7%)	190.89	200.34
Development (4.2%)	296.94	311.64
Self-sustainable (5.7%)	402.99	422.94

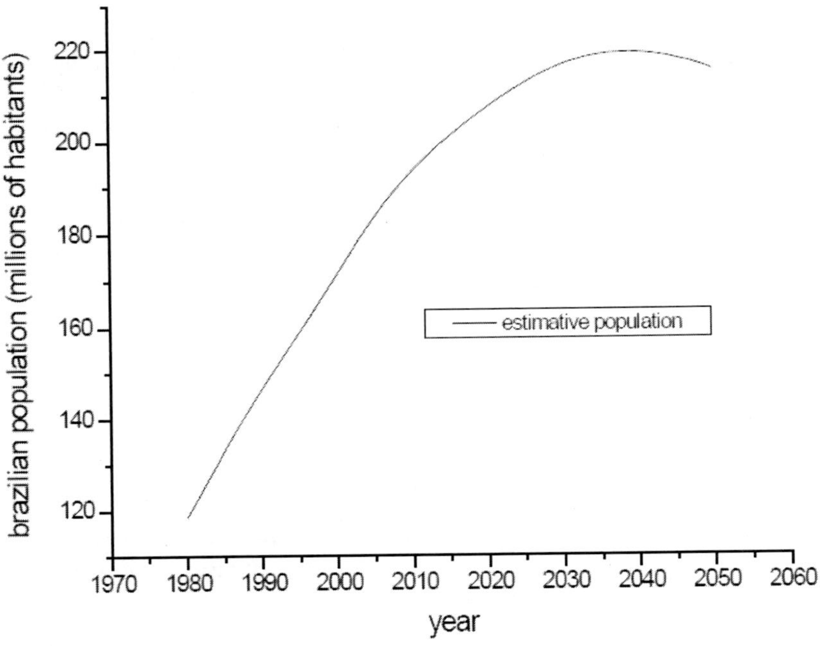

Figure 1. Brazilian population projection, up to 2050 [7].

The parameter energy consumption growth, would give accurate predictions if analyzed for each country, instead of taking the estimates based on global indexes. To put both parameters together, we observe that Iceland has the highest index of energy consumption per capita in the world, but its population is below 5 million people, meaning its contribution to the total world energy consumption is low. Countries with the highest energy consumption per capita are China, the United States, Japan, Russia, India, Canada, South Korea, South Africa, Australia, and the European Union. Taking together population growth and absolute population, these countries are not the first ranked ones (in terms of growth rate, East Timor, with 4.5% by year, is in the first place [1]), but pressure the energy consumption by their high number of inhabitants and also by their high energy consumption per capita. The Brazilian energy consumption per capita can be seen in **Figure 2**. The consumption per capita for the most populated countries is given in **Figure 3**, to acquaint for the differences mentioned in the introduction.

In terms of world energy consumption, it is important to observe that countries of the Europe and North America have lower population growth rates, but their energy consumption is about 1 order of magnitude higher than the mean world energy consumption. This is shown in **Figure 4** and **Figure 5**.

In a recent paper from Tomabechi [6], it was estimated that when the world population reaches $\mathbf{N_{hm}}$ = 10 billion inhabitants, around the next 30 to 50 years, according to the current global mean population growth of 81 million people/year, the energy demand will be $\mathbf{C_m}$ = 2 ZJ/year (where 1 Z = 10^{21}). This number was defined, according to the authors, based on data about the energy resource consumption of developed countries, extended to all the countries of the world. Thus, the estimated energy world consumption per inhabitant ("per capita"), $\mathbf{C_{pc}}$, at the time of $\mathbf{N_{hm}}$, would be:

$$C_{pc} = C_m / N_{hm} = \left[2 \times 10^{21}\,(\mathrm{ZJ/year}) / 10^{10}\,(\mathrm{inhabitants})\right] = \mathbf{200\ BJ/(inhabitant \cdot year)} \quad (1)$$

where BJ = 10^9 J = billion Joules, for a hypothetical scenario which predicts equality in the future energy consumption pattern for all the world's population. It is shown in the following graphs that the developed countries are in general the largest energy consumers, and having the highest energy per capita consumption rates, just to confirm that global indexes do not represent local realities.

According to the indexes presented by Tomabechi [6], to start the projections of **Nhm**, since nuclear power accounts for approximately 5.4% of the total energy consumption in the world [1], the demand for this resource would be:

$$C_{pcnuke} = 10.8\ \mathrm{BJ/(inhabitant\ year)} \quad (1')$$

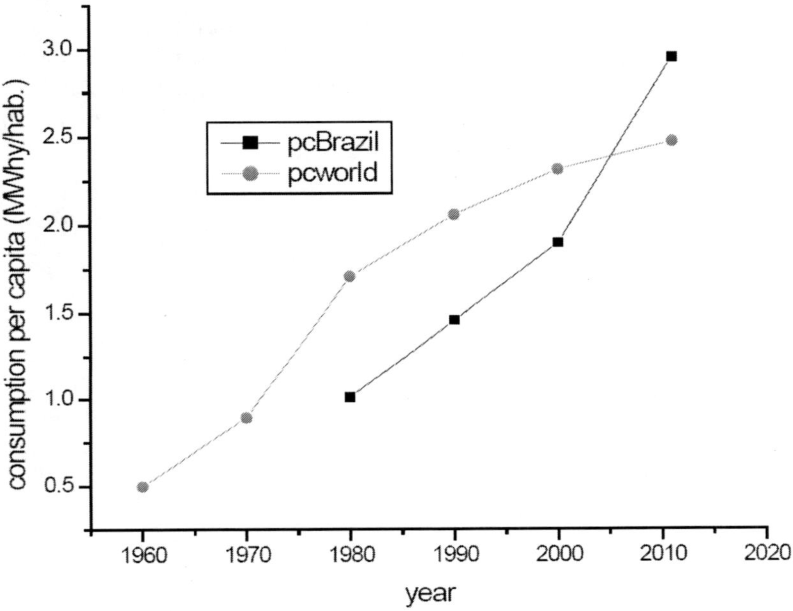

Figure 2. Consumption per capita, Brazil.

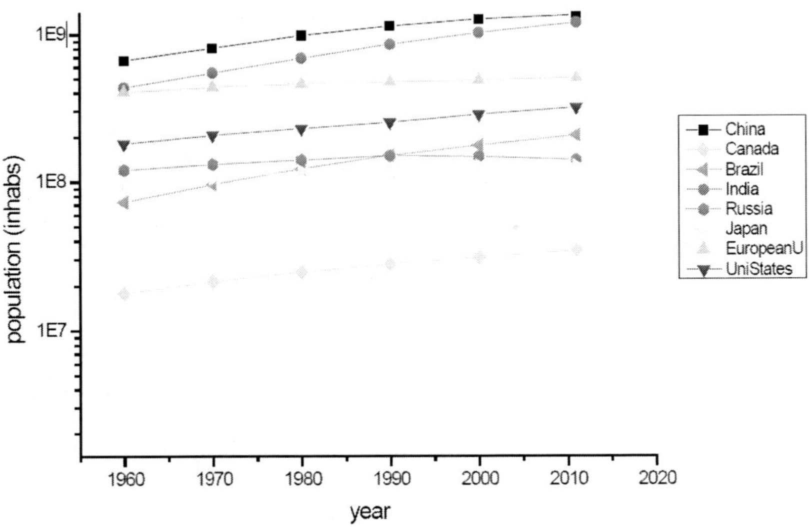

Figure 3. Population growth of the most populated countries.

From the IBGE data [7], in 25 years, the Brazilian population $\mathbf{N_{hBr}}$ will reach its maximum at 219 M inhabitants (**Figure 1**, where 1M = 106), at about the same time the world population is estimated to reach $\mathbf{N_{hm}}$. It can be predicted that, taking into account the above hypothesis, the Brazilian energy consumption per year, $\mathbf{C_{Br}}$, will be:

$$C_{Br} = Cpc \cdot N_{hBr} = 200 \times 10^{9}\,(\text{J/inhabitant year}) \times 219 \times 10^{6}\,(\text{inhabitant}) = \mathbf{0.0438\ \ ZJ/year} \quad (2)$$

corresponding to 2.19% of the global energy needs.

From an estimate based on the data of the 2005 Brazilian National Energy Balance presented in [15], assuming a constant growth rate of 0.67% per year in relation to the total energy consumption, and 0.64% relating to the electricity consumption, some estimates can be carried for the total energy consumption per inhabitant in Brazil, $\mathbf{C_{pcBrtotal}}$.

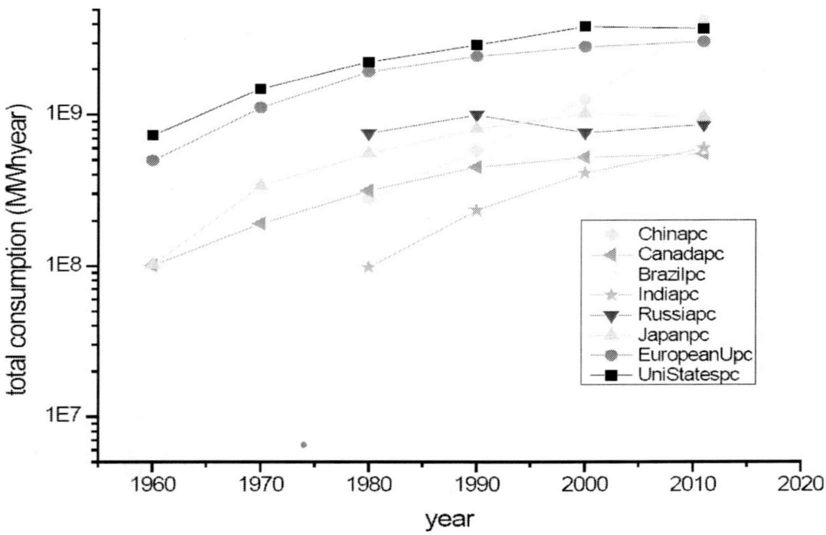

Figure 4. For the same countries of **Figure 3**, total energy consumption.

First, for the total energy consumption, the result is, for 2030:

$$\mathbf{C_{pcBrtotal}} = \left[9.518 \times 10^{-3}\,(\text{ZJ/year})/216{,}010{,}430(\text{inhabitants})\right] = 44.1\ \text{BJ}/(\text{inhabitant year}) \quad (3)$$

and for 2039:

$$C_{pcBrtotal} = \left[1.0175 \times 10^{-2} \left(ZJ/year\right)/219,124,700\left(inhabitants\right)\right] = 46.4\ BJ/\left(inhabitant\ year\right) \quad (4)$$

both results nearly 4.5 times lower than the value of C_{pc}, predicted according to Tomabechi methodology [6], and to the relation in Equation (2). This result shows that the differences can be very high if the estimates are taken from global indexes, which are average values, justifying the use of domestic indexes for the estimates. The goal is not to criticize or favor any methodology, but just to point out that there is a need to present them comparatively, if working together on the reduction or sharing pollution/clean energies deployment costs.

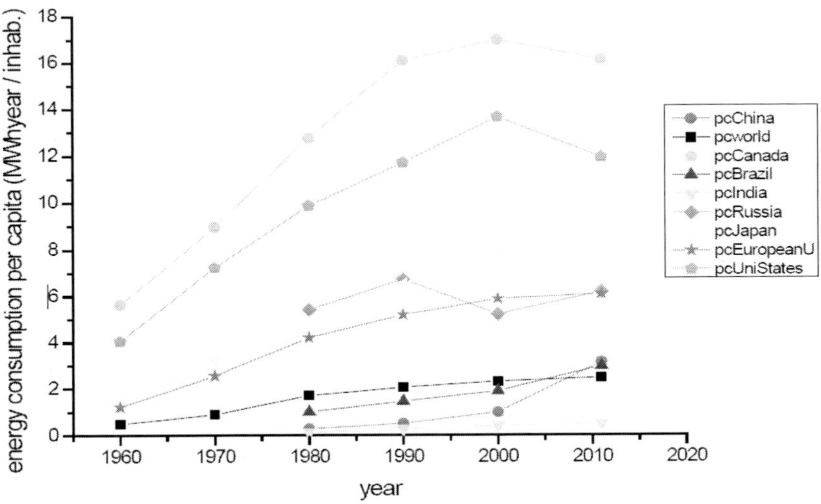

Figure 5. For the same countries, energy consumption per capita.

Taking into account a Brazilian scenario provided by EPE/Eletronuclear [15] about the future participation of the nuclear power in the energy matrix in Brazil, and the INB [4] projections for the Brazilian uranium supplies, its future contribution to the energy demand can be estimated, and the results compared to those from Equation (2).

According to the scenario devised in the EPE document [15], nuclear energy participation in the national electricity demand would grow from the current 2.5% (from a current national total of 78 GW) to 2.7%. Thus, it will be necessary one more nuclear power plant with capacity of 100 to 300 MW and the nuclear matrix will grow from 1950 MW to a maximum of 3550 MW, considering also the construction of Angra III. This work also stated that, in a so called simple development scenario, the estimated participation of nuclear would be 4.2%, comprising the construction of one more nuclear power plant 1300 MW power, (together with Angra III) and two other nuclear power plants of 300 MW each, adding 3200 MW to the current 1950 MW, and making up a total of 5150 MW. Finally, in a so called self-sustained development scenario, more interesting to the safe future supply of energy for the next generations, an increase of 5.7% would be supported by the construction of two new nuclear power plants 1300 MW power each (together with Angra III), and four more new modular stations 300 MW power each. The electricity generation capacity via nuclear power would grow from the current 1950 MW to 6950 MW.

Thus, the participation in the energy consumption in Brazil per inhabitant in relation to nuclear energy can be estimated according to those three scenarios, and the results are shown in **Table 3**, where MJ is equivalent to millions of joules.

Using the assumptions described above for the consumption rates for the total energy and for the energy from electricity, and the values obtained by the relations (3) and (4), we can obtain the electricity consumption values projected for 2030 (16.04%) and 2039 (15.88%). Thus, for 2030, the result of the Brazilian projected electricity consumption demand, $\mathbf{C_{pcBrel}}$ would be:

$$C_{pcBrel} = \left[9.518\times10^{-3}\,(\mathrm{ZJ/year})/216{,}010{,}430\,(\mathrm{inhabitants})\right]\times0.164 = 7.07\,(\mathrm{BJ/inhabitant\ year}) \quad (5)$$

and for 2039:

$$C_{pcBrel} = \left[1.0175\times10^{-2}\,(\mathrm{ZJ/year})/219{,}124{,}700\,(\mathrm{inhabitants})\right]\times0.158 = 7.37\,(\mathrm{BJ/inhabitant\ year}) \quad (6)$$

Considering the self-sustained scenario, electricity generation via nuclear power in the years of 2030 and 2039 in Brazil would be approximately 9×10^{-5} ZJ/year, corresponding to 7.1 GW of projected power.

5. CONCLUSIONS

Taking the current indexes for the world electricity consumption into consideration, pushed up mainly by the developed and highly industrialized countries, and assuming that they are the same for the developing countries, by the time world's population reaches 10^{10} inhabitants, calculations show that the power consumption projections for Brazil, for the adopted time horizon and when using global indexes, become overestimated, when compared with the results calculated by our national indexes. According to our studies, power consumption estimates using global indexes becomes approximately 4.5 times higher than the estimates based on national indexes. Global indexes, as taken by several works and particularly the work of Tomabechi [6] analyzed here, are not reflecting the national realities in terms of energy consumption and demand, as is the case of Brazil.

Data here used for the predictions are subjected to changes and to the accuracy of the governmental agencies at the time of their publication. In the Brazilian case, the recent discovered pre-salt oil deposits offer a possibility for a different solution to a possible energy crisis, when compared, for example, to the United States. Together with China and the European Union, they are responsible for almost 65% of the world's energy consumption, having also combined factors like large population and large energy consumption per capita. Both factors lead to high levels of emissions and to the fast depletion of the current natural resources.

However, despite the great potential of the Brazilian reserves of uranium, water, oil, etc., clean energies must also be taken into consideration in the future, to meet the 2DS requirements. For a non-GHG-emission source, nuclear is the most available candidate. For Brazil, together with the renewables, the nuclear option could be a quick start to solve the problem of emissions due to deforestation, our main source of greenhouse gases.

Construction of nuclear power plants is always a subject for debates and speculation, impacting the forecasts of consumptions of uranium and thorium, by means of a correct campaign for the public acceptance. It was stated above that a way to reach some equilibrium condition could be achieved, however, if countries agree to diffuse their energy technologies for other countries, developing or under-developed, since they are at most low carbon emitters and usually still don't have enough technology development to carry out projects on clean energies. Sharing the technologies could be a fast contribution for the solution to the growing problem of climate changes. Relating to nuclear energy, at least four factors, mentioned is this work, can put Brazil in a comfortable position in front of a possible crisis in the future demand for energy: 1) uranium: since we have enough uranium in the Brazilian territory, and counting with the future improvement of the process of uranium extraction from phosphates; 2) thorium: since Brazil has also enough reserves, for future uses in new technology nuclear power reactors; 3) population growth stability: according to the governmental projections; and 4) nuclear fuel cycle know-how. Climate instabilities can reach us all in a near future, and global warming is one of its possible riggers. Besides, human activities play a very important role in the contribution to the instability of the weather in a global scale. The increase in the demand for energy, in the population growth, mainly in the last 50 years, leading to the increase of the greenhouse gas emissions and to all the climate instabilities we have been facing, should be assessed considering the appropriate indexes. Thus, according to this work, a different criterion must be stated, for the world to face this problem fairly, since a huge amount of investments will be needed to control or to reduce the threat of the so called "non-turning point" phenomena, in terms of the global weather.

Thus, the future transition between the massive global uses of fossil fuels to a massive global use of the alternative sources, supported by our nuclear know-how, could be carried out in such a way to help us in the improvement of the life quality in our country, and as a consequence, in our planet.

REFERENCES

1. http://www.cia.gov/library/publications/the-world-factbook
2. Haratyk, G. and Forsberg, C.W. (2012) Nuclear Renewable Energy System for Hydrogen and Electricity Production. *Nuclear Technology*, **178**, 66-82.
3. Empresa de Pesquisa Energética—EPE (2012) Ministério de Minas e Energia, Balanço Energético Nacional 2012.
4. http://www.inb.gov.br
5. IEA (2012) Tracking Clean Energy Progress—Energy Technology Perspectives 2012 Excerpt as IEA input to the Clean Energy Ministerial. 82 p. http://www. iea.org/publications/freepublications/publication/Tracking_Clean_Energy_Progress.pdf
6. Tomabechi, K. (2010) Energy Resources in the Future. *Energies*, **3**, 686-695.
7. Statistical Series, Instituto Brasileiro de Geografia e Estatística (IBGE). http://www.ibge.gov.br/series_estatisticas/exibedados.php?idnivel=BR&idserie=POP300
8. Piore, A. and Planejando, C.N. (2011) Scientific American Brasil, Edição Especial Energia Nuclear, 7-11.
9. Cochran, T.B., Feiveson, H.A., Mian, Z., Ramana, M.V., Schneider, M. and von Hippel, F.N. (2010) It's Time to Give up on Breeder Reactors. *Bulletin of Atomic Scientists*, **66**, 50-56. http://dx.doi.org/10.2968/066003007
10. World Nuclear Association, WNA (2011) Current Status. http://www.world-nuclear.org/info/reactors.html
11. (2011) Understanding the Roles of Technology Diffusion, Intellectual Property Rights, and Sound Environmental Policy for Climate Change. World Energy Council WEC Energy Sector Environmental Innovation.
12. Cerri, C.C., *et al.* (2009) Brazilian Green-House Gas Emissions, the Importance of Livestock and Agriculture. *Scientia Agricola*, **66**, 831-843. http://dx.doi.org/10.1590/S0103-90162009000600017
13. Kidd, S. (2011) Uranium Supply for the Nuclear Future. *Energy and Environment*, **22**, 61-66.
14. Data to Graphics and Tables Extracted from http://www.nationmaster.com.
15. Geração Termonuclear (2008) Boletim da Empresa de Pesquisa Energética EPE. No. 20080512-7, Ministério de Minas e Energia, Secretaria de Planejamento Estratégico, Maio de 2008, 13.
16. Gonçalves, O.D. (2011) Brasil Ainda Hesita na Área Nuclear. *Scientific American Brasil, Edição Especial Energia Nuclear*, 35-39.

17. Ashley, V., *et al.* (2010) The Commercially Viable ADTR Power Station. *Nuclear Future*, **7**, 41-45.
18. [18] www.epe.gov.br/PNE/20070625_09.pdf
19. [19] Análise Retrospectiva (2008) Boletim da Empresa de Pesquisa Energética EPE. No. 20080512-1, Ministério de Minas e Energia, Secretaria de Planejamento Estratégico, Maio de 2008.

CHAPTER 7

Proposed re-evaluation of the ^{154}Eu thermal (n,γ) capture cross-section based on spent fuel benchmarking studies

Steven E. Skutnik

ABSTRACT

^{154}Eu is a nuclide of considerable importance to both non-destructive measurements of used nuclear fuel assembly burnup as well as for calculating the radiation source term for used fuel storage and transportation. However, recent evidence from code validation studies of spent fuel benchmarks have revealed evidence of a systemic bias in predicted ^{154}Eu inventories when using ENDF/B-VII.0 and ENDF/B-VII.1 nuclear data libraries, wherein ^{154}Eu is consistently over-predicted on the order of 10% or more. Further, this bias is found to correlate with sample burnup, resulting in a larger departure from experimental measurements for higher sample burnups.

In this paper, the bias in ^{154}Eu is characterized across eleven spent fuel destructive assay benchmarks from five different assemblies. Based on these studies, possible amendments to the ENDF/B-VII.0 and VII.1 evaluations of the $^{154}\mathrm{Eu}(n,\gamma)^{155}\mathrm{Eu}$ are explored. By amending the location of the first resolved resonance for the ^{154}Eu radiative capture cross-section (centered at 0.195 eV in ENDF/B-VII.0 and VII.1) to 0.188 eV and adjusting the neutron capture width proportional to $\sqrt{1/E}$, the amended cross-section evaluation was found to reduce the bias in predicted

^{154}Eu inventories by approximately 5–7%. While the amended capture cross-section still results in a residual over-prediction of ^{154}Eu (ranging from 2% to 9%), the effect is substantially attenuated compared with the nominal ENDF/B-VII.0 and VII.1 evaluations.

Keywords: Eu-154, Nuclear data, ENDF, Burnup, Nuclear fuel depletion

1. INTRODUCTION

Passive gamma spectroscopy is a well-established technique burnup evaluation of spent nuclear fuel assemblies (Phillips *et al.*, 1980 ; Phillips, 1991), used for a number of burnup credit (Lebrun and Bignan, 2001 ; Nakahara *et al.*, 2002) and international safeguards applications (Rinard and Bosler, 1988; Willman *et al.*, 2006; Favalli *et al.*, 2016 ; Gauld and Francis, 2010). Burnup measurement via passive gamma spectroscopy relies on the fact that key indicator nuclides (such as ^{137}Cs, ^{134}Cs, ^{154}Eu, and others) show well-established correlations to fuel assembly burnup (for example, showing a near-linear correlation to burnup in the case of ^{137}Cs). ^{154}Eu in particular is a frequently-used burnup indicator nuclide in conjunction with ^{137}Cs, given its relatively long half-life (8.59 years) and strong gamma signatures (996.3 keV, 1,004.8 keV, and 1,274.4 keV). ^{154}Eu has a vanishingly small direct fission yield, its decay chain being blocked by the stable isotope ^{154}Sm. Rather, it is produced almost exclusively via neutron capture reactions with ^{153}Eu (whose direct thermal fission yield from ^{239}Pu is an order of magnitude higher than that of ^{235}U), making the ^{154}Eu inventory proportional to the square of the neutron fluence (Nakahara et al., 2002). The ^{154}Eu activity is frequently measured as the ratio of $\frac{^{154}Eu}{^{137}Cs}$ thereby eliminating the need to perform geometry corrections to measured activities; meanwhile, because of its comparatively long half-life, ^{154}Eu is applicable to measurements of longer-cooled assemblies (i.e., >5000 days post-discharge) (Lebrun and Bignan, 2001 ; Nakahara *et al.*, 2002).

Because of the fact that ^{154}Eu has a non-trivial thermal neutron absorption cross-section, an accurate accounting for the $^{154}Eu(n,\gamma)^{155}Eu$ radiative capture pathway is essential to use of ^{154}Eu for spent fuel NDA and source term calculations. However, a series of recent computational benchmarks to spent fuel destructive assay studies (Shoman and Skutnik, 2016; Skutnik and Shoman, 2016 ; Gauld *et al.*, 2016) using ENDF/B-VII.0 and VII.1 cross-sections has uncovered evidence of systemic biases in the predictions of ^{154}Eu in depletion studies, showing consistent over-predictions in the ^{154}Eu inventory across multiple studies. For example, recent benchmarks of the

Three Mile Island Unit 1 assembly by ORNL found an average ratio of calculated-to-experimental (C/E) for ^{154}Eu across five measured samples of 1.117, well above both the 1-sigma experimental uncertainty of 1.0% and C/E values for other recovered species (Gauld et al., 2016). Recent benchmarks of a CANDU 28-element lattice model for SCALE uncovered similar biases, reporting a C/E for ^{154}Eu of 1.162, again well outside the experimental uncertainty of 5.0% (Shoman and Skutnik, 2016). The consistency of these over-predictions therefore implies a potential bias in the nuclear data, specifically in the thermal region ($E \leq 0.0253$ eV) for the $^{154}Eu(n,\gamma)^{155}Eu$ channel, as previously identified in Shoman and Skutnik (2016). The consequences of a bias in the nuclear data would manifest in computational simulations of spent fuel source terms used for spent fuel instrument development (Galloway et al., 2012), producing over-estimates of the assembly burnup for estimates relying on ^{154}Eu activities (including $\frac{^{154}Eu}{^{137}Cs}$ ratios), thus making the matter a concern for both the reactor physics and nuclear instrumentation communities.

In the most recent Evaluated Nuclear Data File (ENDF) releases (ENDF-B/VII.0, and subsequently ENDF-B/VII.1), the $^{154}Eu(n,\gamma)^{155}Eu$ thermal capture cross-section (σth, reported for $E = 0.0253$ eV and extrapolated to lower energies via the $1/v$ relationship) was updated based on evaluations compiled by Mughabghab (Chadwick, 2011 ; Chadwick *et al.*, 2006), indicating an evaluated cross-section of 1348.5 barns, 36.6% lower than the prior reported evaluation of 1840 barns in ENDF/VI.8 (Cross-Section Evaluation Working Group, 1998). Given the limited production pathways for ^{154}Eu, this cross-section in particular appears to be responsible for the observed over-predictions in ^{154}Eu inventories in lattice physics depletion studies.

This paper presents evidence of this bias across several spent fuel benchmarking studies, drawing from spent fuel benchmark studies conducted from a CANDU 28-element pressurized heavy water reactor (PHWR) assembly well as ten samples obtained from four pressurized water reactor (PWR) assembly types. Working from lattice physics simulations of these destructive assay benchmarks, two proposed modifications to the ENDF/VII.0 and VII.1 thermal-region capture cross-section evaluation for ^{154}Eu are presented, along with their respective impacts on the calculated inventories of Eu species as well as other major constituent species of interest.

2. EVALUATION OF THE ^{154}EU(N,Γ)^{155}EU CAPTURE CROSS-SECTION

In order to understand the cause of discrepancies for Eu isotopes in reactor simulations, it is first important to understand the production and loss channels for Eu species, which can be represented by the general expression in Eq. (1).

$$\frac{dN_i}{dt} = \overbrace{\sum_{m=1}^{M} Y_m\lambda_m N_m}^{\text{(A)}} + \underbrace{\phi\sigma_{i-1}^{a} N_{i-1}}_{\text{(B)}} - \overbrace{\lambda_i N_i}^{\text{(C)}} - \underbrace{\phi\sigma_i^{a} N_i}_{\text{(D)}} + \overbrace{Y_i\Sigma_f\phi}^{\text{(E)}} + \underbrace{\sum_{m\neq i} N_m\sigma_{m\to i}\phi}_{\text{(F)}} \quad (1)$$

where

Term A = Production of nuclide *i* from *M* parent isotope decays

Term B = Production of nuclide *i* neutron absorption

Term C = Loss of nuclide *i* to decay

Term D = Loss of nuclide *i* to neutron absorption

Term E = Direct production of nuclide *i* from fission

Term F = Other neutron-induced reactions (e.g., (n,p), (n,2n), etc.)

The dominant terms controlling the ^{154}Eu and ^{155}Eu inventories over irradiation timescales are (B) and (D). The direct fission yields of ^{154}Eu and ^{155}Eu are both vanishingly small (see Table 1), meaning the predominant production channels for both species is either through decay of parent species or through successive neutron capture reactions, i.e. ^{153}Eu(n,γ)^{154}Eu and ^{154}Eu(n,γ)^{155}Eu. Because β^- decay stops at ^{154}Sm, the accumulation of ^{154}Eu from the decay of parent species is likewise blocked, therefore implying that the predominant production pathway is almost exclusively through the ^{153}Eu(n,γ)^{154}Eu channel, while due to its longer half-life (8.59 years), its primary destruction channel is through the subsequent capture and conversion to ^{155}Eu. In as much, ^{154}Eu inventories in irradiated fuel can primarily be understood through these two pathways.

Table 1. Direct and accumulated yields for select isotopes of Eu from 0.0253 eV fission of ^{235}U and ^{239}Pu; data from ENDF-B/VII.1 fission yield sublibrary (Chadwick, 2011).

Isotope	^{235}U thermal fission		^{239}Pu thermal fission	
	Direct yield	Accumulated yield	Direct yield	Accumulated yield
^{153}Eu	$2.32984 \cdot 10^{-9}$	$4.43181 \cdot 10^{-5}$	$2.50997 \cdot 10^{-8}$	$3.61255 \cdot 10^{-3}$
^{154}Eu	$9.69931 \cdot 10^{-10}$	$1.93986 \cdot 10^{-9}$	$1.39998 \cdot 10^{-7}$	$2.79997 \cdot 10^{-7}$
^{155}Eu	$2.62981 \cdot 10^{-8}$	$3.21357 \cdot 10^{-4}$	$1.92998 \cdot 10^{-6}$	$1.65722 \cdot 10^{-3}$

2.1. Eu radiative capture cross-section evaluations

A comparison of the different evaluations for the production and destruction pathways for ^{154}Eu, i.e., ^{15} Eu(n,γ)^{154}Eu and ^{154} Eu(n,γ)^{155}Eu, respectively, as seen in Fig. 1 ; Fig. 2 reveal several important trends. First, little change is observed in the production pathway for ^{154}Eu between ENDF/VI.8 and VII.0. While small changes to the thermal capture cross-section of ^{153}Eu are present in the ENDF/B-VII.1 evaluation, this change was included on the basis of new capture cross-section studies (Basunia et al., 2014); the effect of this change is evaluated within this paper. Nonetheless, it alone is insufficient to explain the differences observed in the ^{154} Eu inventories.

Turning to the removal channel, a more stark difference is observable between the ENDF/B-VI.8 evaluation and the subsequent VII.0 and VII.1 evaluations, wherein the latter evaluations are based upon a 2200 m/s cross-section of 1348.5 barns, 36.6% lower than the VI.8 evaluation of 1840±90 barns. Likewise considered for comparison is the JEFF-3.2 evaluation, which adopts a similar thermal-region resonance as ENDF/B-VII.0 and VII.1, while retaining the structure of the ENDF/B-VI.8 evaluation for the fast neutron (E>1 MeV) region. Likewise, both ENDF/B-VII.1 and JEFF-3.2 contain minor differences compared to ENDF/B-VII.0 for the resonance region (between 20 and 25 eV). Both of these differences are more clearly illustrated in Fig. 3, which shows the ratios of these evaluations to ENDF/B-VI.8 and VII.0, respectively.

Turning back to the differences in the thermal-region behavior of ENDF/B-VII.0, VII.1, and JEFF-3.2 compared to ENDF/B-VI.8, the ENDF/B-VI.8 evaluation for ^{154}Eu (compiled by R. Q. Wright of ORNL) cites work by Sekine et. al. as the basis for its thermal capture cross-section (Sekine et al., 1987). The ENDF/B-VII.0 evaluation (compiled by Mughabghab) refers to the Atlas of Neutron Resonances as the basis for its thermal capture cross-section (Mughabghab, 2006); however, Mughabghab's compilation makes no mention of the prior evaluation by Sekine et al. (1987), exclusively citing older studies. In private correspondence with Mughabghab regarding the VII.0 and VII.1 evaluations (Mughabghab, 2016), Mughabghab points out that Sekine et. al. assume a Wescott factor g_W (defined as the ratio of departure from the proportionality of the cross-section to $1/v$ in the thermal region) of unity, where (Sekine *et al.*, 1987 ; Mughabghab, 2006):

$$\hat{\sigma} = \sigma_0(g_w + r \cdot s) \quad (2)$$

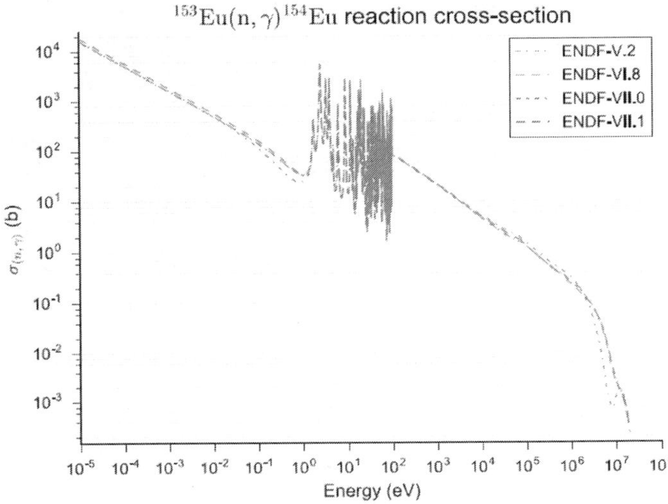

(a) ^{153}Eu(n,γ)^{154}Eu cross-section

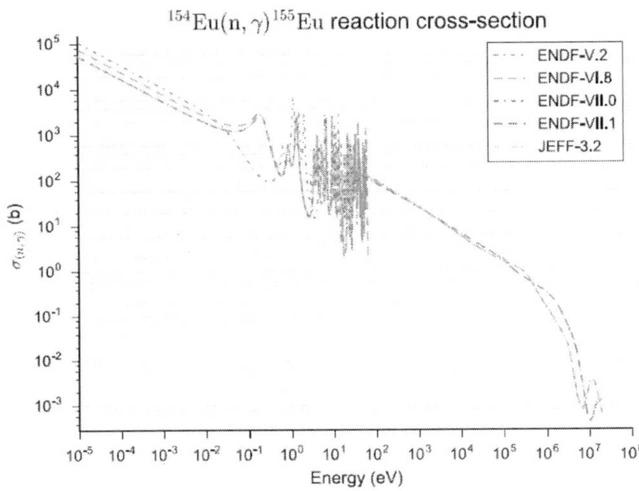

(b) ^{154}Eu(n,γ)^{155}Eu cross-section

Figure. 1. Radiative capture (n,γ) cross-sections for several ENDF library evaluations for ^{153}Eu (left) and ^{154}Eu (right) (Chadwick *et al.*, 2006 ; Cross-Section Evaluation Working Group, 1998).

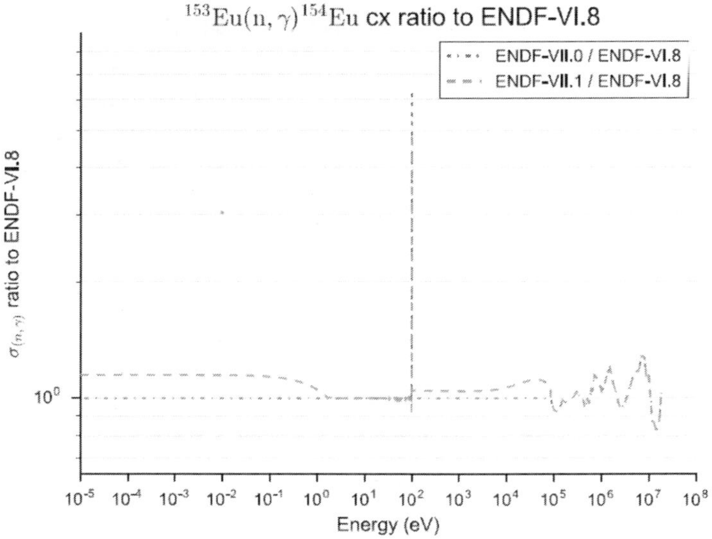

Figure. 2. ^{153}Eu(n,γ)^{154}Eu cross-section ratio to ENDF/B-VI.8 for ENDF/B-VII.0 and VII.1 evaluations.

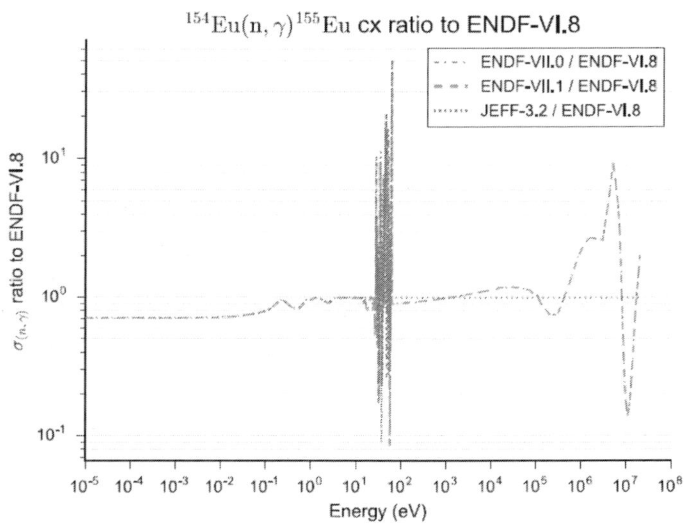

(a) Relative capture ratios to ENDF/B-VI.8

Figure. 3. Radiative capture cross-section ratios for ^{154}Eu(n,γ)^{155}Eu for ENDF/B-VII.0, VII.1, and JEFF-3.2 evaluations.

Here, $\hat{\sigma}$ is the "effective" cross-section measured in the irradiation system (based on the shape of the neutron flux profile), r is defined as the fraction of epithermal neutrons in the flux spectrum, and s is defined as (Mughabghab, 2006):

$$s = \sqrt{\frac{4}{\pi} \cdot \frac{T}{T_0}} \cdot \left(\frac{I'_0}{\sigma_0}\right) \tag{3}$$

where I'_0 is the difference of the resonance integral and its $1/v$ component (referred to as the "reduced resonance integral") and T is the temperature of the Maxwellian distribution that describes the neutron energy spectrum (Mughabghab, 2006).

Mughabghab points out that the assumption of a Wescott factor $g_w = 1$ is incorrect for ^{154}Eu, indicating that a correct value calculated from the resonance parameters would be $g_w = 1.237$ (Mughabghab, 2016), further noting that the published value in the Atlas of Neutron Resonances of $g_w = 0.8979$ is incorrect (Mughabghab, 2006). Therefore, the cross-section reported by Sekine et. al. would not be a true 2200 m/s cross section but rather a Maxwellian-weighted value. As a result, per Mughabghab a corrected 2200 m/s cross-section based on Sekine et al.'s measurements would be 1488 ± 73 barns (Mughabghab, 2016). For comparison, the historical evaluations of the ^{154}Eu thermal cross-sections are presented as Table 2.

Table 2. Measured and evaluated 2200 m/s capture cross-sections (σth) for ^{154}Eu.

Evaluator	Type	σ_{th} (bn)	Uncertainty (bn)
Hayden et al. (1949)	Experimental	880	—
Vertebnyj et al. (1982)	Experimental	1250	160
Sekine et al. (1987)	Experimental	1840	90
Mughabghab (2006)	Calculated	1340	130

Overall, given the relatively high capture cross-section in the thermal region, combined with the typically elevated thermal neutron population in light water reactors (and especially pronounced in heavy-water reactors; see Fig. 4), it is likely that changes to the thermal-region capture cross-section explain much of the biases observed in depletion benchmark studies. This will be further supported by proposed re-evaluations presented in this paper, based on modifications of the ENDF/VII.0 and VII.1

cross-sections to conform to the thermal-region shape of the VI.8 evaluation. Additional evaluations employing the JEFF-3.2 cross-sections serves as a useful cross-check as well; given many shared features of this evaluation between ENDF/B-VII.1 and VI.8, it can be thought of as a "hybrid" evaluation that allows for separate consideration of effects of the nuclear data evaluation. For both the ENDF and JEFF evaluations, proposed amendments to the thermal region resolved resonance are evaluated to examine its effect on spent fuel validation benchmark calculations.

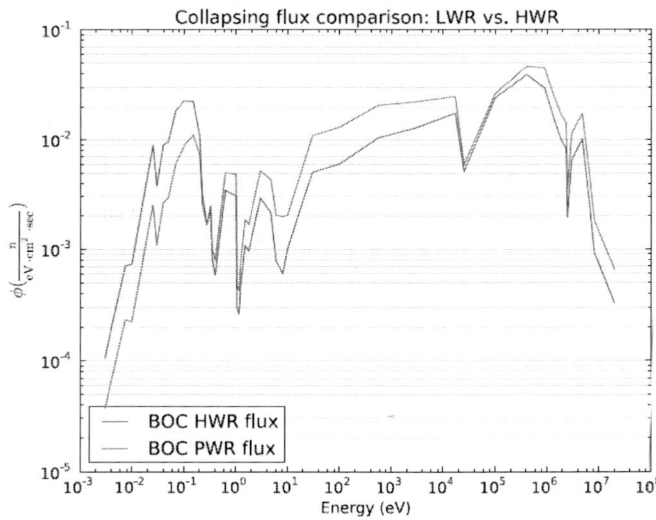

Figure. 4. Distribution of the 49-group collapsing flux used at the beginning-of-core for a typical PWR assembly (Westinghouse 17 × 17) versus a CANDU PHWR (28-element) assembly.

2.2. Proposed cross-section re-evaluations

Given the preceding discussion, two cross-section evaluations will be considered for the ENDF/VII.0 and VII.1 evaluations (wherein the primary difference between the two for this case is in the $^{153}Eu(n,\gamma)^{154}Eu$ production channel): a modification to the ENDF/VII.0 $^{154}Eu(n,\gamma)^{155}Eu$ based on the thermal-region behavior observed in the ENDF/VI.8 evaluation (hereafter referred to as "Wright") and modifications as proposed by Mughabghab (referred to as "Mughabghab"). The changes to the ENDF/B-VII.0/ VII.1 resolved resonance parameters are detailed in Table 3. Here, $|J|$ is the total angular momentum (spin) of the resonance state, Γ is the resonance total width (measured at the resonance energy), Γ_n is the resonance neutron width (evaluated at the resonance energy), and

Γ_γ is the radiation width of the resonance (Cross Sections Evaluation Working Group, 2012).

The "Wright" evaluation consists simply of restoring the resolved resonance region parameters from the ENDF/B-VI.8 evaluation, thereby restoring the thermal-region behavior (subsequently producing a thermal capture cross-section of 1840 barns). This includes both the addition of a bound state at −0.71 eV as well as modifications to the first resolved resonance at 0.190 eV, giving a 2200 m / s cross-section of about 1590 barns. The "Mughabghab" evaluation does not include this bound state but instead shifts the first resolved resonance from 0.195 eV to 0.188 eV, based on results reported by Anufriev et al. (1979), adjusting the neutron width (Γ_n) by a factor of $\sqrt{\frac{0.195}{0.188}} = 1.018$ (i.e., respecting the fact that cross-sections in the thermal region are inherently proportional to $\sqrt{E}$). The proposed modifications yield a 2200 m / s cross-section of approximately 1498 barns, relatively close to the Wescott-corrected Sekine value.

Applying the proposed modifications to the ENDF/B-VII.1 resolved resonance region evaluation (File 2, illustrated in Table 3), new AMPX master format continuous energy (CE) libraries were generated for use with depletion calculations (Wiarda et al., 2016). For further comparison, a modified version of the JEFF-3.2 evaluation based on the Mughabghab amendment to the first thermal resonance is also evaluated. Given that the ENDF/B-VII.1 and JEFF-3.2 unmodified libraries are identical for this region, the modified treatments are likewise identical; the only differences for these evaluations pertain to the resonance region (especially over 100 eV) and the fast region of the cross-section (refer to Fig. 3). The original ENDF/B-VII.1 and modified evaluations used in this study are shown as Fig. 5.

While TRITON/NEWT is a multi-group lattice physics treatment, CE libraries were necessary for this evaluation due to the resonance self-shielding calculation performed by the CENTRM code which calculates pointwise cross-sections used for resonance self-shielding calculations. CENTRM then uses these pointwise cross-sections are then used to calculate problem-specific multi-group cross-sections used in the lattice physics transport calculation (Williams, 2011). Thus, in the context of this study, the modified CE libraries are used by CENTRM to generate updated multi-group cross-sections which reflect the new thermal-region evaluations, where the development of a CE library is necessary to propagate the cross-section changes through the neutron transport and depletion calculation. Modified AMPX CE libraries were generated for each of the amended resolved resonance treatments presented in Table 3; these were then used with TRITON/NEWT to produce depletion using each proposed cross-section treatment.

Table 3. Proposed modifications to the ENDF/VII.0/ VII.1 $^{154}Eu(n, \gamma)^{155}Eu$ resolved resonance parameters, based on the ENDF/B-VI.8 thermal-region behavior ("Wright") and changes proposed by Mughabghab ("Mughabghab"). Note that the changes based on ENDF/B-VI.8 ("Wright") introduces an additional bound state at −0.71 eV not present in the VII.0 or VII.1 evaluations.

Evaluation	E (eV)	J	Γ	Γ_n	Γ_γ
Wright	−0.7100000	2.5	0.1264160	4.160000E−4	0.1260000
	0.1900000	3.5	0.1600595	5.950000E−5	0.1600000
Mughabghab	0.1880000	2.5	0.1500731	7.307365E−5	0.1500000

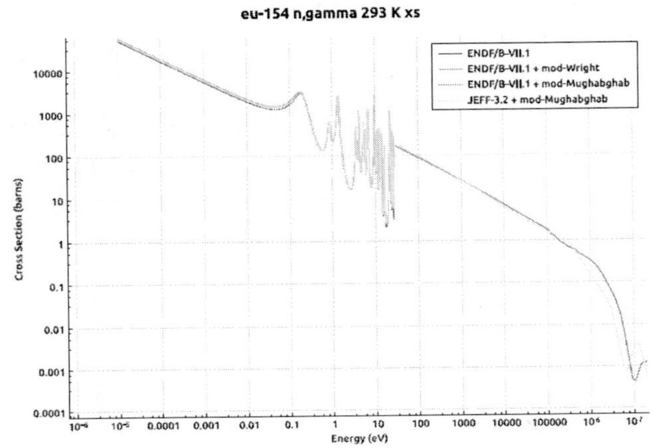

(a) Original and modified ^{154}Eu capture cross-section evaluations

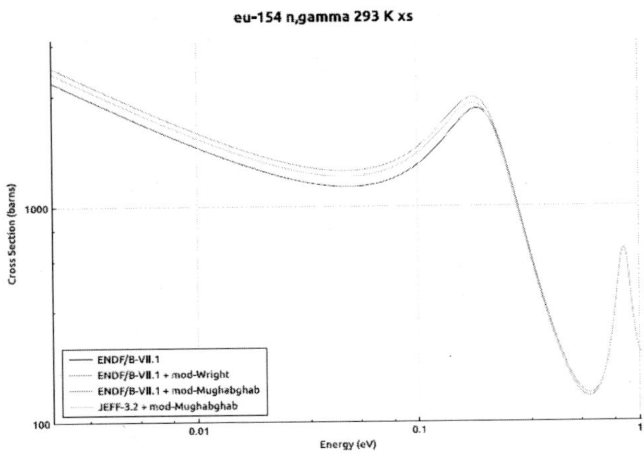

(b) Thermal resonance region of ^{154}Eu with original evaluation and proposed modifications

Figure. 5. Original ENDF/B-VII.0/ VII.1 $^{154}Eu(n,\gamma)^{155}Eu$ evaluation, along with proposed modifications based on restoring ENDF/B-VI.8 thermal-region levels (“Wright”) and revisions proposed by Mughabghab (“Mughabghab”) as evaluated in an AMPX master continuous energy (CE) data library. A modified version of the JEFF-3.2 evaluation (also modified per Mughabghab) is also presented for comparison. The proposed revisions exclusively impact the first resolved resonance and thermal energy regime ($E \leqslant 1$ eV).

3. SPENT FUEL BENCHMARKS

For the purposes of this study, eleven samples from five fuel assemblies will be used: one sample from fuel bundle 19558C, a 28-element CANDU PHWR assembly irradiated in the Pickering-A reactor and evaluated by Atomic Energy of Canada Limited (AECL) (Tait *et al.*, 1995; Gauld and Litwin, 1995 ; Gauld *et al.*, 1995); five samples taken from Three Mile Island Unit 1 (Gauld et al., 2016), a Babcock & Willcox 15×15 assembly; three samples measured from a Combustion Engineering 14×14 assembly from Calvert Cliffs (Ilas et al., 2010a); one sample measured from an 18×18 assembly from the Gemeinschaftskernkraftwerk Unit II reactor in Germany (measured as part of the REBUS campaign) (Ilas et al., 2010b); and finally one sample measured from a 15×15 assembly from the Gösgen reactor in Switzerland (measured as part of the ARIANE campaign) (Ilas et al., 2010b). For all benchmarks, the SCALE 6.2 release was employed with the TRITON/NEWT discrete ordinates transport sequence (DeHart and Bowman, 2011).

3.1. CANDU 28-element PHWR assembly

Fuel bundle 19958C (irradiated in the Pickering-A reactor) is a 28-element CANDU PHWR assembly employing natural uranium fuel elements with Zircaloy-4 cladding, employing a heavy-water (D_2O) moderator, with the fuel lattice depicted as Fig. 6, with characteristic dimensions outlined in Table 4. The nominal assembly power history provided in Gauld and Litwin (1995) is reproduced here as Table 5. For this benchmark, each ring was treated as a separate fuel material with its own Dancoff Factor (Table 6), with the assembly total power history being set to the nominal power history provided in Gauld and Litwin (1995).

In the destructive chemical assay benchmark outlined in Tait *et al.*, 1995; Gauld *et al.*, 1995 ; Gauld and Litwin, 1995, a single outer-ring fuel element was assayed to provide benchmarking data for lattice physics depletion studies. The lattice used for the purposes of this depletion study is the same as that as evaluated in Shoman and Skutnik (2016); thus, for a more extended discussion of this validation effort, including modeling details, the reader is directed to this paper.

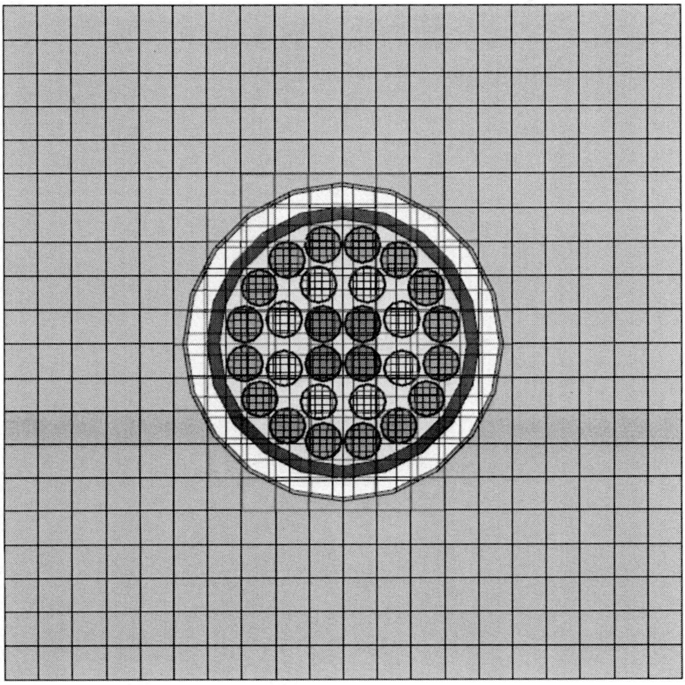

Figure. 6. CANDU 28-element lattice model used for depletion benchmark studies. Depletion benchmark studies are based upon destructive chemical assay of an outer-ring fuel element.

3.2. B&W 15×15 PWR assembly: TMI-1 NJ05YU

Benchmarks of TMI-1 assembly NJ05YU consisted of five samples spanning two pins: samples C2D1 and C2D2 taken from pin H6 (an interior pin), and samples A1, B1, and B2 taken from pin D5, located next to a control rod guide tube, corresponding to the samples studied in Gauld et al. (2016). The lattice model used for this benchmarking study is illustrated as Fig. 7, with the design parameters used for modeling given as Table 7. For the first irradiation cycle (Cycle 9), sixteen Al_4-B_4C control rods were present in the guide tubes; these were removed in the second irradiation cycle (Cycle 10). This removal was modeled using TRITON's material swap feature (new to SCALE 6.2), with the control rod material swapped for an equivalent volume of moderator.

Table 4. Characteristics dimensions of fuel bundle 19958C, a 28-element CANDU PHWR fuel assembly irradiated in the Pickering-A reactor (reproduced from Gauld and Litwin (1995)).

Fuel material	Natural UO_2	Number of fuel pins	28
Fuel density (g/cm^3)	10.6	Inner fuel ring radius (cm)	1.175
Fuel temperature (K)	1003	Middle fuel ring outer radius (cm)	2.685
Fuel element radius (cm)	0.7120	Outer fuel ring radius (cm)	4.229
Cladding material	Zircaloy-4	Fuel channel square pitch (cm)	28.575
Cladding outer radius (cm)	0.7605		
Pressure tube	Zr-Nb 2.5%	**Calandria tube**	Zircaloy-4
Inner radius (cm)	5.1815	Inner radius (cm)	6.5405
Outer radius (cm)	5.6985	Outer radius (cm)	6.6955
Moderator	D_2O	**Coolant**	D_2O
Temperature (K)	336	Temperature (K)	545
Atom purity $\left(\frac{D}{D+H}\right)$	0.9991	Atom purity $\left(\frac{D}{D+H}\right)$	0.9975
Density (g/cm^3)	1.0838	Density (g/cm^3)	0.8445

Table 5. Power history used for modeling the Pickering-A CANDU 28-element bundle (19958C) for the nominal assembly power and the calculated outer element power (Gauld *et al.*, 1995 ; Gauld and Litwin, 1995).

Irradiation (days)	Assembly power ($\frac{MW}{MTU}$)	Cooling time (days)
31.33	28.41	3.25
15.38	26.26	5.61
66.67	28.15	0.00
66.67	26.71	9.00
64.55	26.07	15.0
70.78	24.18	5590.0

Table 6. Calculated Dancoff Factors used for CANDU 28-element assembly calculations (Shoman and Skutnik, 2016).

Position/Ring	Dancoff factor
Inner ring	0.714
Middle ring	0.707
Outer ring	0.447

Because the purpose of this study is to evaluate the impact of changes to the ^{154}Eu cross-section, the modeled pins take into account as many irradiation history details as is feasible. This includes incorporation of the detailed power and fuel temperature history (presented in Table 8) along with moderator characteristics such as the soluble boron letdown curve and moderator density (presented in Table 9). Additionally, to capture local spectrum effects, adjacent fuel pins to the assayed samples were also modeled as separate depletion materials.

3.3. CE 14×14 PWR assembly: Calvert Cliffs D047/rod MKP-109

Calvert Cliffs assembly D047 was a Combustion Engineering 14×14 (CE 14×14) PWR assembly measured as part of a U.S. campaign designed to measure moderate-burnup assemblies representative of many assemblies currently kept in on-site storage (Ilas et al., 2010a). The Calvert Cliffs measurements consisted of three samples taken from rod MKP-109, an interior rod (illustrated in Fig. 8), covering a sample burnup range of 27.4 $\frac{GWd}{MTU}$ to 44.3 $\frac{GWd}{MTU}$. Assembly design parameters used for modeling as given as Table 10. As seen in Fig. 9, assembly D047 is modeled as a half-lattice model with reflective boundary conditions owing to symmetry; the TRITON lattice model was constructed based on sample inputs provided in Ilas et al. (2010a).

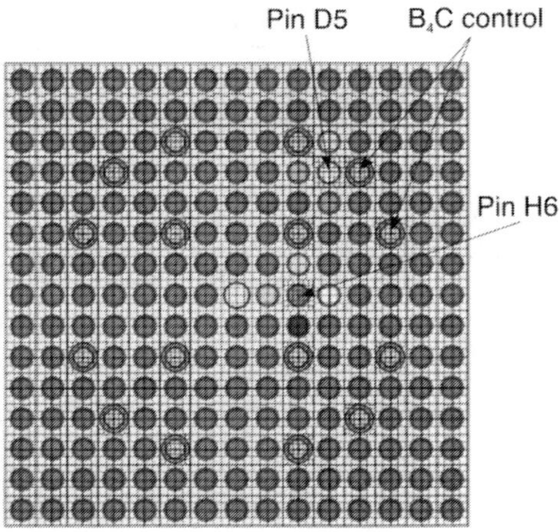

Figure. 7. TMI-1 NJ05YU fuel assembly lattice model used for benchmarking studies; samples were taken from pins H6 (samples C2D1, C2D2; purple pin) and D5 (samples A1, B1, and B2; yellow pin). Note that B_4C control rods were swapped for moderator materials for Cycle 10. (For interpretation of the references to color in this figure legend, the reader is referred to the web version of this article.)

Assembly D047 was irradiated over the course of four cycles, although simplifying the analysis, there were no burnable poisons or gadolinia (Gd_2O_3-UO_2) rods in the assembly. Like the TMI-1 NJ05YU assembly, for Calvert Cliffs assembly D047 the detailed moderator density and soluble boron concentrations are well-known (shown as Table 11); however only cycle-average powers and fuel temperatures were available (given in Table 12). (Note that detailed linear heat generation rate histories were available for the axial locations corresponding to the measured samples, however these were not translated into corresponding equivalent detailed power histories (Ilas *et al.*, 2010a; Radulescu *et al.*, 2010 ; Hermann *et al.*, 1995). This appears to have had a minimal effect on the model fidelity however, as will be illustrated in the following sections.) The cycle powers presented in Table 12 are corrected from their nominal values published in Ilas *et al.*, 2010a; Radulescu *et al.*, 2010 ; Hermann *et al.*, 1995 to match inventories of ^{148}Nd, used as strong indicator of the total number of fissions (and therefore actual fuel burnup) (ASTM International, 2012 ; Suyama and Mochizuki, 2006).

Table 7. TMI-1 NJ05YU assembly data used for modeling (reprinted from Gauld et al. (2016), with permission from Elsevier).

Assembly data			
Rod pitch (cm)	1.44272	Assembly pitch (cm)	21.81098
Number of fuel rods	208	Number of guide tubes	16
		Number of instrument tubes	1
Fuel rod data		**Burnable poison rod (BPR) data**	
Fuel material type	UO_2	Number of poison rods inserted[a]	16
Fuel pellet density (g/cm^3)	10.196	Absorber material	Al_2O_3-B_4C
Fuel pellet diameter (cm)	0.9362	B_4C content (wt%)	1.7
Enrichment (wt% ^{235}U)	4.013	Absorber density (g/cm^3)	3.7
Clad material	Zircaloy-4	Absorber diameter (cm)	0.8636
Clad inner diameter (cm)	0.95758	Cladding material	Zircaloy-4
Clad outer diameter (cm)	1.0922	Clad inner diameter (cm)	0.9144
Average clad temperature (K)	640	Clad outer diameter (cm)	1.0922
Initial uranium composition (wt%)		**Guide/instrument tube data**	
^{234}U	0.040	Guide/instrument tube material	Zircaloy-4
^{235}U	4.013	Guide tube inner diameter (cm)	1.26492
^{238}U	95.947	Guide tube outer diameter (cm)	1.3462
		Instrument tube inner diameter (cm)	1.12014
		Instrument tube outer diameter (cm)	1.25222

[a] BPRs only present during cycle 9.

Table 8. Specific power and fuel temperature history for TMI-1 NJ05YU samples (reprinted from Gauld et al. (2016), with permission from Elsevier). Note that powers are burnup-corrected to match measured ^{148}Nd inventories. (EFPD = effective full-power days).

Cycle	Time (EFPD)	Specific power (MWt/tU)					Fuel temperature (K)				
		AG616			AG536		AG616			AG536	
		A1	B1	B2	C2D1	C2D2	A1	B1	B2	C2D1	C2D2
9	0.0										
	74.2	30.580	45.667	42.156	59.593	57.447	960.5	1084.7	1078.9	1105.7	1098.3
	144.1	33.309	46.386	43.398	61.186	59.904	961.5	1052.2	1051.2	1069.1	1062.9
	214	34.811	46.759	44.038	49.444	48.519	946.9	1019.4	1020.5	1034.4	1029.4
	284.9	36.595	46.968	44.593	51.461	50.779	952.7	992.9	997.5	1003.2	998.2
	349.7	38.559	46.643	44.621	48.619	48.235	937.6	967.5	972.7	976.0	971.2
	425	39.506	45.929	44.103	46.401	45.959	931.7	939.4	946.0	948.9	944.2
	483.9	40.747	45.266	43.651	44.420	44.163	920.8	920.4	926.4	925.5	920.6
	549.2	41.338	44.218	42.708	42.577	42.217	908.9	903.1	909.2	907.8	903.4
	608	41.700	44.348	42.575	42.919	42.398	918.9	905.2	910.5	900.2	889.6
	639.4	46.740	47.275	45.834	44.289	46.160	750.5	725.3	731.5	810.4	791.0
10	0.0										
	68	29.608	42.294	39.052	47.189	45.013	796.1	858.9	851.5	871.0	871.3
	131.8	30.270	41.396	38.583	37.149	35.898	808.4	849.5	847.4	856.6	853.5
	209	31.505	40.445	38.109	34.614	34.028	809.1	841.1	841.9	844.7	838.8
	272.1	32.566	39.837	37.861	33.132	32.902	809.7	834.9	837.4	835.8	829.3
	347.4	33.557	39.289	37.578	32.086	32.032	812.3	828.0	831.5	828.5	822.1
	416.4	34.376	38.776	37.247	31.160	31.244	814.8	822.2	826.6	821.7	815.2
	486.4	34.969	38.252	36.843	30.327	30.460	816.3	815.7	820.6	815.3	808.5
	556.3	35.346	37.664	36.340	29.549	29.608	817.3	808.6	813.9	808.4	801.6
	626.1	35.799	37.120	35.851	28.837	28.852	816.5	803.8	808.7	802.7	796.2
	660.3	36.045	36.720	35.509	28.359	28.383	812.8	802.2	806.5	799.9	794.4

Table 9. Moderator density and soluble boron letdown timetable for TMI-1 NJ05YU samples; reprinted from Gauld et al. (2016) with permission from Elsevier. (EFPD = Effective full-power days).

Cycle	Time (EFPD)	Boron concentration (ppm)	Moderator density (g/cm^3)				
			AG616			AG536	
			A1	B1	B2	C2D1	C2D2
9	0	1670	0.6759	0.6935	0.6876	0.7057	0.6965
	74.2	1481	0.6759	0.6935	0.6876	0.7057	0.6965
	144.1	1342	0.6759	0.6905	0.6876	0.7057	0.6965
	214	1175	0.6731	0.6905	0.6876	0.7057	0.6965
	284.9	990	0.6759	0.6905	0.6876	0.7057	0.6965
	349.7	772	0.6759	0.6935	0.6876	0.7057	0.6965
	425	545	0.6759	0.6935	0.6876	0.7057	0.6965
	483.9	352	0.6759	0.6935	0.6935	0.7057	0.6965
	549.2	134	0.6788	0.6935	0.6905	0.7057	0.6996
	608	13	0.6846	0.7026	0.6965	0.7152	0.7057
	639.4	2	0.7026	0.7088	0.7088	0.7152	0.7120
10	0	1800	0.6846	0.6996	0.6965	0.7120	0.7026
	68	1649	0.6876	0.6996	0.6965	0.7120	0.7057
	131.8	1521	0.6876	0.7026	0.6965	0.7120	0.7057
	209	1322	0.6876	0.7026	0.6965	0.7120	0.7057
	272.1	1140	0.6876	0.7026	0.6996	0.7120	0.7057
	347.4	918	0.6876	0.7026	0.6996	0.7120	0.7057
	416.4	718	0.6876	0.7026	0.6996	0.7120	0.7057
	486.4	506	0.6876	0.7026	0.6996	0.7120	0.7057
	556.3	298	0.6905	0.7026	0.6996	0.7120	0.7057
	626.1	103	0.6905	0.7026	0.6996	0.7120	0.7057
	660.3	1.8	0.6905	0.7026	0.6996	0.7120	0.7057

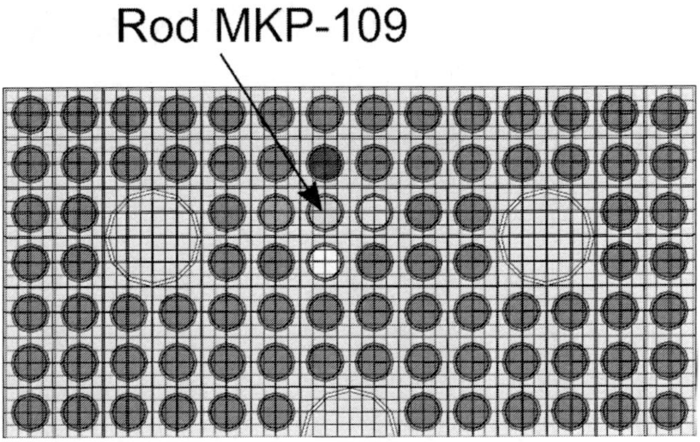

Figure. 8. Calvert Cliffs assembly D047 half-lattice (reflected) model used for benchmarking studies, developed from dimensions, irradiation history, and sample input published in Ilas et al. (2010a); samples were taken from rod MKP-109.

3.4. REBUS *18×18* PWR assembly: GKN-II

The REBUS program consisted of an international collaboration of institutes from Belgium, France, Germany, Japan, and the United States, focused on code validation measurements used for burnup credit in spent fuel storage applications, employing critical experiments conducted on spent fuel rod segments (Ilas et al., 2010b). One of these segments, a rod from assembly 419 of the Gemeinschaftskernkraftwerk Unit II (GKN-II) reactor in Germany (an 18×18 PWR assembly employing UO_2 fuel) was measured for its isotopic content. Key characteristics of this assembly include the presence of 12 gadlonia rods with 7.0 wt% Gd_2O_3 content (with an initial uranium enrichment of 2.6 wt%) and 24 guide tubes (Ilas et al., 2010b). The measured rod (M11) was located relatively toward the interior of the assembly, away from flux-distorting features such as gadlonia rods and guide tubes (as seen in Fig. 9), implying minimal irregularities in the flux spectrum that may otherwise produce confounding biases in isotopic inventories for the study conducted herein.

Table 10. Calvert Cliffs assembly D047 design data used for modeling (reproduced from Ilas *et al.*, 2010a ; Hermann *et al.*, 1995).

Assembly data			
Rod pitch (cm)	1.4732	Assembly pitch (cm)	20.78
Number of fuel rods	176	Number of water rods	5
Fuel rod data			
Fuel material type	UO_2		
Fuel pellet density (g/cm^3)	10.045		
Fuel pellet diameter (cm)	0.563		
Enrichment (wt% ^{235}U)	3.038		
Clad material	Zircaloy-4		
Clad inner diameter (cm)	0.95758		
Clad outer diameter (cm)	1.0922		
Average clad temperature (K)	620		
Initial uranium composition (wt%)		**Water rod data**	
^{234}U	0.027	Water rod material	Zircaloy-4
^{235}U	3.038	Water rod inner diameter (cm)	2.268
^{238}U	96.921	Water rod outer diameter (cm)	2.832

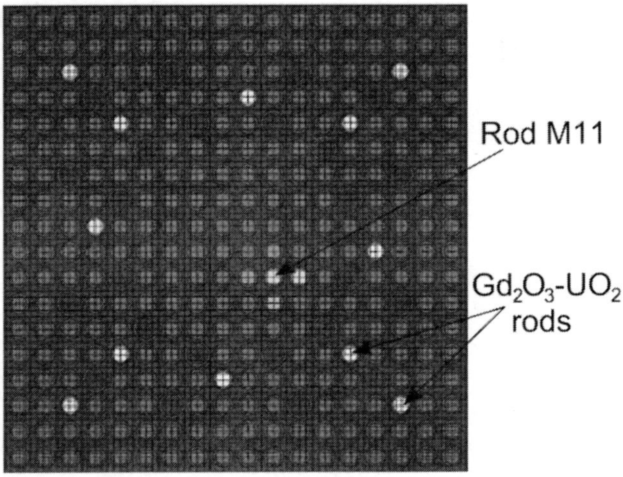

Figure. 9. GKN-II assembly 419 (measured as part of the REBUS campaign) lattice model used for benchmarking studies, developed from dimensions, irradiation history, and sample input published in Ilas et al. (2010b); samples were measured from rod M11.

The characteristic dimensions of the GKN-II assembly are provided as Table 13, reproduced from data available in Ilas et al. (2010b). Additionally, similar to the Calvert Cliffs assembly data, a detailed moderator density and boron letdown history is available (reproduced from Ilas *et al.*, 2010b ; Radulescu *et al.*, 2010), which is given as Table 14; however only cycle-average power, moderator density, and fuel temperature data are available (likewise reproduced from Ilas *et al.*, 2010b ; Radulescu *et al.*, 2010); these parameters are given as Table 15.

3.5. ARIANE *15×15* PWR assembly: Gösgen sample GU1

The ARIANE program was another international collaboration (coordinated by Belgonucleaire and involving participant laboratories and utilities from Belgium, Germany, Japan, Netherlands, Switzerland, the U.K., and the U.S.) focused on improving the database of available isotopic measurements from spent fuel assemblies used for validation of reactor physics and depletion packages (like SCALE) (Ilas et al., 2010b). Three UO_2 samples were measured by the ARIANE program (samples GU1, GU3, and GU4) from the Gösgen reactor located in Switzerland. However, samples GU3 and GU4 had significant irregularities in their irradiation history. For the first two irradiation cycles (cycles 16 and 17), samples GU3 and GU4 were contained in assembly 1601 (with an initial enrichment of 4.1%); on cycle 18, these rods were transfered to a different assembly (assembly 1701, with an initial enrichment of 4.1%), which began cycle 18 with an average burnup of about 20.0 $\frac{\text{GWd}}{\text{MTU}}$ (Ilas et al., 2010b). Because of this substantial anomaly, it was decided not to model these samples given the strongly confounding factors owing to their unique irradiation history.

Sample GU1 from assembly 1240 (illustrated in Fig. 10) was irradiated for four consecutive cycles (cycles 12–15). Following cycle 13, three fuel rods in positions L12, M12, and N12 (the row immediately above sample GU1) were replaced; following cycle 14, the rods at positions N12, K14, and L14 (northeast and southwest of sample GU1 located at M13, respectively) were likewise replaced (Ilas et al., 2010b). Reports from the operator indicate that the burnups of the rods replaced did not differ by more than 3–4% of the original rod burnups, however in the absence of further details it is not possible to explicitly model the effect of this replacement (Ilas et al., 2010b). However, based on the results of nominal evaluations with the ENDF/VII.0 and ENDF/VII.1 libraries (provided in the following section), this does not appear to have produced significantly adverse impacts on isotopic inventories beyond those observed in other assemblies, as evinced both by the isotopes of interest to this study (i.e., Eu) as well as major actinoids (U, Pu) and burnup indicator species (Cs, Nd).

The assembly design parameters used to model the ARIANE GU1 sample are provided as Table 16. As observed in Fig. 10, assembly 1240 was modeled as a quarter-assembly lattice with reflective boundary conditions, given the symmetry of the lattice design. The TRITON input used to create this model was based off of a sample input provided in Ilas et al. (2010b). Finally, the assembly operating history parameters, including the assembly power, soluble boron history, and fuel temperature (each of which were reproduced from Ilas et al. (2010b)) are provided as Table 17. The reader should note that a detailed moderator density history was not available for this model and thus was not explicitly modeled, instead using an averaged value for the entire irradiation history.

Table 11. Moderator density and soluble boron letdown timetable for Calvert Cliffs assembly D047 samples, reproduced from Ilas *et al.*, 2010a; Radulescu *et al.*, 2010 ; Hermann *et al.*, 1995.

Cycle 2		Cycle 3		Cycle 4		Cycle 5	
Time interval (d)	Boron (ppm)	Time interval (d)	Boron (ppm)	Time interval (d)	Boron (ppm)	Time interval (d)	Boron (ppm)
0	663	0	910	0	1033	0	1045
7	654	7.9	883	46.1	960	65	919
30.8	614	14.4	862	24	889	5.5	911
16.3	563	19.7	837	22.6	827	6.6	896
11.4	533	16.8	808	25.7	759	28.6	854
12.5	507	16.3	775	30.2	706	31.2	784
23.7	468	15.4	741	41.2	788	27	715
22.8	418	39.1	684	50.3	720	22.7	655
23.2	368	31.2	611	11	673	27.1	603
8.1	333	31.8	545	32.8	527	55.2	521
31.4	290	31.8	478	23.5	460	20.9	434
34.3	218	44.3	368	29.4	370	41.9	356
16.4	162	25	291	28.1	301	21.6	281
19.2	122	59.1	224	65.4	161	27.6	226
12.8	86	28.9	120	35.7	73	19	173
34.2	36	81.3[a]	83	85[a]	31	61.2	79
1.9	0						
71[a]	0						

[a] Denotes inter-cycle decay interval.

Table 12. ^{148}Nd-corrected cycle powers and fuel temperatures for Calvert Cliffs assembly D047 samples, adapted from Ilas *et al.*, 2010a; Radulescu *et al.*, 2010 ; Hermann *et al.*, 1995.

Cycle	Duration (days)	Down (days)	Cycle power ($\frac{MW}{MTU}$)		
			87–81	87–72	87–63
2	306.0	71.0	20.402	27.632	33.038
3	381.7	81.3	20.319	27.958	32.888
4	466.0	85.0	14.667	19.877	23.788
5	461.1		13.307	18.021	21.535

Table 13. Assembly design parameters for the GKN-II assembly sample measured as part of the REBUS campaign, reproduced from Ilas *et al.*, 2010b ; Radulescu *et al.*, 2010.

Assembly data			
Rod pitch (cm)	1.27	Assembly pitch (cm)	23.116
Number of fuel rods	300	Number of guide tubes	24
Fuel rod data		**Guide tube data**	
Fuel material type	UO_2	Guide tube material	Zircaloy-4
Fuel pellet density (g/cm^3)	10.4	Guide tube inner diameter (cm)	1.11
Fuel pellet diameter (cm)	0.805	Guide tube outer diameter (cm)	1.232
Enrichment (wt% ^{235}U)	3.8		
Clad material	Zircaloy-4	**Gd_2O_3-UO_2 rod data**	
Clad inner diameter (cm)	0.822	Gd_2O_3 content (wt%)	7.0
Clad outer diameter (cm)	0.95	Number of Gd_2O_3-UO_2 rods	12
Average clad temperature (K)	619		
	Initial uranium composition (wt%)		
UO_2 rods		**Gd_2O_3-UO_2 rods**	
^{234}U	0.036	^{234}U	0.0[a]
^{235}U	3.798	^{235}U	2.6
^{236}U	3.798	^{236}U	0.0[a]
^{238}U	96.166	^{238}U	97.4

[a] Per (Ilas et al., 2010b; Radulescu et al., 2010), the initial values of ^{234}U and ^{236}U were unknown for gadolinia-bearing pins and therefore assumed to be zero.

Table 14. Soluble boron history for assembly 419 of the GKN-II sample measured in the REBUS campaign; data reproduced from Ilas *et al.*, 2010b ; Radulescu *et al.*, 2010. Times expressed as cumulative time from the beginning of assembly irradiation.

Cycle 5		Cycle 6		Cycle 7		Cycle 8	
Time (days)	Boron (ppm)	Time (days)	Boron (ppm)	Time (days)	Boron (ppm)	Time (days)	Boron (ppm)
6	965.6	338	1175.9	741.7	1016	1104.5	1228.9
30	876.6	362	1088.9	765.7	926.5	1104.6	1119.9
60	783.2	392	998.8	795.7	833.8	1128.6	1001.3
90	681.8	422	898.8	825.7	732.3	1158.6	874.3
120	583.2	452	800.2	855.7	632.7	1188.6	749.2
150	489.4	482	706.1	885.7	537.4	1218.6	627.3
180	400.9	512	617.3	915.7	447.5	1248.6	509.1
210	308.3	542	529.3	945.7	355.7	1278.6	395.4
240	206.9	572	432	975.7	255	1308.6	282.6
270	99.4	602	323.7	1005.7	148.6	1338.6	169.4
295.4	10.0	632	212.4	1044.6	7.8	1368.6	11.9
310	10.0	662	101.8	1083.6	7.8	1411	11.9
338	1175.9	687	10.0			1445.4	0.0
362	1088.9	718.7	10.0				

Table 15. Cycle-average power, fuel temperature, and moderator temperature for assembly 419 of the GKN-II sample measured in the REBUS campaign; data reproduced from Ilas *et al.*, 2010b ; Radulescu *et al.*, 2010.

Cycle	Duration (EFPD)	Down (days)	Power $\left(\frac{MW}{MTU}\right)$	Fuel temp. (K)	Moderator	
					Temp. (K)	Density $\left(\frac{g}{cm^3}\right)$
5	310.0	22	56.265	1018.0	605.0	0.646
6	386.7	17	47.634	904.3	599.0	0.665
7	347.9	15	40.820	819.7	593.3	0.681
8	346.8	—	11.626	646.1	574.2	0.725

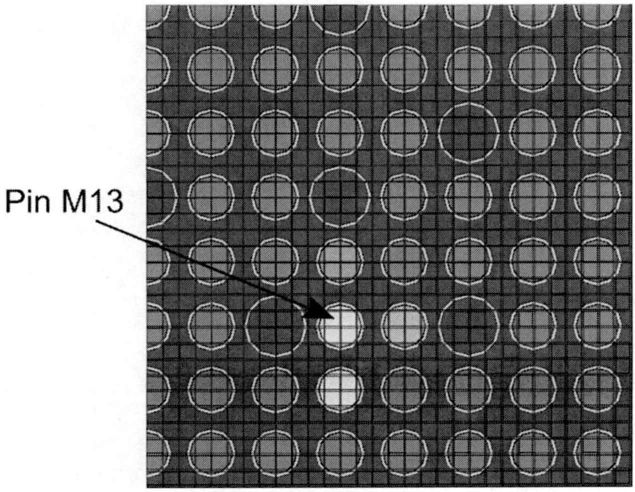

Figure. 10. ARIANE assembly 1240 (sample GU1) quarter-lattice (reflected) model used for benchmarking studies, developed from dimensions, irradiation history, and sample input published in Ilas et al. (2010b); samples were measured from rod M13.

4. DEPLETION BENCHMARKING STUDIES BASED ON THE PROPOSED CROSS-SECTION EVALUATIONS

In the following sections, the results of the isotopic benchmarks of each of the modeled assemblies is presented. Several notable trends common to each of the assemblies are apparent. First, changes to evaluations of the actinoids between ENDF/B-VII.0 and VII.1 lead to several significant changes in calculated actinoid inventories, notably for higher actinoid species (e.g., ^{242}Pu and other transuranic species such as Am and Cm), with a consequence of generally showing superior agreement for difficult to model nuclides such as ^{238}Pu (dependent upon several captures by ^{238}U to produce). As previously noted, the ENDF/B-VII.1 evaluation also contains a minor adjustment to the ^{153}Eu thermal capture cross-section, based on recent measurements (Basunia et al., 2014). As a result, the ENDF/B-VII.1 benchmarks have a higher rate of ^{153}Eu$(n,\gamma)^{154}$Eu captures, resulting in further over-predictions of the ^{154}Eu inventory.

Table 16. Gösgen assembly 1240 (ARIANE sample GU1) assembly design parameters used for modeling (reproduced from Ilas et al. (2010b)).

Assembly data			
Rod pitch (cm)	1.4732	Assembly pitch (cm)	20.78
Number of fuel rods	176	Number of water rods	5
Fuel rod data			
Fuel material type	UO_2		
Fuel pellet density (g/cm^3)	10.045		
Fuel pellet diameter (cm)	0.563		
Enrichment (wt% ^{235}U)	3.038		
Clad material	Zircaloy-4		
Clad inner diameter (cm)	0.95758		
Clad outer diameter (cm)	1.0922		
Average clad temperature (K)	620		
Initial uranium composition (wt%)		**Water rod data**	
^{234}U	0.027	Water rod material	Zircaloy-4
^{235}U	3.038	Water rod inner diameter (cm)	2.268
^{238}U	96.921	Water rod outer diameter (cm)	2.832

Table 17. Gösgen assembly 1240 (ARIANE sample GU1) assembly operating history used for modeling (reproduced from Ilas et al. (2010b)). Cycle powers presented are ^{148}Nd-corrected from Ilas et al. (2010b) to better approximate discharge burnup (ASTM International, 2012 ; Suyama and Mochizuki, 2006). "Effective" down time was calculated in Ilas et al. (2010b) as the sum of the actual down time plus the difference between the actual cycle duration and EFPD.

Cycle	EFPD	Effective down (d)	Soluble boron (ppm)	Power ($\frac{MWd}{MTU}$)	Fuel temperature (K)
12	0	45	1511	0	1151.3
	6		1179	59.19	1171.5
	150		565	59.19	1136.0
	294.9		8	59.19	1078.3
	317		8	53.50	1046.7
13	0	27	1477	0	919.3
	6		1145	47.03	967.7
	150		542	47.03	957.9
	294.9		7	47.03	943.1
	317		7	41.05	842.0
14	0	50	1517	0	888.9
	6		1178	44.74	894.4
	150		549	44.74	854.8
	294.9		5	44.74	841.4
	317		5	32.22	709.8
15	0	11	1594	0	806.6
	6		1243	36.28	829.8
	150		605	36.28	810.6
	294.9		5	36.28	804.0
	317		5	31.57	738.9

The reader should note that in addition to the three nominal nuclear data evaluations presented (ENDF/B-VII.0, VII.1, and JEFF-3.2), four additional evaluations are presented: ENDF/B-VII.0 with modifications based on Wright, and modified versions of ENDF/B-VII.0, ENDF/B-VII.1, and JEFF-3.2 with an amended thermal-region treatment based on Mughabghab's recommendations. (The reader is directed back to Table 3 for details on these amended evaluations.) The ENDF/B-VII.1 case with modifications by Wright was not evaluated due to the fact that this relies on an implicit (incorrect) assumption of a Wescott g_w factor of 1, which as identified by Mughabghab as an erroneous assumption (Mughabghab, 2016). Nevertheless, this evaluation is presented for the ENDF/VII.0 evaluation as a reference point of comparison to bound the possible modifications to the resolved resonance and thermal-region cross-section for ^{154}Eu, especially compared with the modifications proposed by Mughabghab.

When the proposed changes to the ^{154}Eu thermal capture cross-section are evaluated within the benchmarks, the inventories of ^{154}Eu generally come closer in line to experimental measurements, with a corresponding increase in the inventories of ^{155}Eu (also consistently over-predicted). An additional consequence of this change is a corresponding change to the inventories of Gd isotopes (^{154}Gd, produced from the decay of ^{154}Eu; ^{155}Gd, produced by the decay of ^{155}Eu; and ^{156}Gd, produced from the decay of ^{156}Eu). Outside of the Gd isotopes produced as decay products of Eu species, few changes are observed for other measured isotopes between the nominal ENDF/B-VII.0/VII.1 evaluations and the evaluated modifications, as is expected. (For example, actinoid inventories are virtually unchanged by the change to the ^{154}Eu thermal capture cross-section, as are inventories of important burnup indicator isotopes of Cs and Nd.).

Notably, while the overall trends in the predicted inventories of ^{154}Eu show an improved agreement to experimental benchmark values for both evaluations considered (Wright and Mughabghab), the magnitude of the change varies substantially between samples (as shown in the following sections), almost certainly due to local differences in the neutron spectrum shape.

Finally, to aid the reader in reproduction of the studies conducted herein, all models and calculated data used for the benchmarking studies discussed here are available on the author's personal GitHub repository[1] for examination and reuse.

4.1. Pickering-A CANDU 28-element assembly isotopic benchmark

The Pickering-A CANDU 28-element assembly is easily the most prominent example of the effect of the ^{154}Eu thermal capture cross-section, given the substantially higher relative neutron population (see Fig. 4); as a result, it was for this benchmark where the potential nuclear data issue was first discovered (Shoman and Skutnik, 2016). A comparison of the proposed evaluations of the ^{154}Eu capture cross-section are illustrated in Fig. 11. While data for ^{153}Eu is unavailable for this benchmark, several trends are immediately apparent. First, the CANDU assembly shows a significantly elevated sensitivity to thermal-region cross-sections, as is apparent by the change in the ^{154}Eu inventory resulting from the change from the ENDF/B-VII.0 to VII.1 evaluation (wherein the ^{153}Eu thermal capture cross-section increases by about 14.7% (from 312 b to 358 b), driving an observed change in the ^{154}Eu inventory of just over 10%.

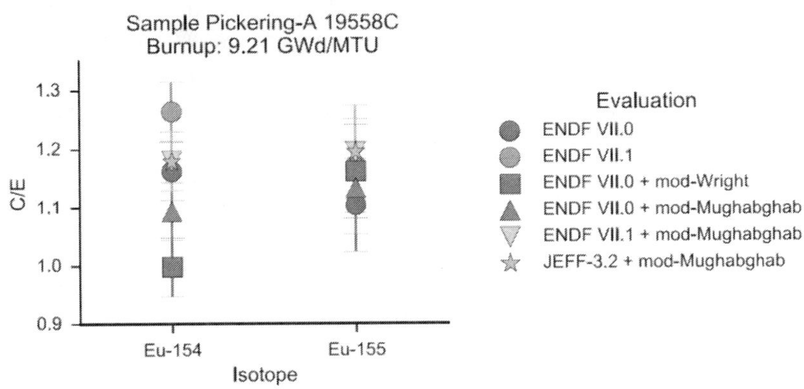

Figure. 11. C/E values for Eu isotopes for Pickering-A assembly 19558C (CANDU 28-element assembly) sample for each tested nuclear data evaluation; C/E values derived from measured isotope data and associated uncertainty ($\sigma_{meas.}$) based on Gauld and Litwin (1995). Error bars indicate relative experimental measurement uncertainty.

The modification to ENDF/VII.0 based on the Wright evaluation (used in ENDF/B-VI.8) brings the measured ^{154}Eu extremely close to the experimentally-measured value (within 0.25%, a change of 16.5%). Similarly, the proposed modifications by Mughabghab with the ENDF/B-VII.0 evaluation likewise corrects the overshoot significantly (by about 6.9%), within two standard deviations of the measurement uncertainty. Meanwhile, while the Mughabghab modification does reduce the overestimation of ^{154}Eu found in the ENDF/B-VII.1 evaluation (by about 8.4%), this modification still substantially over-predicts ^{154}Eu for the ENDF/B-VII.1 evaluation by approximately 18%, well outside the experimental measurement uncertainty bounds. Taken together with the more moderate over-predictions for this evaluation for LWR assemblies, this would appear to indicate that the thermal-region behavior in particular is still at issue, warranting further investigation into either the intensity of the neutron resonance placed at 0.188 eV or the $1/v$ region behavior of the cross-section. Of note is the relative lack of difference between the modified JEFF-3.2 evaluation and the modified ENDF/B-VII.1 evaluation for ^{154}Eu, implying a relative lack of sensitivity of ^{154}Eu inventories to higher-energy neutrons (>100 eV) in this system, likely due to the highly-thermalized spectrum (refer to Fig. 4 for details).

Similarly, for each increase in the ^{154}Eu population (i.e., moving from ENDF/B-VII.0 to VII.1) or increase in the rate of neutron captures by ^{154}Eu (i.e., the Wright and Mughabghab amendments), a corresponding increase in the ^{155}Eu is likewise observed. As expected, the increase in the ^{155}Eu population is directly proportional to the decrease in the ^{154}Eu population for adjustments to the capture cross-section; i.e., the Wright modification results in a 5.8% increase in ^{155}Eu, whereas the Mughabghab modification (for ENDF/B-VII.0) results in a 3.0% increase, directly proportional to the change in ^{154}Eu. Similarly, the change from ENDF/B-VII.0 to VII.1 results in a 6.8% increase in the ^{155}Eu inventory (despite no change to the ^{154}Eu or ^{155}Eu cross-sections). This pattern is relatively consistent between the modifications to the VII.0 and VII.1 evaluation with the Mughabghab amendment (a change of 6.1%). In all cases evaluated, the ^{155}Eu is substantially over-predicted and outside the 8% measurement uncertainty bounds, with the most severe over-prediction occurring for the combination of ENDF/B-VII.1 with the modification to ^{154}Eu proposed by Mughabghab (approximately 19.4%). This overshoot is likewise observed in the TMI-1 NJ05YU assembly samples (where ^{154}Eu over-predictions were likewise most severe), but is not as evident elsewhere. Ultimately, the causes of this overshoot in ^{155}Eu inventories are not immediately clear, warranting further study beyond the scope of this work.

Extended results of the evaluations tested on measured isotopes are presented in Table 18. As discussed previously, the modifications to the ^{154}Eu capture cross-section have almost no discernible effect on other isotopic inventories outside of ^{155}Eu. Additionally, the good agreement of major actinoid species of U and Pu along with minor actinoids ^{237}Np and ^{241}Am and the burnup indicator isotopes of Cs would indicate that the lattice model employed accurately captures the sample irradiation conditions, thus ruling out modeling error as the cause for discrepancies in the isotopic inventories of Eu.

Table 18. Pickering-A assembly 19558C (CANDU 28-element PHWR) sample calculated-to-experimental (measured) ratios for each nuclear data evaluation; C/E values derived from measured and associated uncertainty ($\sigma meas.$) isotope data based on Gauld and Litwin (1995).

Isotope	$\sigma_{meas.}$	C/E						
		ENDF VII.0	ENDF VII.1	ENDF VII.0 + mod-Wright	ENDF VII.0 + mod-Mugh.	ENDF VII.1 + mod-Mugh.	JEFF-3.2	JEFF-3.2 + mod-Mugh.
^{235}U	0.0240	1.0000	0.9940	0.9976	0.9940	0.9940	0.9940	0.9940
^{236}U	0.0370	1.0180	1.0217	1.0212	1.0222	1.0217	1.0217	1.0217
^{238}U	0.0000	1.0000	1.0000	1.0000	1.0000	1.0000	1.0000	1.0000
^{238}Pu	0.0560	0.9514	1.0707	0.9200	0.9364	1.0707	1.0707	1.0707
^{239}Pu	0.0250	0.9978	0.9941	1.0035	0.9963	0.9941	0.9941	0.9941
^{240}Pu	0.0370	1.0041	1.0085	0.9942	1.0065	1.0085	1.0085	1.0085
^{241}Pu	0.0910	0.9929	1.0034	0.9799	0.9966	1.0034	1.0034	1.0034
^{242}Pu	0.0670	1.0163	1.0423	1.0026	1.0256	1.0423	1.0423	1.0423
^{134}Cs	0.0700	0.9654	0.9553	0.9680	0.9714	0.9553	0.9553	0.9553
^{137}Cs	0.0500	0.9681	0.9716	0.9730	0.9727	0.9716	0.9716	0.9716
^{154}Eu	0.0500	1.1623	1.2641	0.9975	1.0936	1.1803	1.2629	1.1787
^{155}Eu	0.0800	1.1033	1.1713	1.1612	1.1331	1.1943	1.1716	1.1946
^{237}Np	0.2000	1.0100	0.9633	0.9789	0.9773	0.9633	0.9633	0.9633
^{241}Am	0.2000	1.0371	1.0435	1.0435	1.0446	1.0435	1.0435	1.0435
^{244}Cm	0.1500	1.1067	1.1650	1.1146	1.1303	1.1650	1.1650	1.1650

4.2. TMI-1 NJ05YU isotopic benchmarks

Fig. 12 illustrates the behavior of several Eu species (^{151}Eu, ^{153}Eu, ^{154}Eu, and ^{155}Eu) for each of the TMI-1 NJ05YU assembly samples. (Not shown is ^{152}Eu, which was substantially under-predicted for all cases, on the order of −90%.) In general, the TMI-1 samples show similar trends as observed to the Pickering-A CANDU assembly samples, wherein ^{154}Eu is consistently over-predicted (by approximately 7–15% for ENDF/B-VII.0), with slightly higher over-predictions (9–16%) for ENDF/B-VII.1. Here however the proposed modification based on the Wright evaluation appears to consistently over-correct ^{154}Eu inventories (by between −6.6 to −0.7%). The ENDF/B-VII.0 evaluation with Mughabghab's proposed changes lowers the overshoot to around 1.2–7.7%. The change from ENDF/B-VII.0 to VII.1 results in a smaller uptick in ^{154}Eu (around 1.4–1.6%), which is consistently observed in the modified (Mughabghab) ENDF/B-VII.1 evaluation (1.6–1.8%). Similarly, the modified JEFF-based evaluations (with a slightly elevated resonance/fast cross-section compared to ENDF) consistently show a reduction of about 0.4% in ^{154}Eu inventories compared to the modified ENDF/B-VII.1 evaluation, implying a moderate sensitivity to this region. Overall the ENDF/B-VII.1 evaluation with the change proposed by Mughabghab still results in a moderate over-prediction of ^{154}Eu (of around 3.0–9.4%), which while a notable improvement over the ENDF/VII.1 nominal case still resides well outside of the experimental uncertainty of 2.0%.

Meanwhile, for ^{155}Eu, similar trends are likewise observed, with moderate over-predictions for ENDF/B-VII.0 (0.7–8.4%), ranging slightly higher for ENDF/B-VII.1 (1.8–9.4%) and highest for the VII.1 evaluations with a modified ^{154}Eu thermal capture cross-section (3.3–11.0%).

Also notable in the TMI-1 samples is in the fact that data on other fission product isotopes of Sm and Gd are available, shown with other selected isotopes as Table 19, Table 20, Table 21, Table 22 ; Table 23. As before, the modifications to the ^{154}Eu capture cross-sections do not appear to have any effects on other isotopic species outside of Gd (plotted as Fig. 13), where noticeable improvements are observed for ^{154}Gd (produced by the decay of ^{154}Eu), while ^{155}Gd (produced from the decay of ^{155}Eu) and ^{156}Gd (produced from captures by ^{155}Eu and subsequent decay of short-lived ^{156}Eu) both show upticks on the order of 3–4%, concordant with an increase in the ^{155}Eu inventory. In general, increases in these latter two isotopes appear to be driven primarily by the change from ENDF/B-VII.0 to VII.1, with the change in the ^{154}Eu showing a noticeable but smaller effect.

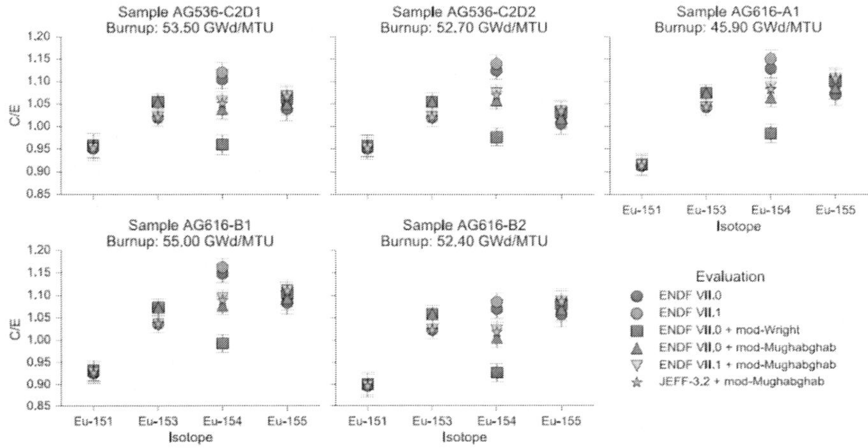

Figure. 12. C/E values for selected Eu isotopes for TMI-1 NJ05YU samples from rods AG536 (pin D5; samples C2D1 and C2D2) and AG616 (pin H6; samples A1, B1, and B2) for each tested nuclear data evaluation; C/E values derived from measured isotope data and associated uncertainty ($\sigma_{meas.}$) based on Gauld et al. (2016). Error bars indicate relative experimental measurement uncertainty.

For the ^{154}Gd inventories, in each evaluation the modifications proposed by Mughabghab to ENDF/B-VII.0 and VII.1 for ^{154}Eu appear to bring the ^{154}Gd closest to experimentally-observed values. While inventories are ^{154}Gd are still generally outside of the 2% measurement uncertainty, the improvements in ^{154}Gd inventories appear to mirror the trends observed in ^{154}Eu.

Finally, as before the detailed isotopic inventories available for comparison in Table 19, Table 20, Table 21, Table 22 ; Table 23, including both major and minor actinoid species (U, Pu, Am, ^{237}Np, and ^{244}Cm) as well as major burnup indicator nuclides (including Nd species as well as Cs) all point to a good model agreement to irradiation conditions for all samples modeled and generally consistent with prior validation studies performed by ORNL (Gauld et al., 2016), again ruling out serious model discrepancies as a potential source of bias in the observed ^{154}Eu inventories.

Table 19. TMI-1 NJ05YU sample AG536-C2D1 calculated-to-experimental (measured) ratios for each nuclear data evaluation; C/E values derived from measured isotope data and associated uncertainty ($\sigma_{meas.}$) based on Gauld et al. (2016).

Isotope	σ_{meas}	C/E						
		ENDF VII.0	ENDF VII.1	ENDF VII.0 + mod-Wright	ENDF VII.0 + mod-Mugh.	ENDF VII.1 + mod-Mugh.	JEFF-3.2	JEFF-3.2 + mod-Mugh.
^{233}U	1.0000	0.0202	0.0181	0.0202	0.0202	0.0181	0.0181	0.0181
^{234}U	0.0240	1.0470	1.0748	1.0470	1.0470	1.0748	1.0748	1.0748
^{235}U	0.0130	0.9488	0.9401	0.9491	0.9488	0.9404	0.9401	0.9404
^{236}U	0.0190	0.9908	0.9901	0.9908	0.9908	0.9901	0.9901	0.9901
^{238}U	0.0100	0.9845	0.9848	0.9845	0.9845	0.9848	0.9848	0.9848
^{238}Pu	0.0240	0.8634	0.9667	0.8634	0.8634	0.9667	0.9667	0.9667
^{239}Pu	0.0150	1.0327	1.0098	1.0330	1.0327	1.0101	1.0098	1.0101
^{240}Pu	0.0180	1.0304	1.0116	1.0304	1.0304	1.0116	1.0116	1.0116
^{241}Pu	0.0200	1.0033	0.9937	1.0038	1.0036	0.9937	0.9937	0.9937
^{242}Pu	0.0170	0.9685	1.0059	0.9683	0.9683	1.0059	1.0059	1.0059
^{237}Np	0.1000	0.7539	0.7454	0.7542	0.7539	0.7454	0.7454	0.7454
^{241}Am	0.1000	1.1585	1.1402	1.1596	1.1596	1.1402	1.1402	1.1402
^{243}Am	0.1000	0.8879	0.7962	0.8879	0.8879	0.7962	0.7962	0.7962
^{244}Cm	0.1000	0.8050	0.8214	0.8052	0.8052	0.8214	0.8214	0.8214
^{133}Cs	0.0130	1.0221	1.0282	1.0221	1.0221	1.0282	1.0282	1.0282
^{134}Cs	0.1000	1.0367	1.0133	1.0370	1.0367	1.0133	1.0133	1.0133
^{135}Cs	0.0140	0.9810	0.9781	0.9813	0.9813	0.9781	0.9781	0.9781
^{137}Cs	0.0140	1.0371	1.0371	1.0371	1.0371	1.0371	1.0371	1.0371
^{142}Nd	0.0200	1.0044	1.0133	1.0044	1.0044	1.0133	1.0133	1.0133
^{143}Nd	0.0200	0.9747	0.9721	0.9747	0.9747	0.9721	0.9721	0.9721
^{144}Nd	0.0180	0.9580	0.9602	0.9571	0.9580	0.9602	0.9602	0.9602
^{145}Nd	0.0200	0.9574	0.9741	0.9574	0.9574	0.9741	0.9741	0.9741
^{146}Nd	0.0200	0.9914	0.9776	0.9914	0.9914	0.9776	0.9776	0.9776
^{148}Nd	0.0210	0.9991	0.9995	0.9991	0.9991	0.9995	0.9995	0.9995
^{150}Nd	0.0190	1.0289	1.0289	1.0289	1.0289	1.0289	1.0289	1.0289
^{147}Sm	0.0160	0.9908	0.9845	0.9908	0.9908	0.9845	0.9845	0.9845
^{148}Sm	0.0150	0.9349	0.9389	0.9349	0.9349	0.9389	0.9389	0.9389
^{149}Sm	0.0200	0.9652	0.9582	0.9658	0.9652	0.9587	0.9582	0.9587
^{150}Sm	0.0150	1.0198	1.0226	1.0198	1.0198	1.0226	1.0226	1.0226
^{151}Sm	0.0200	0.9496	0.9440	0.9496	0.9496	0.9440	0.9440	0.9440
^{152}Sm	0.0170	1.0174	1.0159	1.0174	1.0174	1.0152	1.0159	1.0152
^{154}Sm	0.0170	1.0590	1.0596	1.0590	1.0590	1.0596	1.0596	1.0596
^{151}Eu	0.0270	0.9572	0.9515	0.9577	0.9572	0.9515	0.9515	0.9515
^{152}Eu	0.1100	0.1019	0.0998	0.1020	0.1019	0.0998	0.0998	0.0998
^{153}Eu	0.0190	1.0539	1.0193	1.0539	1.0539	1.0193	1.0193	1.0193
^{154}Eu	0.0220	1.1048	1.1193	0.9602	1.0384	1.0550	1.1172	1.0508
^{155}Eu	0.0250	1.0386	1.0554	1.0633	1.0495	1.0663	1.0554	1.0663
^{152}Gd	0.1600	0.6375	0.6314	0.6378	0.6375	0.6314	0.6314	0.6314
^{154}Gd	0.0180	1.0986	1.1140	0.9584	1.0350	1.0523	1.1124	1.0482
^{155}Gd	0.0180	1.0344	1.0508	1.0586	1.0453	1.0617	1.0508	1.0617
^{156}Gd	0.0160	0.9654	0.9920	0.9945	0.9784	1.0045	0.9920	1.0055
^{157}Gd	0.0490	0.8251	0.8317	0.8402	0.8322	0.8387	0.8322	0.8387
^{158}Gd	0.0170	1.1869	1.1991	1.2004	1.1932	1.2053	1.1991	1.2053
^{160}Gd	0.0210	1.0357	1.0347	1.0357	1.0357	1.0347	1.0347	1.0347

Table 20. TMI-1 NJ05YU sample AG536-C2D2 calculated-to-experimental (measured) ratios for each nuclear data evaluation; measured isotope data and associated uncertainty ($\sigma meas.$) based on Gauld et al. (2016).

Isotope	σ_{meas}	C/E						
		ENDF VII.0	ENDF VII.1	ENDF VII.0 + mod-Wright	ENDF VII.0 + mod-Mugh.	ENDF VII.1 + mod-Mugh.	JEFF-3.2	JEFF-3.2 + mod-Mugh.
^{233}U	1.0000	0.0136	0.0123	0.0137	0.0137	0.0123	0.0123	0.0123
^{234}U	0.0410	1.0178	1.0454	1.0178	1.0178	1.0454	1.0454	1.0454
^{235}U	0.0210	0.9665	0.9574	0.9668	0.9668	0.9577	0.9577	0.9577
^{236}U	0.0220	0.9973	0.9970	0.9973	0.9973	0.9970	0.9970	0.9970
^{238}U	0.0130	0.9870	0.9874	0.9870	0.9870	0.9872	0.9874	0.9872
^{238}Pu	0.0110	0.9442	1.0590	0.9442	0.9442	1.0590	1.0590	1.0590
^{239}Pu	0.0100	1.0745	1.0507	1.0748	1.0745	1.0507	1.0507	1.0507
^{240}Pu	0.0110	1.0697	1.0497	1.0697	1.0697	1.0497	1.0497	1.0497
^{241}Pu	0.0150	1.0551	1.0444	1.0556	1.0554	1.0448	1.0444	1.0448
^{242}Pu	0.0130	0.9941	1.0317	0.9941	0.9941	1.0317	1.0326	1.0317
^{237}Np	0.1000	0.7675	0.7584	0.7677	0.7675	0.7586	0.7584	0.7586
^{241}Am	0.1000	1.1616	1.1423	1.1616	1.1616	1.1423	1.1423	1.1423
^{243}Am	0.1000	0.9041	0.8110	0.9041	0.9041	0.8110	0.8104	0.8104
^{244}Cm	0.1000	0.8423	0.8594	0.8426	0.8423	0.8597	0.8590	0.8592
^{133}Cs	0.0190	1.0345	1.0406	1.0345	1.0345	1.0406	1.0406	1.0406
^{134}Cs	0.1000	0.9452	0.9242	0.9452	0.9452	0.9242	0.9242	0.9242
^{135}Cs	0.0210	1.0032	0.9997	1.0032	1.0032	0.9997	0.9997	0.9997
^{137}Cs	0.0200	1.0485	1.0485	1.0485	1.0485	1.0485	1.0485	1.0485
^{142}Nd	0.0190	1.0098	1.0183	1.0094	1.0094	1.0178	1.0183	1.0178
^{143}Nd	0.0170	0.9808	0.9782	0.9808	0.9808	0.9782	0.9782	0.9782
^{144}Nd	0.0160	0.9552	0.9570	0.9552	0.9552	0.9570	0.9570	0.9570
^{145}Nd	0.0180	0.9577	0.9747	0.9575	0.9577	0.9747	0.9747	0.9747
^{146}Nd	0.0180	0.9957	0.9821	0.9957	0.9957	0.9821	0.9813	0.9813
^{148}Nd	0.0190	1.0017	1.0020	1.0017	1.0017	1.0020	1.0020	1.0020
^{150}Nd	0.0180	1.0199	1.0206	1.0199	1.0199	1.0206	1.0206	1.0206
^{147}Sm	0.0130	0.9879	0.9814	0.9873	0.9879	0.9814	0.9814	0.9814
^{148}Sm	0.0140	0.9464	0.9504	0.9464	0.9464	0.9504	0.9504	0.9504
^{149}Sm	0.0230	0.9971	0.9904	0.9977	0.9977	0.9904	0.9904	0.9904
^{150}Sm	0.0130	1.0230	1.0257	1.0230	1.0230	1.0261	1.0257	1.0261
^{151}Sm	0.0180	0.9795	0.9737	0.9795	0.9795	0.9737	0.9737	0.9737
^{152}Sm	0.0150	1.0014	0.9993	1.0014	1.0014	0.9993	0.9993	0.9993
^{154}Sm	0.0150	1.0729	1.0737	1.0729	1.0729	1.0737	1.0737	1.0737
^{151}Eu	0.0240	0.9568	0.9510	0.9572	0.9572	0.9510	0.9510	0.9515
^{152}Eu	0.0630	0.0229	0.0224	0.0229	0.0229	0.0224	0.0224	0.0224
^{153}Eu	0.0200	1.0544	1.0196	1.0544	1.0544	1.0196	1.0196	1.0196
^{154}Eu	0.0200	1.1242	1.1388	0.9764	1.0574	1.0731	1.1367	1.0689
^{155}Eu	0.0240	1.0066	1.0227	1.0302	1.0179	1.0331	1.0227	1.0340
^{152}Gd	0.0600	0.2207	0.2185	0.2208	0.2207	0.2186	0.2185	0.2186
^{154}Gd	0.0200	1.1240	1.1396	0.9797	1.0585	1.0758	1.1380	1.0719
^{155}Gd	0.0200	1.0378	1.0541	1.0618	1.0494	1.0649	1.0541	1.0656
^{156}Gd	0.0180	0.9778	1.0049	1.0074	0.9911	1.0177	1.0049	1.0187
^{157}Gd	0.0640	0.8176	0.8245	0.8324	0.8240	0.8309	0.8245	0.8309
^{158}Gd	0.0200	1.2092	1.2219	1.2235	1.2154	1.2280	1.2219	1.2287
^{160}Gd	0.0290	1.0430	1.0420	1.0430	1.0430	1.0420	1.0420	1.0420

Table 21. TMI-1 NJ05YU sample AG616-A1 calculated-to-experimental (measured) ratios for each nuclear data evaluation; C/E values derived from measured isotope data and associated uncertainty ($\sigma_{meas.}$) based on Gauld et al. (2016).

Isotope	σ_{meas}	C/E						
		ENDF VII.0	ENDF VII.1	ENDF VII.0 + mod-Wright	ENDF VII.0 + mod-Mugh.	ENDF VII.1 + mod-Mugh.	JEFF-3.2	JEFF-3.2 + mod-Mugh.
^{233}U	1.0000	0.0113	0.0100	0.0113	0.0113	0.0100	0.0100	0.0100
^{234}U	0.0240	1.0385	1.0569	1.0385	1.0385	1.0569	1.0569	1.0569
^{235}U	0.0240	0.9146	0.9105	0.9148	0.9146	0.9105	0.9105	0.9107
^{236}U	0.0240	1.0054	1.0050	1.0054	1.0054	1.0050	1.0050	1.0050
^{238}U	0.0110	0.9921	0.9923	0.9921	0.9921	0.9923	0.9923	0.9923
^{238}Pu	0.0180	0.9278	1.0306	0.9278	0.9278	1.0306	1.0306	1.0306
^{239}Pu	0.0140	1.0220	1.0052	1.0223	1.0220	1.0052	1.0052	1.0052
^{240}Pu	0.0150	1.0594	1.0438	1.0594	1.0594	1.0438	1.0438	1.0438
^{241}Pu	0.0190	1.0209	1.0154	1.0214	1.0214	1.0154	1.0154	1.0154
^{242}Pu	0.0160	1.0337	1.0699	1.0337	1.0337	1.0695	1.0699	1.0699
^{237}Np	0.1000	0.8239	0.8147	0.8239	0.8239	0.8150	0.8147	0.8147
^{241}Am	0.1000	1.0478	1.0356	1.0478	1.0478	1.0356	1.0356	1.0356
^{243}Am	0.1000	1.0164	0.9103	1.0164	1.0164	0.9108	0.9103	0.9103
^{244}Cm	0.1000	0.9644	0.9815	0.9649	0.9644	0.9825	0.9815	0.9820
^{133}Cs	0.0150	1.0558	1.0614	1.0558	1.0558	1.0614	1.0614	1.0614
^{134}Cs	0.1000	0.9908	0.9649	0.9908	0.9908	0.9653	0.9649	0.9649
^{135}Cs	0.0160	1.0082	1.0068	1.0082	1.0082	1.0068	1.0068	1.0068
^{137}Cs	0.0160	1.0733	1.0743	1.0733	1.0733	1.0743	1.0743	1.0743
^{142}Nd	0.0210	1.0451	1.0512	1.0451	1.0451	1.0512	1.0512	1.0512
^{143}Nd	0.0190	0.9714	0.9705	0.9714	0.9714	0.9705	0.9705	0.9705
^{144}Nd	0.0170	0.9846	0.9862	0.9846	0.9846	0.9862	0.9862	0.9862
^{145}Nd	0.0190	0.9718	0.9868	0.9718	0.9718	0.9865	0.9868	0.9868
^{146}Nd	0.0190	1.0104	0.9975	1.0104	1.0104	0.9975	0.9975	0.9975
^{148}Nd	0.0190	1.0158	1.0162	1.0158	1.0158	1.0162	1.0162	1.0162
^{150}Nd	0.0180	1.0387	1.0387	1.0387	1.0387	1.0387	1.0387	1.0387
^{147}Sm	0.0160	0.9994	0.9951	0.9994	0.9994	0.9957	0.9951	0.9951
^{148}Sm	0.0160	0.9578	0.9605	0.9578	0.9578	0.9605	0.9605	0.9605
^{149}Sm	0.0240	0.8905	0.8871	0.8905	0.8905	0.8871	0.8871	0.8871
^{150}Sm	0.0160	1.0411	1.0437	1.0411	1.0411	1.0437	1.0437	1.0437
^{151}Sm	0.0250	0.7796	0.7771	0.7802	0.7796	0.7771	0.7771	0.7771
^{152}Sm	0.0220	1.0262	1.0239	1.0254	1.0262	1.0239	1.0239	1.0239
^{154}Sm	0.0200	1.0598	1.0604	1.0598	1.0598	1.0604	1.0604	1.0604
^{151}Eu	0.0230	0.9163	0.9128	0.9163	0.9163	0.9133	0.9128	0.9133
^{152}Eu	0.0590	0.0277	0.0273	0.0277	0.0277	0.0273	0.0273	0.0273
^{153}Eu	0.0190	1.0736	1.0421	1.0736	1.0736	1.0414	1.0421	1.0421
^{154}Eu	0.0200	1.1281	1.1497	0.9843	1.0629	1.0860	1.1481	1.0820
^{155}Eu	0.0250	1.0706	1.0917	1.1007	1.0847	1.1052	1.0919	1.1057
^{152}Gd	0.0720	0.2191	0.2179	0.2192	0.2191	0.2179	0.2179	0.2179
^{154}Gd	0.0200	1.0795	1.1008	0.9451	1.0185	1.0418	1.0998	1.0374
^{155}Gd	0.0190	1.0508	1.0706	1.0800	1.0640	1.0837	1.0715	1.0847
^{156}Gd	0.0170	1.0145	1.0407	1.0480	1.0291	1.0560	1.0415	1.0567
^{157}Gd	0.0840	0.6695	0.6771	0.6814	0.6748	0.6824	0.6771	0.6824
^{158}Gd	0.0190	1.1965	1.2066	1.2084	1.2022	1.2123	1.2066	1.2123
^{160}Gd	0.0220	1.0089	1.0082	1.0089	1.0089	1.0082	1.0082	1.0082

Table 22. TMI-1 NJ05YU sample AG616-B1 calculated-to-experimental (measured) ratios for each nuclear data evaluation; C/E values derived from measured isotope data and associated uncertainty ($\sigma_{meas.}$) based on Gauld et al. (2016).

Isotope	$\sigma_{meas.}$	C/E						
		ENDF VII.0	ENDF VII.1	ENDF VII.0 + mod-Wright	ENDF VII.0 + mod-Mugh.	ENDF VII.1 + mod-Mugh.	JEFF-3.2	JEFF-3.2 + mod-Mugh.
^{233}U	1.0000	0.0182	0.0166	0.0183	0.0182	0.0165	0.0166	0.0166
^{234}U	0.0240	1.0205	1.0520	1.0205	1.0205	1.0520	1.0520	1.0520
^{235}U	0.0240	0.9071	0.8998	0.9074	0.9071	0.8998	0.8998	0.8998
^{236}U	0.0240	1.0030	1.0019	1.0030	1.0030	1.0019	1.0019	1.0019
^{238}U	0.0110	0.9887	0.9887	0.9885	0.9885	0.9887	0.9887	0.9887
^{238}Pu	0.0180	0.9594	1.0825	0.9600	0.9594	1.0825	1.0825	1.0825
^{239}Pu	0.0140	1.0415	1.0214	1.0419	1.0419	1.0217	1.0214	1.0217
^{240}Pu	0.0150	1.0627	1.0445	1.0627	1.0627	1.0445	1.0445	1.0445
^{241}Pu	0.0190	1.0308	1.0215	1.0311	1.0308	1.0215	1.0215	1.0215
^{242}Pu	0.0160	1.0182	1.0565	1.0182	1.0182	1.0565	1.0565	1.0565
^{237}Np	0.1000	0.7779	0.7677	0.7781	0.7781	0.7679	0.7677	0.7679
^{241}Am	0.1000	1.0757	1.0589	1.0757	1.0757	1.0589	1.0589	1.0589
^{243}Am	0.1000	0.9404	0.8424	0.9404	0.9404	0.8430	0.8424	0.8424
^{244}Cm	0.1000	0.8939	0.9107	0.8941	0.8939	0.9113	0.9107	0.9109
^{133}Cs	0.0150	1.0428	1.0494	1.0428	1.0428	1.0494	1.0494	1.0494
^{134}Cs	0.1000	1.0244	1.0000	1.0244	1.0244	1.0000	1.0000	1.0000
^{135}Cs	0.0160	1.0028	1.0002	1.0028	1.0028	1.0002	1.0002	1.0002
^{137}Cs	0.0160	1.0665	1.0665	1.0657	1.0665	1.0665	1.0665	1.0665
^{142}Nd	0.0210	1.0603	1.0682	1.0599	1.0603	1.0682	1.0682	1.0682
^{143}Nd	0.0190	0.9759	0.9733	0.9759	0.9759	0.9733	0.9733	0.9733
^{144}Nd	0.0170	0.9799	0.9816	0.9799	0.9799	0.9816	0.9816	0.9816
^{145}Nd	0.0190	0.9667	0.9848	0.9657	0.9657	0.9838	0.9848	0.9848
^{146}Nd	0.0190	1.0116	0.9967	1.0116	1.0116	0.9967	0.9967	0.9967
^{148}Nd	0.0190	1.0137	1.0143	1.0137	1.0137	1.0143	1.0143	1.0143
^{150}Nd	0.0180	1.0361	1.0367	1.0361	1.0361	1.0367	1.0367	1.0367
^{147}Sm	0.0160	0.9953	0.9891	0.9944	0.9944	0.9891	0.9891	0.9891
^{148}Sm	0.0160	0.9681	0.9720	0.9688	0.9681	0.9720	0.9720	0.9720
^{149}Sm	0.0240	1.0337	1.0279	1.0342	1.0337	1.0285	1.0279	1.0285
^{150}Sm	0.0160	1.0461	1.0486	1.0461	1.0461	1.0486	1.0486	1.0486
^{151}Sm	0.0250	0.9512	0.9462	0.9519	0.9512	0.9462	0.9462	0.9462
^{152}Sm	0.0220	1.0236	1.0215	1.0229	1.0229	1.0208	1.0215	1.0208
^{154}Sm	0.0200	1.1019	1.1026	1.1019	1.1019	1.1026	1.1026	1.1026
^{151}Eu	0.0230	0.9313	0.9261	0.9318	0.9313	0.9266	0.9261	0.9266
^{152}Eu	0.0590	0.0170	0.0166	0.0170	0.0170	0.0167	0.0166	0.0167
^{153}Eu	0.0190	1.0735	1.0368	1.0735	1.0735	1.0362	1.0368	1.0368
^{154}Eu	0.0200	1.1486	1.1622	0.9926	1.0769	1.0936	1.1611	1.0894
^{155}Eu	0.0250	1.0841	1.1006	1.1054	1.0938	1.1103	1.1006	1.1103
^{152}Gd	0.0720	0.2022	0.2004	0.2023	0.2022	0.2004	0.2004	0.2004
^{154}Gd	0.0200	1.1185	1.1333	0.9696	1.0505	1.0677	1.1322	1.0636
^{155}Gd	0.0190	1.0594	1.0752	1.0805	1.0692	1.0850	1.0759	1.0850
^{156}Gd	0.0170	1.0239	1.0511	1.0530	1.0375	1.0638	1.0511	1.0648
^{157}Gd	0.0840	0.9810	0.9908	0.9973	0.9886	0.9984	0.9913	0.9989
^{158}Gd	0.0190	1.2610	1.2752	1.2758	1.2675	1.2823	1.2752	1.2823
^{160}Gd	0.0220	1.0709	1.0709	1.0719	1.0719	1.0709	1.0709	1.0709

Table 23. TMI-1 NJ05YU sample AG616-B2 calculated-to-experimental (measured) ratios for each nuclear data evaluation; C/E values derived from measured isotope data based on Gauld et al. (2016).

Isotope	σ_{meas}	C/E						
		ENDF VII.0	ENDF VII.1	ENDF VII.0 + mod-Wright	ENDF VII.0 + mod-Mugh.	ENDF VII.1 + mod-Mugh.	JEFF-3.2	JEFF-3.2 + mod-Mugh.
^{233}U	1.0000	0.0143	0.0129	0.0143	0.0143	0.0129	0.0129	0.0129
^{234}U	0.2500	1.0285	1.0561	1.0285	1.0285	1.0561	1.0561	1.0561
^{235}U	0.0180	0.8712	0.8650	0.8715	0.8712	0.8653	0.8653	0.8653
^{236}U	0.0220	0.9981	0.9973	0.9981	0.9981	0.9973	0.9973	0.9973
^{238}U	0.0160	0.9890	0.9892	0.9890	0.9890	0.9892	0.9892	0.9892
^{238}Pu	0.0290	0.9468	1.0631	0.9468	0.9468	1.0631	1.0631	1.0631
^{239}Pu	0.0160	0.9682	0.9502	0.9685	0.9682	0.9502	0.9502	0.9502
^{240}Pu	0.0170	1.0158	0.9987	1.0158	1.0158	0.9987	0.9987	0.9987
^{241}Pu	0.0250	0.9735	0.9657	0.9740	0.9738	0.9659	0.9657	0.9659
^{242}Pu	0.0200	1.0179	1.0548	1.0179	1.0179	1.0548	1.0548	1.0548
^{237}Np	0.1000	0.7730	0.7633	0.7730	0.7730	0.7635	0.7633	0.7633
^{241}Am	0.1000	1.0477	1.0327	1.0477	1.0477	1.0327	1.0327	1.0327
^{243}Am	0.1000	0.9443	0.8467	0.9443	0.9443	0.8467	0.8460	0.8460
^{244}Cm	0.1000	0.8693	0.8860	0.8696	0.8696	0.8862	0.8855	0.8857
^{133}Cs	0.0200	1.0490	1.0552	1.0490	1.0490	1.0552	1.0552	1.0552
^{134}Cs	0.1000	1.0893	1.0630	1.0896	1.0893	1.0630	1.0626	1.0630
^{135}Cs	0.0220	0.9688	0.9664	0.9688	0.9688	0.9664	0.9664	0.9664
^{137}Cs	0.0210	1.0576	1.0584	1.0576	1.0576	1.0584	1.0584	1.0584
^{142}Nd	0.0170	1.0504	1.0574	1.0499	1.0504	1.0574	1.0574	1.0574
^{143}Nd	0.0160	0.9603	0.9585	0.9603	0.9603	0.9585	0.9585	0.9585
^{144}Nd	0.0140	0.9840	0.9859	0.9840	0.9840	0.9859	0.9859	0.9859
^{145}Nd	0.0160	0.9681	0.9849	0.9681	0.9681	0.9849	0.9854	0.9854
^{146}Nd	0.0150	1.0079	0.9939	1.0079	1.0079	0.9939	0.9930	0.9930
^{148}Nd	0.0170	1.0110	1.0114	1.0110	1.0110	1.0114	1.0114	1.0114
^{150}Nd	0.0160	1.0344	1.0352	1.0344	1.0344	1.0352	1.0352	1.0352
^{147}Sm	0.0170	0.9952	0.9902	0.9952	0.9952	0.9902	0.9902	0.9902
^{148}Sm	0.0170	0.9376	0.9405	0.9376	0.9376	0.9405	0.9405	0.9405
^{149}Sm	0.0210	0.9960	0.9909	0.9960	0.9960	0.9909	0.9909	0.9909
^{150}Sm	0.0170	1.0276	1.0307	1.0276	1.0276	1.0307	1.0307	1.0307
^{151}Sm	0.0240	0.8775	0.8734	0.8781	0.8775	0.8734	0.8734	0.8734
^{152}Sm	0.0180	1.0360	1.0339	1.0353	1.0353	1.0339	1.0339	1.0339
^{154}Sm	0.0180	1.0734	1.0740	1.0734	1.0734	1.0740	1.0740	1.0740
^{151}Eu	0.0270	0.9003	0.8961	0.9007	0.9007	0.8961	0.8961	0.8961
^{152}Eu	0.0420	0.0404	0.0396	0.0404	0.0404	0.0397	0.0397	0.0397
^{153}Eu	0.0200	1.0584	1.0232	1.0584	1.0584	1.0232	1.0232	1.0232
^{154}Eu	0.0210	1.0700	1.0855	0.9269	1.0048	1.0221	1.0834	1.0181
^{155}Eu	0.0290	1.0582	1.0753	1.0813	1.0693	1.0863	1.0753	1.0863
^{152}Gd	0.0980	0.3445	0.3416	0.3447	0.3445	0.3416	0.3416	0.3416
^{154}Gd	0.0290	1.0462	1.0626	0.9093	0.9836	1.0019	1.0609	0.9978
^{155}Gd	0.0190	1.0466	1.0640	1.0703	1.0577	1.0750	1.0640	1.0750
^{156}Gd	0.0160	0.9990	1.0259	1.0290	1.0124	1.0388	1.0259	1.0399
^{157}Gd	0.0760	0.6752	0.6825	0.6870	0.6805	0.6878	0.6825	0.6878
^{158}Gd	0.0180	1.2053	1.2182	1.2189	1.2113	1.2245	1.2182	1.2245
^{160}Gd	0.0240	1.0397	1.0397	1.0397	1.0397	1.0397	1.0397	1.0397

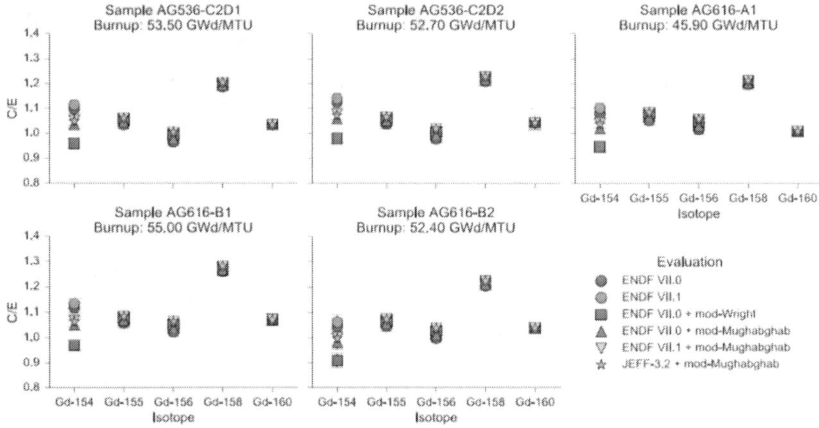

Figure. 13. C/E values for selected Gd isotopes for TMI-1 NJ05YU samples from rods AG536 (pin D5; samples C2D1 and C2D2) and AG616 (pin H6; samples A1, B1, and B2) for each tested nuclear data evaluation; C/E values derived from measured isotope data and associated uncertainty ($\sigma_{meas.}$) based on Gauld et al. (2016). Error bars indicate relative experimental measurement uncertainty.

4.3. Calvert Cliffs D047/MKP-109 isotopic benchmarks

Fig. 14 shows the effect of the tested evaluations on selected Eu isotopes for the Calvert Cliffs assembly D047 samples. In general, the Calvert Cliffs samples show an overall lower initial bias in ^{154}Eu compared to the TMI-1 samples, in part likely owing to the lower average burnup of these samples (ranging from 27.4 to 44.3 $\frac{GWd}{MTU}$, compared with 45.9 to 55.9 $\frac{GWd}{MTU}$ for TMI-1 NJ05YU). Because ^{154}Eu tracks closely with total fuel burnup, a bias in the capture cross-section for ^{154}Eu should be expected to manifest more strongly for higher discharge burnups. Within the Calvert Cliffs samples, this appears to be true, where for the nominal ENDF/B-VII.0 and VII.1 evaluations, the over-predictions in ^{154}Eu are greatest for sample 87–63 (44.3 $\frac{GWd}{MTU}$ burnup) at 10.0% and 11.9% ($\pm 7.0\%$ measurement uncertainty), respectively; followed by sample 87–72 (discharge burnup: 37.1 $\frac{GWd}{MTU}$) with an over-prediction of 5.4% and 8.2% ($\pm 6.4\%$ measurement uncertainty), respectively; and finally, sample 87–81 (discharge burnup: 27.4 $\frac{GWd}{MTU}$), with an over-prediction for ^{154}Eu of 4.0% and 7.6% ($\pm 8.8\%$ measurement uncertainty), respectively. Given the relatively wide measurement uncertainties, it is difficult to draw a definitive conclusion of a bias in the ^{154}Eu inventory from these samples, given the proximity of the calculated inventories to the experimental uncertainty bounds. Similarly, the modified JEFF-3.2 evaluation gives ^{154}Eu inventories 0.3–0.4% closer to the measured values, consistent with findings for the TMI-1 samples. While this is a relatively small difference, it is notable in its consistency across evaluations.

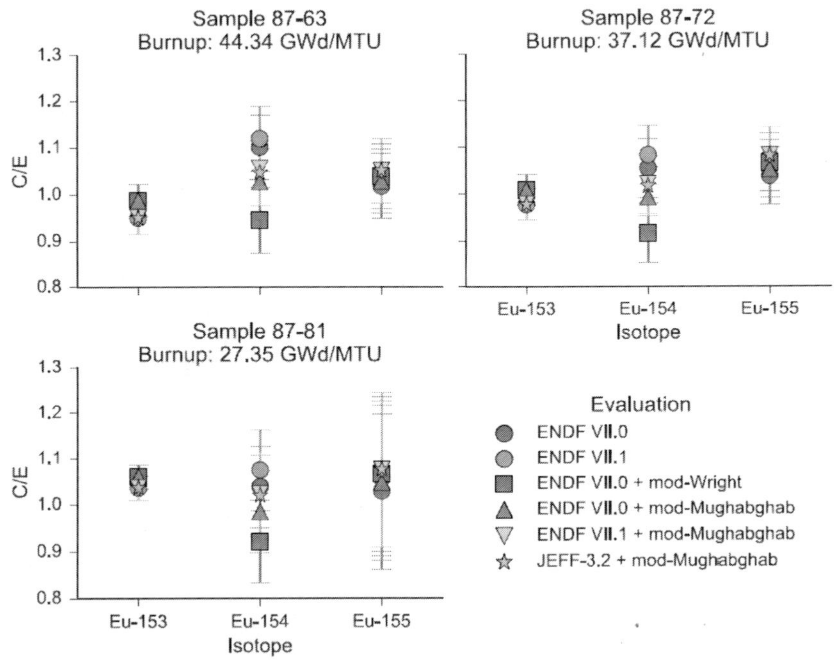

Figure. 14. C/E values for selected Eu isotopes for Calvert Cliffs samples 87–63, 87–72, and 87–71 for each tested nuclear data evaluation; C/E values derived from measured isotope data and associated uncertainty ($\sigma meas.$) based on Ilas et al. (2010a). Error bars indicate relative experimental measurement uncertainty.

By contrast, the Wright treatment of the ^{154}Eu cross-section strongly under-predicts ^{154}Eu inventories for nearly all of the Calvert Cliffs samples, coming close to the lower experimental uncertainty bounds for sample 87–63 (at -5.5%) and well below them for lower-burnup samples 87–72 and 87–81 (at -8.5% and -7.9%, respectively). This finding suggests that the treatment adapted from the Wright evaluation may in fact be too severe of an increase in the ^{154}Eu thermal capture cross-section, with the adjustment proposed by Mughabghab giving closer values for this particular case.

In contrast to the TMI-1 samples, the ENDF/B-VII.1 evaluation with the Mughabghab modification to ^{154}Eu show very good agreement to experimental measurement bounds, with samples 87–72 and 87–81 both well within the measurement uncertainty for this evaluation; sample 87–63 (the highest burnup sample) shows a slightly elevated concentration of ^{154}Eu at 5.5%, but still well within experimental measurement bounds. The inventories of ^{155}Eu likewise show excellent agreement for the modified ENDF/B-VII.1 evaluation, although all samples do show a slightly greater deviation from experimental values compared to the nominal cases. The ^{155}Eu concentration for sample 87–72 now lies slightly outside the measurement uncertainty at 8.0% difference, consistent with prior observations and again potentially indicating further study is needed as to the drivers of the ^{155}Eu concentration.

As with the TMI-1 samples, selected isotopes of Gd are also evaluated for the Calvert Cliffs samples and plotted as Fig. 15. Here the most significant changes are again to the inventory of ^{154}Gd, due to the elevated removal of ^{154}Eu via capture in the proposed modified evaluations, with a decreases of 8.0% from the ENDF/B-VII.1 evaluation for samples 87–63 and 87–72; in both of these cases however, inventories of ^{154}Gd are substantially over-predicted even with the modified ENDF/B-VII.1 treatment (by approximately 29.5% and 33.3%, respectively). This finding is consistent with prior ORNL validation studies, which saw similar over-predictions in the ^{154}Eu inventories (and subsequent under-predictions of ^{155}Gd of roughly similar magnitude), indicating a potential problem with the measurement data itself (Ilas et al., 2010a). For sample 87–81, ^{154}Gd is consistently under-predicted, showing a slight decrease (-4.4%) upon adoption of the modified ^{154}Eu cross-section compared to the nominal ENDF/B-VII.1 case, again consistent with prior findings (Ilas et al., 2010a).

Detailed isotopic comparisons are available for each evaluation for the Calvert Cliffs samples as Table 24, Table 25 ; Table 26. The reader will again observe very good agreement for the major actinoid and burnup indicator species (e.g., Cs, Nd), indicating a good model agreement to the irradiation conditions. As before, the modifications to the ^{154}Eu capture cross-section produce few discernible impacts on other isotopic species outside of decay products of Eu (i.e., isotopes of Gd).

4.4. REBUS/GKN-II isotopic benchmark

Fig. 16 illustrates the impact of the tested nuclear data evaluations for the REBUS GKN-II sample. Notably, the GKN-II sample shows similar levels of bias in the ^{154}Eu inventory as the TMI-1 NJ05YU samples (13.3% and 14.5% for the ENDF/B-VII.0 and VII.1 evaluations), given the relatively similar discharge burnup of the REBUS sample to the TMI-1 samples, well outside the ±1.7% measurement uncertainty bounds. Likewise, similar to prior samples evaluated, the modified ENDF/VII.1 evaluation (based on Mughabghab's proposed changes to the resolved resonance) brings the ^{154}Eu inventory into closer agreement to experimentally-measured values (8.0%, which is still well outside the experimental uncertainty of 2.0%). Likewise, the modified JEFF-3.2 evaluation gives an inventory for ^{154}Eu about 0.44% closer to the experimental value compared to ENDF/B-VII.1.

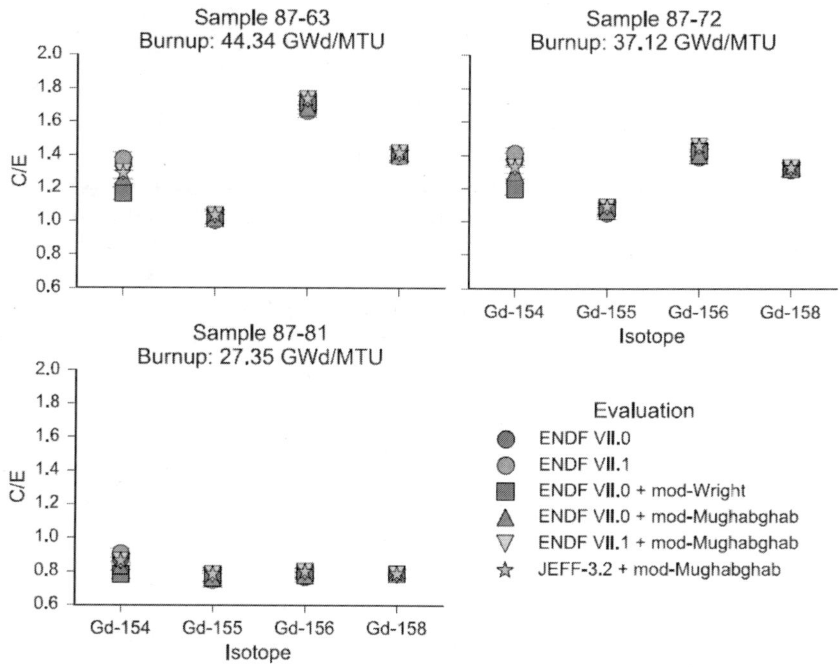

Figure. 15. C/E values for selected Gd isotopes for Calvert Cliffs samples 87–63, 87–72, and 87–71 for each tested nuclear data evaluation; C/E values derived from measured isotope data and associated uncertainty ($\sigma meas.$) based on Ilas et al. (2010a). Error bars indicate relative experimental measurement uncertainty.

Table 24. Calvert Cliffs sample MKP-87–63 calculated-to-experimental (measured) ratios for each nuclear data evaluation; C/E values derived from measured isotope data and associated uncertainty ($\sigma_{meas.}$) based on Ilas et al. (2010a).

Isotope	σ_{meas}	C/E						
		ENDF VII.0	ENDF VII.1	ENDF VII.0 + mod-Wright	ENDF VII.0 + mod-Mugh.	ENDF VII.1 + mod-Mugh.	JEFF-3.2	JEFF-3.2 + mod-Mugh.
^{234}U	0.0230	1.0368	1.0515	1.0368	1.0368	1.0515	1.0515	1.0515
^{235}U	0.0230	1.0323	1.0199	1.0323	1.0323	1.0199	1.0199	1.0199
^{236}U	0.0230	1.0191	1.0191	1.0191	1.0191	1.0191	1.0191	1.0191
^{238}U	0.0230	0.9989	0.9989	0.9989	0.9989	0.9989	0.9989	0.9989
^{238}Pu	0.0230	0.9049	1.0262	0.9049	0.9049	1.0262	1.0262	1.0262
^{239}Pu	0.0230	1.0586	1.0364	1.0586	1.0586	1.0364	1.0364	1.0364
^{240}Pu	0.0230	1.0278	1.0069	1.0278	1.0278	1.0069	1.0069	1.0069
^{241}Pu	0.0230	1.0000	0.9914	1.0000	1.0000	0.9914	0.9914	0.9914
^{242}Pu	0.0230	0.9265	0.9591	0.9265	0.9265	0.9591	0.9591	0.9591
^{237}Np	0.0250	1.0282	1.0151	1.0282	1.0282	1.0151	1.0151	1.0151
^{241}Am	0.0520	0.9329	0.9051	0.9329	0.9329	0.9051	0.9051	0.9051
^{133}Cs	–	1.0071	1.0142	1.0071	1.0071	1.0142	1.0142	1.0142
^{134}Cs	–	0.8529	0.8353	0.8529	0.8529	0.8353	0.8353	0.8353
^{135}Cs	0.1410	1.0328	1.0287	1.0328	1.0328	1.0287	1.0287	1.0287
^{137}Cs	0.0380	0.9718	0.9718	0.9718	0.9718	0.9718	0.9718	0.9718
^{143}Nd	0.0190	1.0150	1.0115	1.0150	1.0150	1.0115	1.0115	1.0115
^{144}Nd	–	0.9731	0.9785	0.9731	0.9731	0.9785	0.9785	0.9785
^{145}Nd	0.0190	0.9692	0.9870	0.9692	0.9692	0.9870	0.9870	0.9870
^{146}Nd	–	1.0042	0.9894	1.0042	1.0042	0.9894	0.9894	0.9894
^{148}Nd	–	1.0000	1.0000	1.0000	1.0000	1.0000	1.0000	1.0000
^{150}Nd	–	1.0254	1.0254	1.0254	1.0254	1.0254	1.0254	1.0254
^{147}Sm	0.0380	0.9610	0.9545	0.9610	0.9610	0.9545	0.9545	0.9545
^{148}Sm	0.0380	0.9375	0.9420	0.9375	0.9375	0.9420	0.9420	0.9420
^{149}Sm	0.0760	0.5417	0.5373	0.5417	0.5417	0.5395	0.5373	0.5395
^{150}Sm	0.0380	1.0165	1.0192	1.0165	1.0165	1.0192	1.0192	1.0192
^{151}Sm	0.0510	0.9533	0.9439	0.9533	0.9533	0.9439	0.9439	0.9439
^{152}Sm	0.0430	0.9699	0.9699	0.9699	0.9699	0.9699	0.9699	0.9699
^{154}Sm	0.0380	0.9467	0.9467	0.9467	0.9467	0.9467	0.9467	0.9467
^{151}Eu	0.0340	0.3090	0.3090	0.3090	0.3090	0.3090	0.3090	0.3090
^{152}Eu	0.0340	0.0130	0.0127	0.0130	0.0130	0.0127	0.0127	0.0127
^{153}Eu	0.0340	0.9877	0.9506	0.9877	0.9877	0.9506	0.9506	0.9506
^{154}Eu	0.0700	1.1009	1.1193	0.9450	1.0275	1.0550	1.1193	1.0459
^{155}Eu	0.0690	1.0168	1.0391	1.0391	1.0279	1.0503	1.0391	1.0503
^{154}Gd	0.0390	1.3450	1.3750	1.1650	1.2650	1.2950	1.3700	1.2900
^{155}Gd	0.0390	1.0010	1.0241	1.0241	1.0141	1.0341	1.0241	1.0341
^{156}Gd	0.0390	1.6614	1.7140	1.7140	1.6824	1.7350	1.7140	1.7350
^{158}Gd	0.0390	1.3932	1.4078	1.4078	1.3981	1.4126	1.4078	1.4126

Table 25. Calvert Cliffs sample MKP-87–72 calculated-to-experimental (measured) ratios for each nuclear data evaluation; C/E values derived from measured isotope data and associated uncertainty ($\sigma_{meas.}$) based on Ilas et al. (2010a).

Isotope	$\sigma_{meas.}$	C/E						
		ENDF VII.0	ENDF VII.1	ENDF VII.0 + mod-Wright	ENDF VII.0 + mod-Mugh.	ENDF VII.1 + mod-Mugh.	JEFF-3.2	JEFF-3.2 + mod-Mugh.
^{234}U	0.0230	0.9874	0.9874	0.9874	0.9874	0.9874	0.9874	0.9874
^{235}U	0.0230	1.0051	0.9966	1.0051	1.0051	0.9983	0.9983	0.9983
^{236}U	0.0230	1.0225	1.0250	1.0225	1.0225	1.0250	1.0250	1.0250
^{238}U	0.0230	0.9968	0.9968	0.9968	0.9968	0.9968	0.9968	0.9968
^{238}Pu	0.0230	0.9019	1.0140	0.9019	0.9019	1.0140	1.0140	1.0140
^{239}Pu	0.0230	1.0303	1.0101	1.0303	1.0303	1.0121	1.0101	1.0121
^{240}Pu	0.0230	1.0157	1.0000	1.0157	1.0157	1.0000	1.0000	1.0000
^{241}Pu	0.0230	0.9765	0.9696	0.9765	0.9765	0.9696	0.9696	0.9696
^{242}Pu	0.0230	0.9418	0.9724	0.9418	0.9418	0.9724	0.9724	0.9724
^{237}Np	0.0250	1.1139	1.0990	1.1139	1.1139	1.0990	1.0990	1.0990
^{241}Am	0.0520	0.9126	0.8895	0.9126	0.9126	0.8895	0.8895	0.8895
^{133}Cs	—	1.0000	1.0000	1.0000	1.0000	1.0000	1.0000	1.0000
^{134}Cs	—	0.9031	0.8811	0.9031	0.9031	0.8811	0.8811	0.8811
^{135}Cs	0.1410	1.0330	1.0308	1.0330	1.0330	1.0308	1.0308	1.0308
^{137}Cs	0.0380	0.9831	0.9831	0.9831	0.9831	0.9831	0.9831	0.9831
^{143}Nd	0.0190	1.0025	1.0000	1.0025	1.0025	1.0000	1.0000	1.0000
^{144}Nd	—	0.9803	0.9803	0.9803	0.9803	0.9803	0.9803	0.9803
^{145}Nd	0.0190	0.9717	0.9852	0.9717	0.9717	0.9852	0.9852	0.9852
^{146}Nd	—	1.0026	0.9897	1.0026	1.0026	0.9897	0.9897	0.9897
^{148}Nd	—	1.0000	1.0000	1.0000	1.0000	1.0000	1.0000	1.0000
^{150}Nd	—	1.0205	1.0205	1.0205	1.0205	1.0205	1.0205	1.0205
^{147}Sm	0.0310	1.0444	1.0407	1.0444	1.0444	1.0407	1.0407	1.0407
^{148}Sm	0.0300	0.9876	0.9938	0.9876	0.9876	0.9938	0.9938	0.9938
^{149}Sm	0.1210	1.2162	1.2108	1.2162	1.2162	1.2108	1.2108	1.2108
^{150}Sm	0.0300	1.0483	1.0517	1.0483	1.0483	1.0517	1.0517	1.0517
^{151}Sm	0.0370	0.9692	0.9649	0.9702	0.9692	0.9660	0.9649	0.9660
^{152}Sm	0.0330	1.0636	1.0636	1.0636	1.0636	1.0636	1.0636	1.0636
^{154}Sm	0.0300	1.0866	1.0866	1.0866	1.0866	1.0866	1.0866	1.0866
^{151}Eu	0.0330	0.9124	0.9086	0.9133	0.9124	0.9086	0.9086	0.9086
^{152}Eu	0.0330	0.1141	0.1125	0.1143	0.1141	0.1125	0.1125	0.1125
^{153}Eu	0.0330	1.0077	0.9769	1.0077	1.0077	0.9769	0.9769	0.9769
^{154}Eu	0.0640	1.0537	1.0822	0.9155	0.9920	1.0194	1.0811	1.0160
^{155}Eu	0.0620	1.0365	1.0657	1.0657	1.0511	1.0803	1.0657	1.0803
^{154}Gd	0.0360	1.3800	1.4133	1.2000	1.3000	1.3333	1.4133	1.3333
^{155}Gd	0.0360	1.0499	1.0783	1.0837	1.0661	1.0931	1.0783	1.0931
^{156}Gd	0.0360	1.3915	1.4325	1.4325	1.4052	1.4598	1.4325	1.4598
^{158}Gd	0.0360	1.3154	1.3221	1.3221	1.3221	1.3289	1.3221	1.3289

Table 26. Calvert Cliffs sample MKP-87–81 calculated-to-experimental (measured) ratios for each nuclear data evaluation; C/E values derived from measured isotope data and associated uncertainty ($\sigma_{meas.}$) based on Ilas et al. (2010a).

Isotope	$\sigma_{meas.}$	C/E						
		ENDF VII.0	ENDF VII.1	ENDF VII.0 + mod-Wright	ENDF VII.0 + mod-Mugh.	ENDF VII.1 + mod-Mugh.	JEFF-3.2	JEFF-3.2 + mod-Mugh.
^{234}U	0.0230	0.9945	0.9945	0.9945	0.9945	0.9945	0.9945	0.9945
^{235}U	0.0230	0.9979	0.9948	0.9979	0.9979	0.9948	0.9948	0.9948
^{236}U	0.0230	1.0169	1.0169	1.0169	1.0169	1.0169	1.0169	1.0169
^{238}U	0.0230	0.9937	0.9948	0.9937	0.9937	0.9948	0.9948	0.9948
^{238}Pu	0.0230	0.8957	0.9826	0.8957	0.8957	0.9826	0.9826	0.9826
^{239}Pu	0.0230	1.0248	1.0104	1.0248	1.0248	1.0104	1.0104	1.0104
^{240}Pu	0.0230	1.0103	0.9949	1.0103	1.0103	0.9949	0.9949	0.9949
^{241}Pu	0.0230	0.9677	0.9651	0.9677	0.9677	0.9651	0.9651	0.9651
^{242}Pu	0.0230	0.9421	0.9665	0.9421	0.9421	0.9665	0.9665	0.9665
^{237}Np	0.0250	1.0395	1.0263	1.0395	1.0395	1.0263	1.0263	1.0263
^{241}Am	0.0520	0.9716	0.9539	0.9716	0.9716	0.9539	0.9539	0.9539
^{133}Cs	—	0.9959	0.9990	0.9959	0.9959	0.9990	0.9990	0.9990
^{134}Cs	—	1.0000	0.9735	1.0000	1.0000	0.9735	0.9735	0.9735
^{135}Cs	0.1410	1.0466	1.0466	1.0466	1.0466	1.0466	1.0466	1.0466
^{137}Cs	0.0380	0.9805	0.9805	0.9805	0.9805	0.9805	0.9805	0.9805
^{143}Nd	0.0190	1.0000	1.0000	1.0000	1.0000	1.0000	1.0000	1.0000
^{144}Nd	—	0.9813	0.9813	0.9813	0.9813	0.9813	0.9813	0.9813
^{145}Nd	0.0190	0.9793	0.9896	0.9793	0.9793	0.9896	0.9896	0.9896
^{146}Nd	—	1.0054	0.9946	1.0054	1.0054	0.9946	0.9946	0.9946
^{148}Nd	—	1.0000	1.0000	1.0000	1.0000	1.0000	1.0000	1.0000
^{150}Nd	—	1.0142	1.0142	1.0142	1.0142	1.0142	1.0142	1.0142
^{147}Sm	0.0380	1.0652	1.0652	1.0652	1.0652	1.0652	1.0652	1.0652
^{148}Sm	0.0260	0.9790	0.9820	0.9790	0.9790	0.9820	0.9820	0.9820
^{149}Sm	0.2010	0.7138	0.7103	0.7138	0.7138	0.7103	0.7103	0.7103
^{150}Sm	0.0460	1.0478	1.0478	1.0478	1.0478	1.0478	1.0478	1.0478
^{151}Sm	0.3850	1.1169	1.1142	1.1169	1.1169	1.1142	1.1142	1.1142
^{152}Sm	0.0370	1.0591	1.0569	1.0591	1.0591	1.0569	1.0569	1.0569
^{154}Sm	0.0600	0.8925	0.8925	0.8925	0.8925	0.8925	0.8925	0.8925
^{151}Eu	0.0990	0.4447	0.4441	0.4447	0.4447	0.4441	0.4441	0.4441
^{152}Eu	1.0000	0.0842	0.0835	0.0842	0.0842	0.0836	0.0836	0.0836
^{153}Eu	0.0260	1.0610	1.0364	1.0610	1.0610	1.0364	1.0364	1.0364
^{154}Eu	0.0880	1.0395	1.0757	0.9211	0.9852	1.0230	1.0740	1.0197
^{155}Eu	0.1680	1.0288	1.0577	1.0673	1.0481	1.0769	1.0577	1.0769
^{154}Gd	0.0280	0.8759	0.9051	0.7810	0.8321	0.8613	0.9051	0.8613
^{155}Gd	0.0420	0.7456	0.7680	0.7792	0.7600	0.7824	0.7680	0.7840
^{156}Gd	0.0260	0.7645	0.7865	0.7928	0.7771	0.7991	0.7865	0.7991
^{158}Gd	0.0280	0.7857	0.7857	0.7857	0.7857	0.7857	0.7857	0.7857
^{160}Gd	0.1210	0.5248	0.5241	0.5248	0.5248	0.5241	0.5241	0.5241

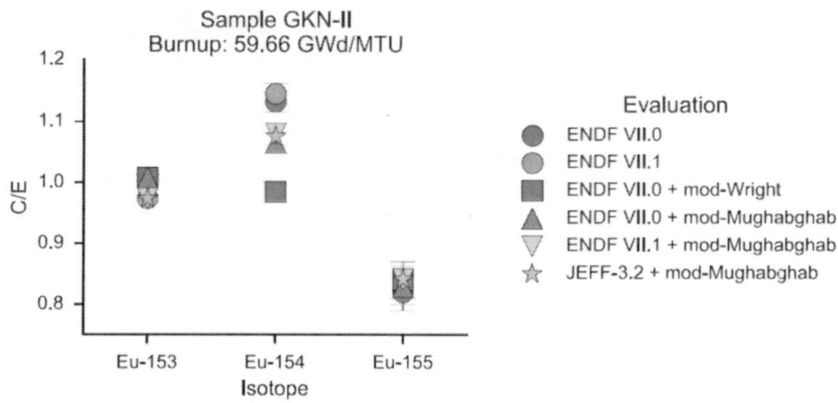

Figure. 16. REBUS GKN-II sample C/E values for Eu isotopes for each tested nuclear data evaluation; C/E values derived from measured and associated uncertainty ($\sigma meas.$) isotope data based on Ilas et al. (2010b).

Unlike the TMI-1 and Calvert Cliffs samples however, another significant feature of this benchmark is in the fact that ^{155}Eu is substantially under-predicted (by -18.0% and -16.8% for ENDF/B-VII.0 and VII.1, respectively); this under-prediction likewise persists for the modified ^{154}Eu treatment (-15.9%). The precise cause of this under-prediction is unclear, especially in light of an overall trend in over-predicting ^{155}Eu inventories, however previous SCALE benchmarks of the REBUS sample found even greater discrepancies when using ENDF/B-V cross-sections (approximately -42%) (Ilas et al., 2010b). Thus, the discrepancy here may be due to other factors, including possible measurement bias in the sample or due to modeling features not captured.

A detailed isotopic benchmark for the REBUS sample is presented in Table 27. In general the modeled sample shows very good agreement overall with experimental measurements, although several major plutonium isotopes (^{239}Pu and ^{241}Pu) lie outside the experimental uncertainty bounds, a possible indicator of factors influencing the neutron energy spectrum shape. However, given that the GKN-II sample pin was located relatively far from guide tubes and Gd_2O_3-UO_2 absorber pins, it is unlikely that local effects would be responsible for particular distortions to the flux shape. By contrast, major burnup indicators ^{148}Nd and ^{137}Cs show very good agreement with measured values, thus indicating a good correspondence between the model burnup and sample discharge burnup.

Table 27. REBUS GKN-II sample calculated-to-experimental (measured) ratios for each nuclear data evaluation; C/E values derived from measured and associated uncertainty ($\sigma_{meas.}$) isotope data based on Ilas et al. (2010b). Error bars indicate relative experimental measurement uncertainty.

Isotope	$\sigma_{meas.}$	C/E						
		ENDF VII.0	ENDF VII.1	ENDF VII.0 + mod-Wright	ENDF VII.0 + mod-Mugh.	ENDF VII.1 + mod-Mugh.	JEFF-3.2	JEFF-3.2 + mod-Mugh.
^{234}U	0.0250	1.2006	1.2198	1.2006	1.2006	1.2198	6172.8395	1.2198
^{235}U	0.0050	1.0800	1.0685	1.0806	1.0806	1.0685	179.8561	1.0685
^{236}U	0.0050	0.9995	0.9986	0.9995	0.9995	0.9979	172.1170	0.9979
^{238}U	0.0250	1.0000	1.0000	1.0000	1.0000	1.0000	1.0000	1.0000
^{237}Np	0.1000	1.2367	1.2215	1.2367	1.2367	1.2220	0.0000	1.2220
^{238}Pu	0.0160	0.9138	1.0267	0.9138	0.9138	1.0267	0.0000	1.0267
^{239}Pu	0.0040	1.1246	1.0987	1.1251	1.1246	1.0987	0.0000	1.0987
^{240}Pu	0.0040	1.0550	1.0344	1.0550	1.0550	1.0344	0.0000	1.0344
^{241}Pu	0.0040	1.0533	1.0419	1.0533	1.0533	1.0419	0.0000	1.0419
^{242}Pu	0.0040	0.9331	0.9701	0.9331	0.9331	0.9701	0.0000	0.9701
^{241}Am	0.0180	1.3526	1.3219	1.3532	1.3526	1.3225	0.0000	1.3225
^{242m}Am	0.0550	0.9218	0.9706	0.9224	0.9218	0.9712	0.0000	0.9712
^{243}Am	0.0180	1.2833	1.1496	1.2833	1.2833	1.1496	0.0000	1.1496
^{242}Cm	0.1600	0.9413	0.9915	0.9419	0.9413	0.9921	0.0000	0.9921
^{243}Cm	0.1000	0.8718	1.1185	0.8718	0.8718	1.1188	0.0000	1.1188
^{244}Cm	0.0130	0.9340	0.9528	0.9340	0.9340	0.9528	0.0000	0.9528
^{245}Cm	0.0280	1.1056	1.0701	1.1062	1.1062	1.0701	0.0000	1.0701
^{133}Cs	0.0130	1.0575	1.0632	1.0569	1.0575	1.0632	0.0000	1.0632
^{135}Cs	0.0130	1.0451	1.0406	1.0451	1.0451	1.0411	0.0000	1.0411
^{137}Cs	0.0130	0.9846	0.9846	0.9846	0.9846	0.9846	0.0000	0.9846
^{144}Ce	0.0500	0.9594	0.9596	0.9594	0.9594	0.9596	0.0000	0.9596
^{142}Nd	0.0050	0.8938	0.9014	0.8938	0.8938	0.9014	0.0000	0.9014
^{143}Nd	0.0040	1.0465	1.0422	1.0465	1.0465	1.0422	0.0000	1.0422
^{144}Nd	0.0040	0.9796	0.9820	0.9796	0.9796	0.9820	0.0000	0.9820
^{145}Nd	0.0040	0.9935	1.0120	0.9935	0.9935	1.0111	0.0000	1.0111
^{146}Nd	0.0040	1.0172	1.0024	1.0172	1.0172	1.0024	0.0000	1.0024
^{148}Nd	0.0040	1.0156	1.0158	1.0156	1.0156	1.0158	0.0000	1.0158
^{150}Nd	0.0040	1.0206	1.0203	1.0206	1.0206	1.0203	0.0000	1.0203
^{147}Sm	0.0050	1.0111	1.0034	1.0111	1.0111	1.0034	0.0000	1.0034
^{148}Sm	0.0050	0.9460	0.9498	0.9460	0.9460	0.9498	0.0000	0.9498
^{149}Sm	0.0110	1.0880	1.0768	1.0884	1.0880	1.0772	0.0000	1.0768
^{150}Sm	0.0050	1.0154	1.0183	1.0154	1.0154	1.0183	0.0000	1.0183
^{151}Sm	0.0050	1.0342	1.0271	1.0348	1.0342	1.0271	0.0000	1.0271
^{152}Sm	0.0050	0.9768	0.9743	0.9768	0.9768	0.9743	0.0000	0.9743
^{154}Sm	0.0050	1.0260	1.0261	1.0260	1.0260	1.0261	0.0000	1.0261
^{153}Eu	0.0050	1.0077	0.9736	1.0077	1.0077	0.9736	0.0000	0.9736
^{154}Eu	0.0170	1.1320	1.1448	0.9844	1.0652	1.0800	0.0000	1.0756
^{155}Eu	0.0300	0.8200	0.8319	0.8400	0.8293	0.8407	0.0000	0.8412
^{155}Gd	0.0250	0.9664	0.9800	0.9891	0.9773	0.9900	0.0000	0.9909

4.5. ARIANE/Gösgen GU1 isotopic benchmark

ARIANE sample GU1 was the last sample tested in this benchmark study; a comparison between measured and calculated values for isotopes of Eu for the evaluations considered is presented as Fig. 17. Sample GU1 presents a similar pattern as previous samples with respect to ^{154}Eu behavior, including a similar level of bias (12.8% and 13.9% above experimental measurements for ENDF/B-VII.0 and VII.1, again well outside the ±2.0% measurement uncertainty). Likewise similar to other samples, the modified ENDF/B-VII.1 treatment for ^{154}Eu reduces this bias to around 6.3% above the experimental value; still well outside of the measurement uncertainty but a significant correction. The modified JEFF-3.2 evaluation gives a ^{154}Eu inventory 0.34% closer to the measured value compared to the modified ENDF/B-VII.1 evaluation, consistent with prior samples. Finally, as before the modification to the VII.0 cross-sections based on Wright again appear to over-correct, leading to an under-calculation by -4.0%.

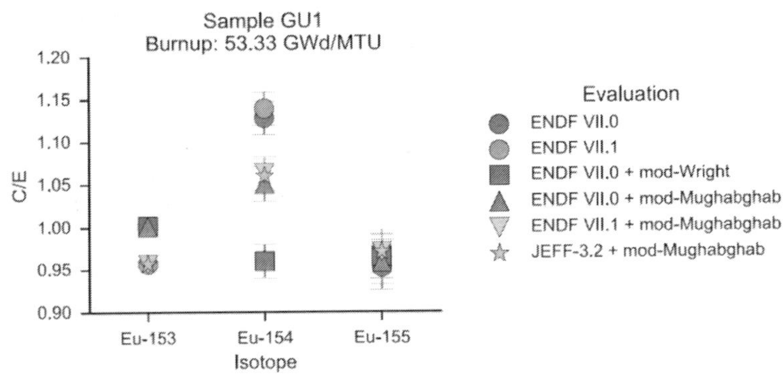

Figure. 17. ARIANE sample GU1 C/E values for Eu isotopes for each tested nuclear data evaluation; C/E values derived from measured and associated uncertainty ($\sigma_{meas.}$) isotope data based on Ilas et al. (2010b).

By contrast, the ^{155}Eu inventory is much closer to the measured values, under-predicted by 4.8% and 3.5% for ENDF/B-VII.0 and VII.1, respectively, with a measurement uncertainty of ±2.6%. Similar to before, the increased capture by ^{154}Eu in the modified evaluations drives up ^{155}Eu inventories, bringing them to an under-calculation of -2.9% for the modified ENDF/B-VII.1 treatment, just outside the 1-σ measurement uncertainty. It is not immediately clear why the GU1 sample exhibits better agreement for ^{155}Eu compared with other samples, especially in light of the rather unique history of the assembly itself (with multiple pins replaced between cycles).

Table 28 gives a detailed account of the correspondence between the modeled and measured isotopic values for the GU1 sample. In general, the model shows very good agreement for nearly all actinoid species, although ^{235}U is substantially over-predicted (by about 10.4% in the ENDF/B-VII.1-based evaluations). The major fission product burnup indicators of Cs and Nd likewise show excellent agreement to measured values, indicating a good match between modeled and measured burnups.

4.6. Trends in Eu inventories with sample burnup

With the existence of a bias in calculated ^{154}Eu robustly confirmed across multiple spent fuel benchmarks, it is important to understand whether trends exist with sample burnup. The existence of an upward trend in departures from experimental values with increasing burnup would be strong evidence of a bias in the nuclear data used to calculate ^{154}Eu inventories. First, looking to ^{153}Eu (Fig. 18), no clear trend is evident with the C/E ratio as a function of burnup. (Note that the CANDU assembly is excluded from this trends analysis, given the significantly different spectral shape of this sample compared to the PWR samples analyzed). With the ^{153}Eu samples, the only clear effect persistent between samples is the impact of the modification of the thermal capture cross-section from ENDF/B-VII.0 to VII.1, consistently lowering the calculated ^{153}Eu inventory across samples. The lack of a trend in burnup for ^{153}Eu is important however for ruling out possible contributors to burnup-dependent behavior in ^{154}Eu inventories.

Fig. 19 illustrates the trend in C/E ratios for ^{154}Eu as a function of sample burnup. Unlike ^{153}Eu, a clear trend with respect to burnup is evident among samples, showing a consistently greater departure from experimentally-measured values of ^{154}Eu with increasing burnup. This is further evident in the linear regression fits to C/E data as a function of burnup (presented as dashed lines in Fig. 19). Here, a larger slope in the regression fit would denote a more significant bias effect with burnup. These slope coefficients from the linear regression fits are presented as Table 29. One will observe that the lowest slope (i.e., smallest effect with burnup) occurs with the Mughabghab modification to ENDF/B-VII.1 (with the largest bias observed in ENDF/B-VII.0, likely implying an under-estimation effect of ^{153}Eu), along with the Mughabghab modification to the JEFF-3.2 evaluation. Overall the amendment proposed by Mughabghab likewise appears to show the closest correspondence to experimentally-measured values for all samples over the burnup range considered, with the JEFF-3.2-based evaluations showing slightly closer agreement (about 0.4–0.5%) to measured ^{154}Eu inventories compared to ENDF.

Table 28. ARIANE/ Gösgen sample GU1 calculated-to-experimental (measured) ratios for each nuclear data evaluation; C/E values derived from measured and associated uncertainty ($\sigma meas.$) isotope data based on Ilas et al. (2010b).

Isotope	$\sigma_{meas.}$	C/E						
		ENDF VII.0	ENDF VII.1	ENDF VII.0 + mod-Wrghr	ENDF VII.0 + mod-Mugh.	ENDF VII.1 + mod-Mugh.	JEFF-3.2	JEFF-3.2 + mod-Mugh.
^{234}U	0.0250	1.1908	1.2033	1.1908	1.1908	1.2033	1.2033	1.2033
^{235}U	0.0100	1.1213	1.1033	1.1223	1.1213	1.1038	1.1033	1.1038
^{236}U	0.0030	1.0153	1.0139	1.0153	1.0153	1.0139	1.0139	1.0139
^{238}U	0.0020	0.9978	0.9984	0.9978	0.9978	0.9984	0.9984	0.9984
^{238}Pu	0.0150	0.9037	1.0573	0.9040	0.9040	1.0573	1.0573	1.0573
^{239}Pu	0.0030	1.0534	1.0305	1.0534	1.0534	1.0307	1.0305	1.0307
^{240}Pu	0.0030	1.0274	1.0066	1.0274	1.0274	1.0066	1.0066	1.0066
^{241}Pu	0.0030	1.0139	1.0014	1.0146	1.0139	1.0014	1.0014	1.0014
^{242}Pu	0.0030	0.9116	0.9458	0.9116	0.9116	0.9458	0.9458	0.9458
^{237}Np	0.0300	0.0000	0.0000	0.0000	0.0000	0.0000	0.0000	0.0000
^{241}Am	0.0180	1.0988	1.0556	1.0976	1.0972	1.0560	1.0560	1.0560
^{242m}Am	0.0530	1.0349	1.0431	1.0498	1.0491	1.0431	1.0431	1.0431
^{243}Am	0.0180	1.1104	0.9573	1.0648	1.0648	0.9573	0.9573	0.9573
^{242}Cm	0.0360	0.9553	0.9689	0.9557	0.9553	0.9693	0.9689	0.9693
^{243}Cm	0.3680	2.7553	3.4467	2.7568	2.7553	3.4497	3.4467	3.4497
^{244}Cm	0.0150	0.9262	0.9475	0.9266	0.9262	0.9480	0.9475	0.9480
^{245}Cm	0.0300	1.1097	1.0783	1.1109	1.1103	1.0783	1.0783	1.0783
^{246}Cm	0.1010	0.8894	0.8703	0.8894	0.8894	0.8703	0.8703	0.8703
^{133}Cs	0.0250	1.0436	1.0500	1.0436	1.0436	1.0500	1.0500	1.0500
^{134}Cs	0.0250	1.0231	1.0028	1.0231	1.0231	1.0028	1.0028	1.0028
^{135}Cs	0.0250	0.9882	0.9839	0.9884	0.9884	0.9839	0.9839	0.9839
^{137}Cs	0.0250	0.9897	0.9901	0.9897	0.9897	0.9901	0.9901	0.9901
^{142}Nd	0.0500	0.9842	0.9941	0.9842	0.9842	0.9941	0.9941	0.9941
^{143}Nd	0.0030	1.0614	1.0558	1.0620	1.0620	1.0558	1.0558	1.0558
^{144}Nd	0.0030	0.9837	0.9863	0.9837	0.9837	0.9863	0.9863	0.9863
^{145}Nd	0.0030	0.9942	1.0173	0.9942	0.9942	1.0173	1.0173	1.0173
^{146}Nd	0.0030	1.0135	0.9947	1.0135	1.0135	0.9947	0.9947	0.9947
^{148}Nd	0.0030	1.0005	1.0005	1.0005	1.0005	1.0005	1.0005	1.0005
^{150}Nd	0.0030	0.9861	0.9864	0.9861	0.9861	0.9864	0.9864	0.9864
^{147}Sm	0.0030	1.0095	1.0009	1.0095	1.0095	1.0009	1.0009	1.0009
^{148}Sm	0.0030	0.9540	0.9577	0.9540	0.9540	0.9577	0.9577	0.9577
^{149}Sm	0.0110	0.9174	0.9113	0.9177	0.9174	0.9116	0.9113	0.9116
^{150}Sm	0.0030	1.0252	1.0272	1.0252	1.0252	1.0272	1.0272	1.0272
^{151}Sm	0.0040	0.9838	0.9785	0.9838	0.9838	0.9785	0.9785	0.9785
^{152}Sm	0.0030	0.9512	0.9488	0.9506	0.9506	0.9488	0.9488	0.9488
^{154}Sm	0.0030	0.9913	0.9927	0.9913	0.9913	0.9927	0.9927	0.9927
^{151}Eu	0.0110	0.4146	0.4124	0.4149	0.4148	0.4124	0.4124	0.4124
^{153}Eu	0.0030	1.0014	0.9567	1.0014	1.0014	0.9567	0.9567	0.9567
^{154}Eu	0.0200	1.1283	1.1391	0.9599	1.0509	1.0630	1.1376	1.0596
^{155}Eu	0.0260	0.9522	0.9646	0.9655	0.9584	0.9708	0.9646	0.9708
^{155}Gd	0.0250	1.0178	1.0306	1.0306	1.0242	1.0369	1.0306	1.0369

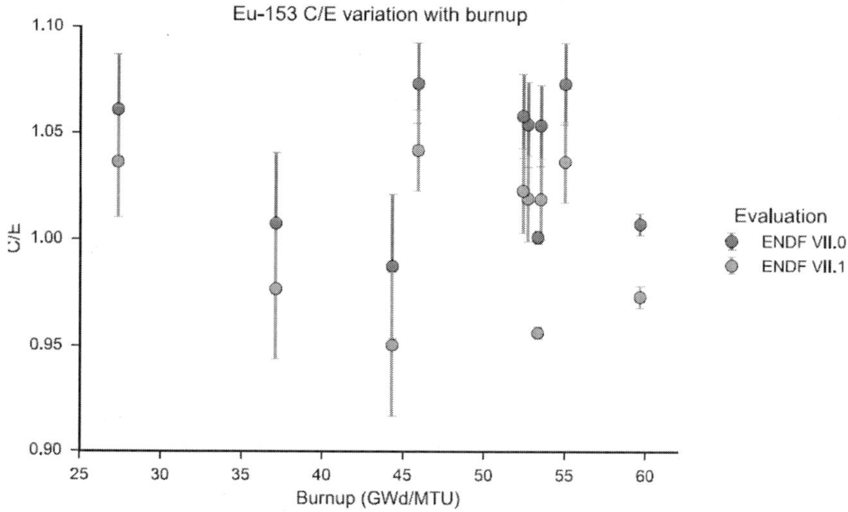

Figure. 18. ^{153}Eu C/E values as a function of burnup for all tested PWR samples. Only ENDF/B-VII.0 and VII.1 evaluations are compared due to the fact that all changes in the ^{153}Eu capture cross-section in all evaluation sets tested are contained within these two evaluations.

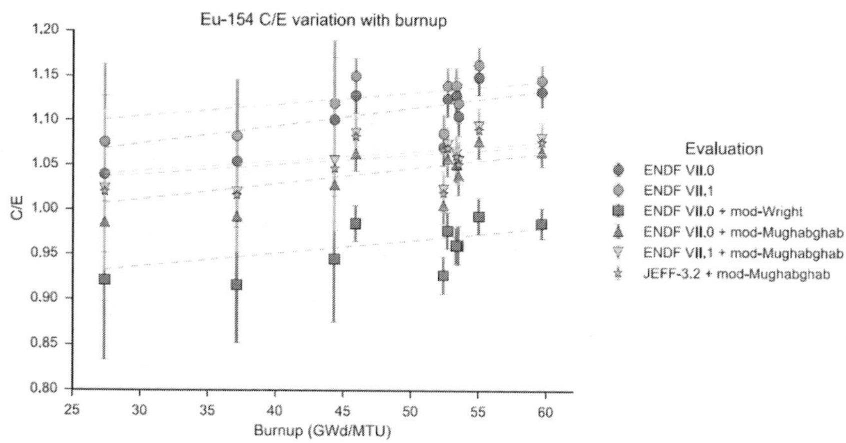

Figure. 19. ^{154}Eu C/E values as a function of burnup for all tested PWR samples. Dashed lines indicate a linear regression fit to series data.

Table 29. Linear regression parameters for C/E as a function of burnup, where $C/E = m \cdot BU + k$, where m is the slope parameter, BU is the sample burnup (in $\frac{GWd}{MTU}$), and k is a constant.

Evaluation	Slope (m)	Intercept (k)	Weighted R^2
ENDF/B-VII.0	0.0020	0.9492	0.4188
ENDF/B-VII.1	0.0013	1.0100	0.2879
ENDF/B-VII.0 + mod-Wright	0.0014	0.8940	0.3442
ENDF/B-VII.0 + mod-Mughabghab	0.0017	0.9599	0.3989
ENDF/B-VII.1 + mod-Mughabghab	0.0011	1.0110	0.2419
JEFF-3.2 + mod-Mughabghab	0.0011	1.0075	0.2506

Notably however, the evaluation based on Wright appears to under-estimate ^{154}Eu, particularly at lower burnups (implying that the capture cross-section is too high). Similarly, the variation in C/E shows a discernible relationship to burnup for the ENDF/B-VII.0 evaluation (uncertainty-weighted $R^2=0.4188$), with the weakest relationship manifested by modified cross-section evaluations, thereby likewise strongly supporting the premise that a bias in the nuclear data is responsible for the observed discrepancies.^{155}Eu (shown in Fig. 20) does not manifest the same observable relationship to burnup as ^{154}Eu, although the effect of both increasing the rate of capture reactions by ^{154}Eu (through the amended cross-section evaluations) as well as driving up inventories of ^{154}Eu through captures by ^{153}Eu (via the ENDF/B-VII.1 evaluation) is clear across all samples measured.

Finally, an evaluation of C/E trends is presented for ^{154}Gd and ^{155}Gd as Fig. 21 ; Fig. 22. Here, for ^{154}Gd, the modifications proposed by Mughabghab in almost every case bring the inventories of ^{154}Gd closer to the respective measured value, given the removal of ^{154}Eu (thereby decreasing decay into ^{154}Gd). For ^{155}Gd, the effect is notably smaller, although like for ^{155}Eu, elevated captures by ^{154}Eu likewise drive increased inventories of ^{155}Gd (via decay from ^{155}Eu). As with ^{155}Eu, no discernible burnup trend emerges with ^{154}Gd and ^{155}Gd as a function of burnup.

4.7. Criticality (k_∞) impacts

In addition to evaluating changes in isotopic inventories, also evaluated as part of the change in the cross-section evaluation was the change in k_∞ for each modeled assembly. ^{154}Eu and ^{155}Eu both have substantial absorption cross-sections (with the latter being a key burnup credit nuclide for spent fuel storage). However, overall the effects on k_∞ were found to be trivial due to the change in the cross-section evaluation, largely balanced out over the irradiation cycle. Overall for the CANDU PHWR, the reactivity differences due to the modified ^{154}Eu evaluations were found to be <10 pcm; for PWR assemblies, the reactivity differences arising from the modified capture cross-section were <30 pcm. Furthermore, cross-section changes introduced in ENDF/B-VII.1 overwhelm any differences due to the change in the ^{154}Eu capture cross-section, ranging from +178 pcm at the beginning of assembly irradiation to +112 pcm at the end of cycle for PWR lattices (TMI-1 sample C2D2) and -357 pcm for fresh assemblies to -40 pcm at assembly discharge for CANDU PHWR assemblies. A summary of these effects is laid out in Table 30.

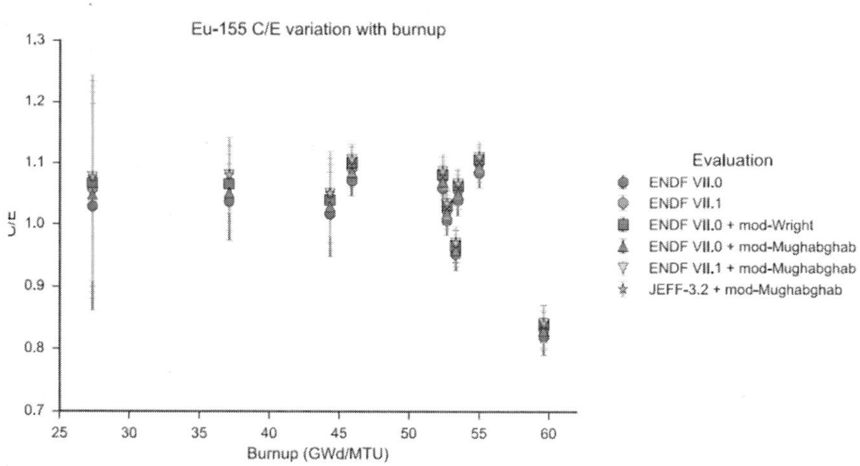

Figure. 20. ^{155}Eu C/E values as a function of burnup for all tested PWR samples. Dashed lines indicate an uncertainty-weighted linear regression fit to series data.

5. CONCLUSIONS

In this paper, I have presented evidence of bias in the ENDF/B-VII.0 and ENDF/B-VII.1 thermal capture cross-sections for ^{154}Eu, using the results of spent fuel benchmarking studies. Working from these studies, I have shown that these nuclear data evaluations result in systemic over-predictions in the inventory of ^{154}Eu. To correct this, two proposed amendments to the evaluation were considered: the first, based on the ENDF/VI.8 evaluation performed by R. Q. Wright, and the second, proposed by S. F. Mughabghab (the original author of the VII.0 and VII.1 evaluations). By developing a modified AMPX master cross-section library for each of these evaluations, it was possible test the effect of each of these evaluations on eleven samples from five different spent fuel assemblies. The results of these studies indicate that the amended ^{154}Eu capture cross-section proposed by Mughabghab, which shifts the location of the first resolved resonance to 0.188 eV (and the neutron capture width proportionally) yields substantially improved correspondence to experimentally-measured values for most samples measured.

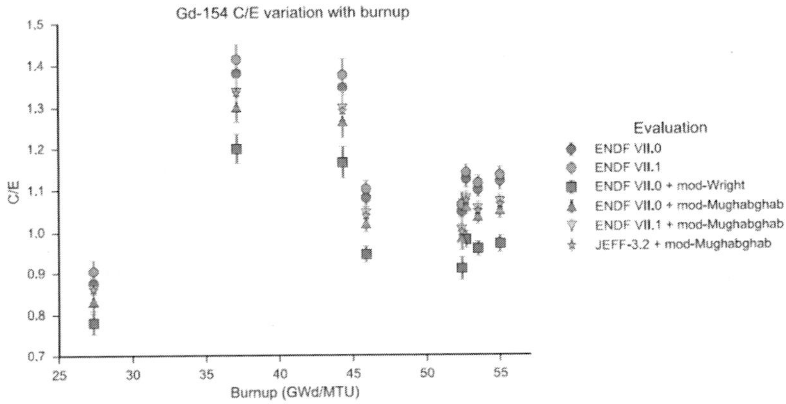

Figure. 21. ^{154}Gd C/E values as a function of burnup for all tested PWR samples.

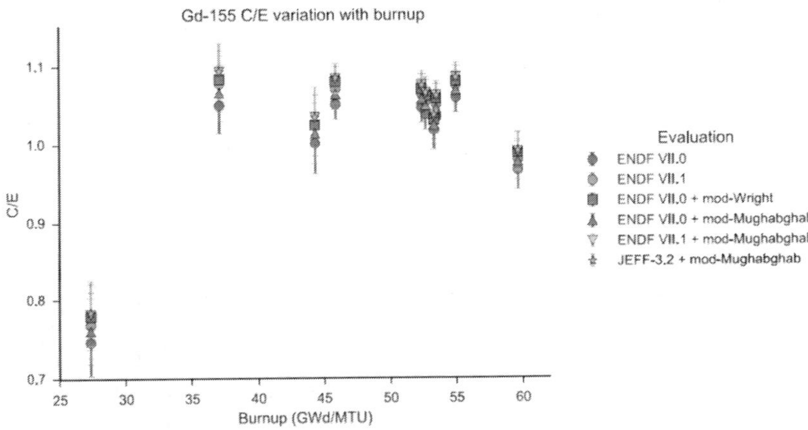

Figure. 22. ^{155}Gd C/E values as a function of burnup for all tested PWR samples.

Table 30. Δk_∞ (in pcm) from ENDF/B-VII.0 and VII.1 nominal cases for each tested sample, in terms of absolute maximum change and lifetime-averaged values, along with the change in calculated multiplication factor due to employing ENDF/B-VII.1 nuclear data evaluations over ENDF/B-VII.0.

Assembly	Sample	ENDF/B-VII.0				ENDF/B-VII.1		JEFF-3.2		ENDF	
		Wright		Mughabghab		Mughabghab		Mughabghab		VII.0—VII.1	
		|Max|	Avg.	|Max|	Avg.	|Max|	Avg.	|Max|	Avg.	|Max|	Avg.
TMI-1 NJ05YU	C2D1	−27	−17	−12	−8	−12	−7	−13	−8	215	180
	C2D2	−25	−15	−12	−7	−12	−7	−10	−5	215	179
	A1	−27	−14	−10	−5	−15	−8	−13	−6	234	201
	B1	−26	−16	−12	−7	−14	−9	−12	−7	228	191
	B2	−25	−15	−11	−7	−11	−7	−9	−5	227	192
Calvert Cliffs	87-63	−23	−13	−11	−6	−11	−6	−8	−4	214	129
	87-72	−23	−12	−10	−6	−11	−6	−8	−4	206	138
	87-81	−24	−10	−11	−5	−11	−5	−8	−3	192	153
REBUS	GKN-II	−25	−14	−12	−6	−12	−6	−11	−6	261	112
ARIANE	GU1	−23	−13	−10	−6	−10	−6	−9	−4	180	121
Pickering-A		−18	−7	−5	−2	−5	−2	−6	−2	−357	−185

It is important to note however that the ENDF/B-VII.1 evaluation with Mughabghab's proposed amendment still substantially over-predicts ^{154}Eu inventories (especially for highly-thermalized systems like CANDU), although less severely than the original evaluations. Conversely, the modification based on Wright appears to over-correct for the rate of neutron captures, leading to under-predictions of the ^{154}Eu inventory in many cases. Based on this, it is likely that the true $^{154}\mathrm{Eu}(n,\gamma)^{155}\mathrm{Eu}$ capture cross-section lies somewhere between the values indicated by the Mughabghab and Wright evaluations, potentially warranting further study in the next evaluation. This conclusion is further supported by evaluations employing a modified version of the JEFF-3.2 library, which adopts much of the same structure as ENDF/B-VII.1 with the exception of the resonance region over 100 eV and the fast (>1 MeV) region, which employ capture cross-sections identical to the ENDF/B-VI.8 evaluation. Inventories of ^{154}Eu were found to be consistently about 0.3–0.4% closer to experimental values with the JEFF-based evaluations (employing the same thermal-region structure as ENDF).

Overall however, the modified ^{154}Eu cross-section results in substantially improved calculations of ^{154}Eu, reducing the over-prediction in ^{154}Eu inventories by approximately 5–10% on average. Further, the modified capture cross-section appears to reduce systemic biases introduced as a function of sample discharge burnup. The increased capture rate by ^{154}Eu does however drive up over-predictions of ^{155}Eu, which appears to be also consistently over-predicted across nearly all samples evaluated. On the balance however, the effects of this change are negligible in terms of the assembly reactivity at discharge burnups and are thus most likely to manifest in terms of reducing error in simulations used to model spent fuel burnup measurements as well as slight reductions to calculated radiation source terms for spent fuel storage and transportation. Key outcomes of adopting a modified evaluation for the $^{154}\mathrm{Eu}(n,\gamma)^{155}\mathrm{Eu}$ capture cross-section will be in reducing model conservatism for spent fuel source terms in storage while also reducing bias in high-fidelity models used for predicting instrument response in spent fuel NDA applications.

ACKNOWLEDGMENTS

This work was supported by an Nuclear Energy University Programs (NEUP) grant sponsored by the U.S. Department of Energy, Office of Nuclear Energy, award number DE-NE0000737. The author would like to acknowledge the contributions of Nathan Shoman for his assistance in developing the initial TMI-1 lattice model. The author likewise wishes to express his sincere appreciation to Shane Hart, Doro Wiarda, and Mark

Williams of Oak Ridge National Laboratory for their invaluable assistance with the AMPX system in order to prepare the modified AMPX master libraries used for testing the ^{154}Eu thermal capture cross-sections. Finally, the author wishes to thank Ian Gauld (also of ORNL) for his assistance in interpreting the TMI-1 assay data used in this study.

REFERENCE

1. Anufriev, V.A., Babich, S.I., Kolesov, A.G., Nefelov, V.N., Poruchikov, V.A., Safonov, V. A., Chetverikov, A.P., Artamonov, V.S., Ivanov, R.N., Kalebin, S.M., 1979. Measurement of the total neutron cross sections of 153Eu, 154Eu, and 155Eu. Soviet At. Energy 46 (3), 182–185 [Online]. Available: http://dx.doi.org/10.1007/BF01125733.
2. ASTM International, 2012. Standard Test Method for Atom Percent Fission in Uranium and Plutonium Fuel (Neodymium-148 Method). http://www.astm. org/cgi-bin/resolver.cgi?E321, West Conshohocken, PA, Tech. Rep. ASTM E321- 96.
3. Basunia, M.S., Firestone, R.B., Révay, Z., Choi, H.D., Belgya, T., Escher, J.E., Hurt, A.M., Krtic˘ka, M., Szentmiklósi, L., Sleaford, B., Summers, N.C., 2014. Determination of the 151Eu (n,c)152m1,gEu and 153Eu(n,c)154Eu reaction cross sections at thermal neutron energy. Nucl. Data Sheets 119, 88–90.
4. Chadwick, M. et al., 2011. Fission product yields and decay data. Nucl. Data Sheets 112 (12), 2887–2996. special Issue on ENDF/B-VII.1 Library. [Online]. Available: http://www.sciencedirect.com/science/article/pii/S009037521100113X.
5. Chadwick, M., Obloz˘insky´ , P., Herman, M., Greene, N., McKnight, R., Smith, D., Young, P., MacFarlane, R., Hale, G., Frankle, S., Kahler, A., Kawano, T., Little, R., Madland, D., Moller, P., Mosteller, R., Page, P., Talou, P., Trellue, H., White, M., Wilson, W., Arcilla, R., Dunford, C., Mughabghab, S., Pritychenko, B., Rochman, D., Sonzogni, A., Lubitz, C., Trumbull, T., Weinman, J., Brown, D., Cullen, D., Heinrichs, D., McNabb, D., Derrien, H., Dunn, M., Larson, N., Leal, L., Carlson, A., Block, R., Briggs, J., Cheng, E., Huria, H., Zerkle, M., Kozier, K., Courcelle, A., Pronyaev, V., van der Marck, S.,.ENDF/B-VII.0: Next generation evaluated nuclear data library for nuclear science and technology. Nucl. Data Sheets 107 (12), 2931–3060. Evaluated Nuclear Data File ENDF/B-VII.0. [Online]. Available: http://www.sciencedirect.com/science/article/pii/S0090375206000871.
6. Cross-Section Evaluation Working Group, 1998. ENDF/B-VI Summary Documentation. Brookhaven National Laboratory, Tech. Rep. BNL-NCS-17541 (ENDF-201), Data Library ENDF/B-VI, update 1998, by the U.S. National Nuclear Data Center on behalf of the Cross-Section Evaluation Working Group.
7. Cross Sections Evaluation Working Group, 2012. ENDF-6 formats manual, Brookhaven National Laboratory, Upton, NY, October 2012, bNL-90365-2009 Rev.2. [Online]. Available: http://www.nndc.bnl.gov/csewg/docs/endf-manual.pdf.
8. DeHart, M.D., Bowman, S.M., 2011. Reactor physics methods and analysis capabilities in SCALE. Nucl. Technol. 174 (2), 196–213 [Online]. Available: http://dx.doi.org/10.13182/NT174-196.
9. Favalli, A., Vo, D., Grogan, B., Jansson, P., Liljenfeldt, H., Mozin, V., Schwalbach, P., Sjöland, A., Tobin, S.J., Trellue, H., Vaccaro, S., 2016. Determining initial enrichment, burnup, and cooling time of pressurized-water-reactor spent fuel assemblies by analyzing passive gamma spectra measured at the clab interimfuel storage facility in Sweden. Nucl. Instrum. Methods Phys. Res., Sect. A 820, 102–111 [Online]. Available: http://www.sciencedirect.com/science/article/pii/S0168900216002485.
10. Galloway, J.D., Trellue, H.R., Fensin, M.L., Broadhead, B.L., 2012. Design and description of the NGSI spent fuel library with emphasis on the passive gamma signal. J. Nucl. Mater. Manage. XL (3), 25–34. Spring 2012.

11. Gauld, I.C., Francis, M., July 2010. Investigation of passive gamma spectroscopy to verify spent nuclear fuel content. In: Institute of Nuclear Materials Management, 51st Annual Meeting.
12. Gauld, I.C., Litwin, K.A., August 1995. Verification and Validation of the ORIGEN-S Code and Nuclear Data Libraries. Atomic Energy of Canada Limited (AECL), Tech. Rep. RC-1429/ COG-I-95-150. [Online]. Available: http://web.ornl.gov/sci/scale/pubs/ rc1429.pdf.
13. Gauld, I.C., Carlson, P., Litwin, K., October 1995. Production and Validation of ORIGEN-S Cross-Section Libraries for CANDU Reactor Fuel Studies. Whiteshell Laboratories, Tech. Rep. RC-1442; COG-I-95-200.
14. Gauld, I.C., Giaquinto, J.M., Delashmitt, J.S., Hu, J., Ilas, G., Haverlock, T.J., Romano, C., 2016. Re-evaluation of spent nuclear fuel assay data for the Three Mile Island unit 1 reactor and application to code validation. Ann. Nucl. Energy 87 (Part 2), 267–281 [Online]. Available: http://www.sciencedirect.com/science/article/pii/S0306454915004454.
15. Hayden, R.J., Reynolds, J.H., Inghram, M.G., 1949. Reactions induced by slow neutron irradiation of europium. Phys. Rev. 75, 1500–1507 [Online]. Available: http://link.aps.org/doi/10.1103/PhysRev.75.1500.
16. Hermann, O.W., Bowman, S.M., Brady, M.C., Parks, C.V., March 1995. Validation of the scale system for pwr spent fuel isotopic composition analyses. http://web.ornl.gov/sci/scale/pubs/tm12667.pdf, Oak Ridge National Laboratory, Tech. Rep. ORNL/TM-12667. [Online]. Available: http://web.ornl.gov/sci/scale/pubs/tm12667.pdf.
17. Ilas, G., Gauld, I.C., Difilippo, F.C., Emmett, M., February 2010a. Analysis of Experimental Data for High Burnup PWR Spent Fuel Isotopic Validation– Calvert Cliffs, Takahama, and Three Mile Island Reactors. http://www.nrc.gov/docs/ML1009/ML100900227.pdf, U.S. Nuclear Regulatory Commission(NRC), Tech. Rep. NUREG/CR-6968/ ORNL/TM-2008/071. [Online]. Available: http://www.nrc.gov/docs/ML1009/ML100900227.pdf.
18. Ilas, G., Gauld, I.C., Murphy, B.D., February 2010b. Analysis of Experimental Data for High Burnup PWR Spent Fuel Isotopic Validation–ARIANE and REBUS Programs (UO2 Fuel). http://www.nrc.gov/docs/ML1008/ML100850211.pdf, U.S. Nuclear Regulatory Commission (NRC), Tech. Rep. NUREG/CR-6969/ ORNL/TM-2008/072. [Online]. Available: http://www.nrc.gov/docs/ML1008/ML100850211.pdf.
19. Lebrun, A., Bignan, G., 2001. Nondestructive assay of nuclear low-enriched uranium spent fuels for burnup credit application. Nucl. Technol. 135 (3), 216–229.
20. Mughabghab, S.F., 2006. Atlas of Neutron Resonances, Fifth Edition: Resonance Parameters and Thermal Cross Sections. Z=1–100. Elsevier Science.
21. Mughabghab, S., May 2016. Private correspondence.
22. Nakahara, Y., Suyama, K., Suzaki, T., January 2002. Translation of Technical Development on Burn-up Credit for Spent LWR Fuels. Oak Ridge National Laboratory, Tech. Rep. JAERI-Tech 2000-071 (ORNL/TR-2001/01).
23. Phillips, J.R., 1991. Irradiated fuel measurements. In: Passive Nondestructive Analysis of Nuclear Materials. Los Alamos National Laboratory, Los Alamos, NM.
24. Phillips, J.R., Halbig, J.K., Lee, D.M., Beach, S.E., Bement, T.R., Dermendjiev, E., Hatcher, C.R., Kaieda, K., Medina, E.G., 1980. Application of non-destructive gamma-ray and neutron techniques for the safeguarding of irradiation fuel materials. Los Alamos National Laboratory, Tech. Rep. LA-8212. [Online]. Available: http://www.osti.gov/scitech/servlets/purl/5277679-FVEIuh.
25. Radulescu, G., Gauld, I.C., Ilas, G., March 2010. SCALE 5.1 Predictions of PWR Spent Nuclear Fuel Isotopic Compositions. http://info.ornl.gov/sites/ publications/files/Pub23359.pdf, Oak Ridge National Laboratory, Tech. Rep. ORNL/TM-2010/44. [Online]. Available: http://info.ornl.gov/sites/publications/files/Pub23359.pdf.
26. Rinard, P.M., Bosler, G.E., 1988. Safeguarding LWR spent fuel with the Fork detector. Los Alamos National Laboratory, Tech. Rep. Report LA-11096-MS.
27. Sekine, T., Ichikawa, S.-I., Baba, S., 1987. Triple neutron capture of 153Eu in a reactor: the cross sections of 154Eu and 155Eu. Int. J. Radiat. Appl. Instrum. Part A. Appl. Radiat. Isotopes 38 (7), 513–516.

28. Shoman, N.T., Skutnik, S.E., 2016. Development of modern CANDU PHWR crosssection libraries for SCALE. Nucl. Eng. Des. 302 (Part A), 56–67 [Online]. Available: http://www.sciencedirect.com/science/article/pii/S0029549316300255.
29. Skutnik, S.E., Shoman, N.T., May 2016. Re-evaluation of the 154Eu thermal capture cross-section based on spent fuel benchmarking studies. In: PHYSOR 2016: Unifying Theory and Experiments in the 21st Century, Sun Valley, ID, pp. 633–633.
30. Suyama, K., Mochizuki, H., 2006. Corrections to the 148Nd method of evaluation of burnup for the PIE samples from Mihama-3 and Genkai-1 reactors. Ann. Nucl. Energy 33, 335–352.
31. Tait, J., Gauld, I., Kerr, A.H., 1995. Validation of the ORIGEN-S code for predicting radionuclide inventories in used CANDU fuel. J. Nucl. Mater. 223 (2), 109–121 [Online]. Available: http://www.sciencedirect.com/science/article/pii/0022311594006830#.
32. Vertebnyj, V.P., Vorona, P.N., Kal'chenko, A.I., Krivenko, V.G., Pshenichnyj, V.A., 1982. Neutron resonances of the radioactive nuclides Eu-154 and Eu-155. Vop. At. Nauki i Tekhn., Ser.Yadernye Konstanty 5 (49), 16.
33. Wiarda, D., Dunn, M.E., Greene, N.M., Williams, M.L., Celik, C., Petrie, L.M., April 2016. AMPX-6: A Modular Code System for Processing ENDF/B, Oak Ridge, TN.
34. Williams, M.L., 2011. Resonance self-shielding methodologies in SCALE 6. Nucl. Technol. 174 (2), 149–168.
35. Willman, C., Håkansson, A., Osifo, O., Bäcklin, A., Svärd, S.J., 2006. Nondestructive assay of spent nuclear fuel with gamma-ray spectroscopy. Ann. Nucl. Energy 33 (5), 427–438 [Online]. Available: http://www.sciencedirect.com/science/article/pii/ S030645490500 2847.

CHAPTER 8

Comparison of Small Modular Reactor and Large Nuclear Reactor Fuel Cost

Christopher P. Pannier[1]*, Radek Skoda[1,2]

[1]*Department of Nuclear Engineering, Texas A&M University, College Station, USA*

[2]*Czech Technical University, Prague, Czech Republic*

ABSTRACT

Small modular reactors (SMRs) offer simple, standardized, and safe modular designs for new nuclear reactor construction. They are factory built, requiring smaller initial capital investment and facilitating shorter construction times. SMRs also promise competitive economy when compared with the current reactor fleet. Construction cost of a majority of the projects, which are mostly in their design stages, is not publicly available, but variable costs can be determined from fuel enrichment, average burn-up, and plant thermal efficiency, which are public parameters for many near-term SMR projects. The fuel cost of electricity generation for selected SMRs and large reactors is simulated, including calculation of optimal tails assay in the uranium enrichment process. The results are compared between one another and with current generation large reactor designs providing a rough comparison of the long-term economics of a new nuclear reactor project. SMRs are predicted to have higher fuel costs than large reactors. Particularly, integral pressurized water reactors (iPWRs) are shown to have from 15% to 70% higher fuel costs than large light water reactors using 2014 nuclear fuels market data. Fuel cost sensitivities to reactor design parameters are presented.

Keywords: Nuclear Energy, New Nuclear, Nuclear Fuel Cost, Small Modular Reactors, SMR, Light Water Reactors.

1. INTRODUCTION

There are over 60 electricity-generating nuclear reactors under construction around the world as of March 2013 [1]. Of these, only six units are considered small reactors with power ratings below 400 MWe (megawattselectric): two 30 MWe units in Russia, two 100 MWe units in China and two 320 MWe units in Pakistan [2]. In the Americas, Asia, and Europe, reactor vendors have adopted a philosophy of small modular reactor designs to reduce the capital cost and increase the number of suitable sites for new nuclear electric generation projects. These small designs are offered in addition to conventional large reactor designs to suit a variety of utility needs. The two classes of nuclear reactors (small modular and large) are expected to have different overnight, financing, and variable costs.

The development of small electric-generating nuclear reactors was strongly influenced by naval applications. These naval reactors and scaled-down versions of existing large nuclear power plants have inspired current SMR designs which can be broadly classified as integral pressurized water reactors (iPWRs), marine-derivative PWRs, boiling water reactors (BWRs), pressurized heavy water reactors (PHWRs), gas-cooled, liquid metal-cooled, and various non-conventional designs [3] -[5]. Integral pressurized water reactors (iPWRs) integrate the steam generator and pressurizer into the core pressure vessel. SMRs can be divided into two broad groups based on their design and licensing status: those for early deployment based on a proven LWR technology, and those for longer-term deployment, based on other innovative designs. Other authors have written extensively on the potential for and variety of SMR designs [6].

Generally, SMRs offer utilities the opportunity to add nuclear generating capacity with a smaller total project cost than large LWRs. Construction cost is significant to the total cost of any nuclear power project. However, after the construction cost is amortized the fuel, operations and maintenance costs drive the cost of electric generation. In other words, construction cost drives the economics in the short and medium terms, but fuel cost is important for the long term. Economies of scale tend to favor larger plants over small plants on variable costs. While the smaller SMRs may save on construction cost compared to LWRs, their variable cost of electric generation is not openly reported. In this paper, emphasis is placed on assessing the economic viability of a few selected near-term SMR designs.

Through this assessment, areas for economic improvement of SMR nuclear designs will be identified.

Since the fuel of current LWRs is very similar to the fuel of several SMRs derived from them, it is straightforward to compare fuel costs of the two concepts, giving a comparison of variable costs of the two reactor classes. These papers summarize the fuel cost issue and put it into perspective with the initial capital investment, economy of scale and overall cost of construction [7] [8]. The SMR designs that use gas, liquid metal or molten salt coolants have significantly different fuel designs from the low enriched uranium dioxide fuel used in current LWRs and PWR-derived SMR designs. Therefore only the PWR-derived SMR designs are fully suitable for direct SMR to LWR fuel cost comparison.

In this paper, five SMRs based on LWRs were chosen. All SMR reactors in the study are still in pre-build or pre-licensing stages, so the design characteristics are nominal and subject to change by their designers. The five other PWR-derived designs for near term land deployment in this study are: NuScale, mPower, SMART, the Westinghouse SMR, and HI-SMUR. The NuScale power reactor is a 45 MWe reactor based on a PWR design to be installed underground with up to six units sharing a common housing [9]. The B&W mPower reactor was announced to be installed in modules of 125 MWe with a long operating cycle between refuelings [10]. The South Korean SMART (System-integrated Modular Advanced ReacTor) is being developed by KAERI with a thermal power of 330 MWt to be used for both electric generation and water desalination or other process heat applications [11]. The Westinghouse Small Modular Reactor (W-SMR) builds upon the design concepts of the much larger AP1000 reactor currently under construction in China and the United States. The Westinghouse SMR was announced to be a 200 MWe integral pressurized water reactor using a derivative of AP 1000 fuel [12]. Holtec Inherently Safe Modular Underground Reactor, (HI-SMUR) was announced to be a 140 MWe reactor design does not require coolant pumps as it relies on natural convection. It does not depend on off-site power for safe shutdown and thus qualifies as inherently safe. As its name implies, the HI-SMUR is to be built underground [13]. The HI-SMUR is not considered an integral PWR because its steam generators are not integrated into the reactor vessel.

Six well-known light water reactors and one large heavy water reactor were chosen to represent currently available options on the nuclear reactor market. The VVER 1000 class represents the standard Russian pressurized water reactor offered by Atomstroyexport. It is available in several power variants. Many units were built in Russia, India, China, and other countries [14]. The AP 1000 is the Westinghouse pressurized water reactor currently under construction in USA and China [15]. The ABWR is the Japanese boiling water reactor offered by Toshiba and Hitachi and already operated in Japan and Taiwan [16]. The EPR is Areva pressurized water reactor under construction in Finland, France, and China [17]. The US-EPR is the variant for the US market. The US-APWR is a large pressurized water reactor designed by Mitsubishi Heavy Industries [18]. The ESBWR is a boiling water reactor designed by Hitachi and General Electric [19]. Finally, the Enhanced CANDU- 6 is the heavy water designed in Canada [20]. It uses uranium at natural enrichment, but requires heavy water moderator in the core.

2. METHODS

A light water moderated reactor using uranium fuel requires fuel with a ^{235}U concentration above that of natural uranium. The economics of uranium enrichment are well known and are described in literature [21]. To calculate fuel price in a nuclear system using enriched uranium fuel, one must know the price of enriched uranium used to make the fuel. This price depends on the enrichment of the fuel, a reactor design parameter. Furthermore, three market parameters are needed to calculate the price of enriched uranium: the price of U_3O_8 from a uranium mill, the price of conversion of U_3O into the gaseous UF_6 for enrichment, and the price of SWU (separative work units). The number of SWUs required to enrich to a certain level is a measure of the electric energy required to perform the enrichment. All three of these costs are available from the Uranium Consulting Company's uranium price indices [22].

The cost of fabricating nuclear fuel assemblies from enriched uranium is proprietary to the nuclear fuel producer and is strongly dependent on the type of fuel assembly chosen for a reactor. Integral PWR fuel is not expected to vary in the fuel fabrication techniques used for large LWRs. In a 1994 study, fuel fabrication prices are given with a range between $200 and $400 per kg U, with the higher costs for fuel designed to withstand a higher burnup. In that study, a value of $275/kg U is assumed for fabrication cost [23]. A more recent 2010 study by EPRI uses a nominal $200/kg U for PWR fuel fabrication cost [24]. A constant value of $275/kg U will be used for fabrication cost throughout this study.

To specify the cost of mining, conversion and enrichment nuclear fuel per kWh, a simple formula will be used. When the price of nuclear material is known, the fuel cost of electrical energy produced by unit of nuclear fuel can be evaluated as:

$$P = \frac{U}{\eta \cdot B \cdot 240} \quad (1)$$

Where P is fuel cost in ¢/kWh, U is price of enriched uranium product in \$/kg of uranium, η

is plant net efficiency, and B is burn-up in MWd/kg of uranium. The factor of 240 comes from the units conversion of days into hours and \$/MWh into ¢/kWh. For fuel cost minimization as attempted in enrichment plants, the uranium enrichment process must be optimized.

2.1. Uranium Enrichment Optimization

In the first step of the optimization, the optimal depletion rate of the tailing uranium is computed according the following procedure, as shown by Benedict and Pigford in [26]. Separative capacity, *SC*, is defined by the following formula:

$$SC = W(2x_W - 1)\ln\left(\frac{x_W}{1-x_W}\right) + P(2x_P - 1)\ln\left(\frac{x_P}{1-x_P}\right) - F(2x_F - 1)\ln\left(\frac{x_F}{1-x_F}\right) \quad (2)$$

where W stands for waste (tails) rate in kilograms per unit time, P for product rate in kilograms per unit time, F for feed rate in kilograms per unit time. The mass fraction of ^{235}U are x_w as waste enrichment (tails assay), x_p as product enrichment, and x_F as feed uranium enrichment. Separative capacity represents a measure of the rate at which an enrichment cascade performs the separation and is used as a measure of isotope separation work magnitude. In the isotope separation plant, the initial cost of the separation plant is proportional to the separative capacity of the plant. The annual operating costs are proportional to the separative work done per year. Separative work, *SW*, is measure of the amount of separation achieved by a cascade producing M_W kilograms of waste and M_P kilograms of product from M_F kilograms of feed, and is assigned as:

$$SW = M_W(2x_W - 1)\ln\left(\frac{x_W}{1-x_W}\right) + M_P(2x_P - 1)\ln\left(\frac{x_P}{1-x_P}\right) - M_F(2x_F - 1)\ln\left(\frac{x_F}{1-x_F}\right) \quad (3)$$

The annual charges for enrichment plant investment plus annual operating costs exclusive of feed is in \$/year, where *SC* is the annual separative capacity in kilograms of uranium per year, and c_s is the unit cost of separative work in dollars per kilogram of separative work units (\$/kg-SWU). If M_F kg of feed is charged per year at a unit cost of c_F \$/kg, the total annual cost c is:

$$c = SC \cdot c_S + M_F \cdot c_F \tag{4}$$

If P kg of product is made per year, the unit cost of product, c_p , is:

$$c_p = \frac{SC \cdot c_S}{M_F} + \frac{M_F \cdot c_F}{M_P} \tag{5}$$

Applying the conservation of mass equation,

$$M_F = M_P + M_W \tag{6}$$

and the conservation of the ^{235}U mass equation,

$$M_F \cdot x_F = M_P \cdot x_P + M_W \cdot x_W \tag{7}$$

to the enrichment plant gives two mass ratios as functions of the three mass fractions in Equations (8) And (9)

$$\frac{M_F}{M_P} = \frac{x_P - x_W}{x_F - x_W} \tag{8}$$

$$\frac{M_W}{M_P} = \frac{x_P - x_F}{x_F - x_W} \tag{9}$$

The amount of separative work required in the up-stream enriching section is independent of tails composition. But the cost of separative work required in the stripping down-stream section varies from zero when $x_W = x_F$ to infinity when $x_W =$ Conversely, the cost of feed varies from infinity when $x_W = x_F$ to a minimum at $x_W = 0$. There is therefore an optimum tails assay, $x_{W,\text{ optimal}}$ ⍰$(0, x_F)$ at which the sum of the cost of separative work and the cost of natural uranium feed is a minimum. Optimum tails composition occurs when:

$$\left(\frac{\partial c_p}{\partial x_w}\right)_{x_P, x_F} = \tag{10}$$

$$c_P = c_F \frac{x_P - x_W}{x_F - x_W} + c_S \frac{x_P - x_F}{x_F - x_W}(2x_W - 1)\ln\left(\frac{x_W}{1 - x_W}\right)$$
$$+ c_S (2x_P - 1)\ln\left(\frac{x_P}{1 - x_P}\right) - c_S \frac{x_P - x_W}{x_F - x_W}(2x_F - 1)\ln\left(\frac{x_F}{1 - x_F}\right) \quad (11)$$

$$\frac{c_F}{c_S} = (2x_P - 1)\ln\left(\frac{x_F \cdot x_{W,\text{optimal}}}{x_{W,\text{optimal}}(1 - x_F)}\right) + \frac{(x_F - x_{W,\text{optimal}})(1 - 2x_{W,\text{optimal}})}{x_{W,\text{optimal}}(1 - x_{W,\text{optimal}})} \quad (12)$$

2.2. Fuels Market Price Inputs

The enriched uranium cost for the selected power plants, the prices of material, conversion, and enrichment are shown in Table 1. Price input parameters are volatile and their evolution over time is shown in Figure 1 and Figure 2.

For a comparison with operating commercial nuclear power plants, performances and parameters of the selected LWRs and SMRs are shown in Table 2.

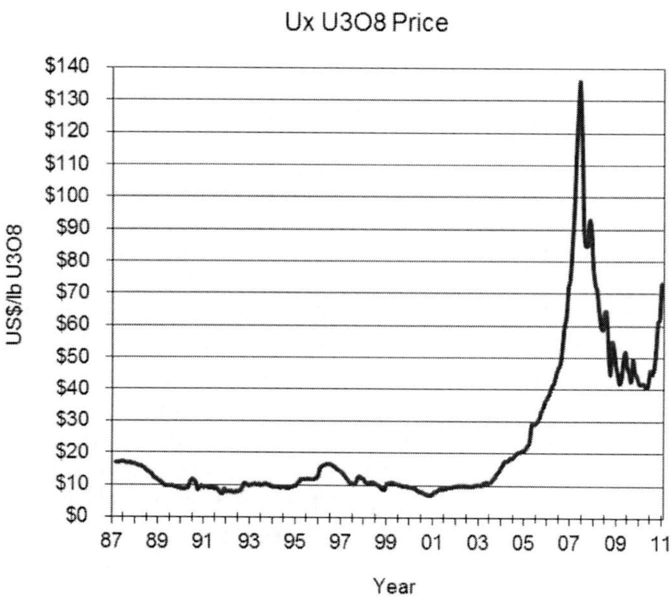

Figure 1. Price of uranium, data from UxC [22]

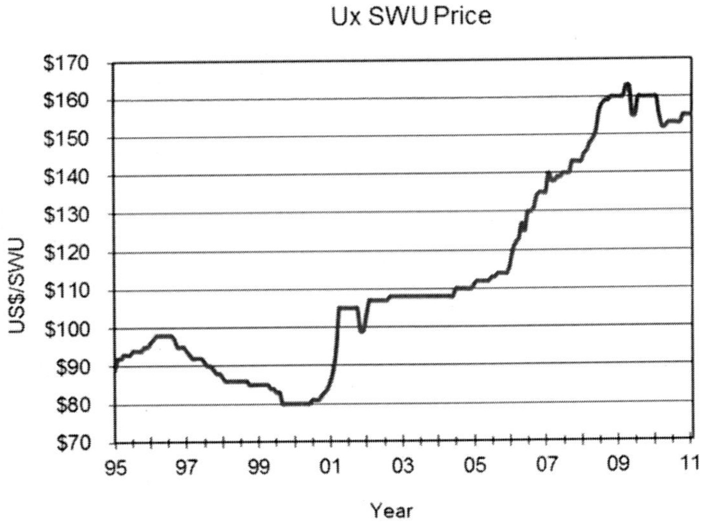

Figure 2. Price of SWU for the enrichment process, data from UxC [22]

Table 1. Mean fuels cycle cost parameters in Spring 2014.

Parameter	Unit	Average Cost
U_3O_8	\$/lb of U_3O_8	35.6
Conversion	\$/kg U in UF_6	8
SWU	\$/SWU	98
Fabrication	\$/kg U in UO_2	275

3. RESULTS

Based on historical enrichment, conversion and uranium costs, a simulation for a given time horizon was performed using correlated simulated input variables. The calculation for all reactors resulted in full distribution of fuel cost and sensitivities to increased burn-ups. The simulated fuel cost for new large LWRs is 3.86 \$/MWh. The simulated iPWR (excludes HI-SMUR) design average fuel costs are between 3.95 and 7.47 \$/MWh with an average for the iPWR reactor class of 5.84 \$/MWh as shown in Table 3. Hence an increase between +15% to +70% can be expected for iPWR SMR fuel cost compared to new large LWR fuel cost.

Table 2. Model inputs for selected nuclear power plant designs.

Reactor design	Average fuel enrichment	Fuel burnup [GWd/tU]	Net electric power [MWe]	Thermal power output [MWt]	Plant net efficiency
SMR					
NuScale	4.95%	50	45	160	28%
SMART	4.88%	60	90	330	27%
mPower	5.00%	40	125	400	31%
HI-SMUR	4.95%	35	140	450	31%
W-SMR	4.95%	54	200	600	33%
Large Reactors					
EC 6	Natural	7.5	745	2084	36%
VVER1000	3.50%	43.4	1000	3000	33%
AP1000	4.55%	60	1100	3400	32%
ABWR	3.70%	45	1315	3811	35%
US-APWR	5.00%	62	1600	4451	36%
US-EPR	5.00%	60	1600	4500	36%
ESBWR	4.20%	50	1561.5	4500	35%

The simulated mean and standard deviation fuel cost for every reactor studied is presented in Table 4. The reactor types are ordered from lowest mean fuel cost to highest. The large light water reactors had the lowest mean fuel cost per unit electric energy produced. In the simulation, the large heavy water moderated reactor had a 42% higher mean fuel cost than large light water reactors. However, it must be remembered that the fabrication cost that includes the processes of cladding, assembling, and shipping new fuel elements was estimated at $275/kg U regardless of the fuel design. The fabrication cost for CANDU fuel may be somewhat different than the fabrication cost for LWR fuel. As shown in Figure 10 and Figure 11, heavy water reactor fuel cost is more dependent on fabrication cost than light water reactor fuel, as is to be expected for a natural enrichment fuel.

Assuming the economies of scale of large reactors (reduced neutron leakage and thermal energy losses) one would expect lower fuel cost for larger SMR units. The results do not follow this trend. Rather, SMR designers often aim for fuel management goals like long times between refueling outages or using reactor heat to desalinate seawater, both of which raise the fuel cost of electricity produced for one design over another. Current fuel management goals for SMRs are not related to reactor electric power rating in the 45 to 300 MW electric power rating range. This may change as SMR designs mature and utilities build and operate SMRs.

Sensitivity analysis directly shows which parameters influence the fuel cost. Hence a sensitivity analysis of mean fuel cost to burnup was performed; the results are shown in Figure 3. From industry history, one can expect that plant owners will try to improve fuel economy by pushing burn-ups higher [26] through core materials modifications after a reactor is built and operating.

Table 3. Mean fuel cost and standard deviation for the compared reactor classes.

Reactor Class	Mean Fuel Cost [US$/MWh]
Large LWR	3.86
Large Reactor	4.09
PWR-Derived SMR	5.84

Table 4. Mean fuel cost and standard deviation for the compared reactor designs in order of increasing mean fuel cost.

Reactor Type	Reactor Classes	Mean Fuel Cost [US$/MWh]	Standard Deviation [US$/MWh]
US-APWR	Large LWR, Large Reactor	3.67	0.116
AP1000	Large LWR, Large Reactor	383	0.120
US-EPR	Large LWR, Large Reactor	3.84	0.121
ABWR	Large LWR, Large Reactor	3.89	0.120
VVER1000	Large LWR, Large Reactor	3.95	0.121
ESBWR	Large LWR, Large Reactor	3.95	0.124
W-SMR	iPWR, PWR-derived, SMR	4.51	0.142
SMART	iPWR, PWR-derived, SMR	4.88	0.154
EC-6	Large Reactor	5.49	0.091
NuScale	iPWR, PWR-derived, SMR	5.76	0.182
mPower	iPWR, PWR-derived, SMR	6.60	0.208
HI-SMUR	PWR-derived, SMR	7.47	0.236

Plant net efficiency is not only design dependent but also strongly site dependent, so the sensitivity was calculated and the results are shown in Figure 4. Similarly, deducing from improvements in thermodynamic efficiency over the last 40 years, one assumes fuel economy will be improved by better net efficiency during plant operations and in newer plant builds. The curves in Figure 4 can also be used to estimate the fuel cost for a site-specific net thermodynamic efficiency.

Higher fuel enrichment on its own, as illustrated on Figure 5, has a negative impact of fuel economy. However, as higher fuel enrichment goes hand in hand with a higher burn-up (other fuel limits permitting), the enrichment dependency must be presented along with the burn-up dependency for a complete picture.

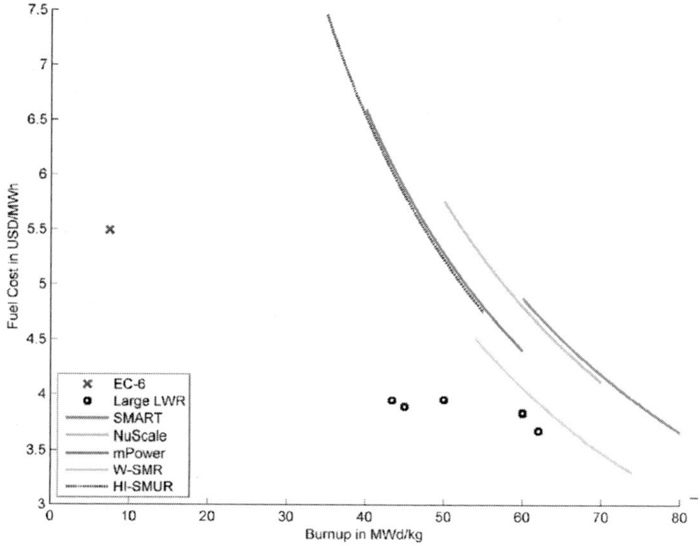

Figure 3. Fuel cost sensitivity to discharge burn-up for PWR-derived SMRs compared to large reactors

Histograms of model simulation outputs are presented for the heavy water reactor, a large light water reactor and two integral pressurized water reactor SMRs in Figures 6-9.

Pie charts of the simulated average cost of each component of fuel cost are given for the heavy water reactor, a large light water reactor and two integral pressurized water reactor SMRs in Figures 10-13. Because the heavy water reactor (Enhanced CANDU 6) uses natural uranium enrichment, it has no separation or enrichment cost.

4. CONCLUSIONS

It is expected that SMRs may be easier to license and find more acceptable by the general public then current LWRs. However, the predicted fuel cost is 15% - 70% higher for integral pressurized water reactor type small and modular reactors than for currently operating or under-construction large reactors. Other studies indicate. that the fixed construction cost of small and large nuclear power stations have a similar behavior [7]. In conclusion, the projected range of the fuel cost when producing electricity from SMRs shows these reactors are more fuel-wise expensive than current large LWRs.

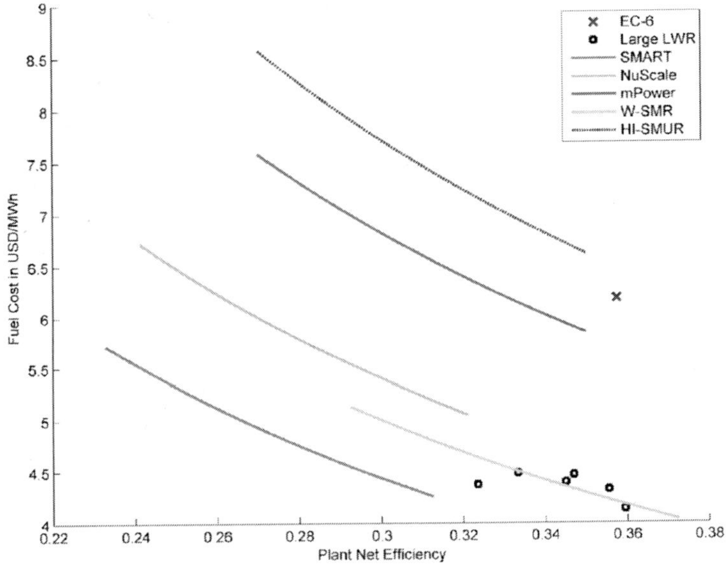

Figure 4. Fuel cost sensitivity to plant net efficiency for PWR-derived SMRs compared to large reactors.

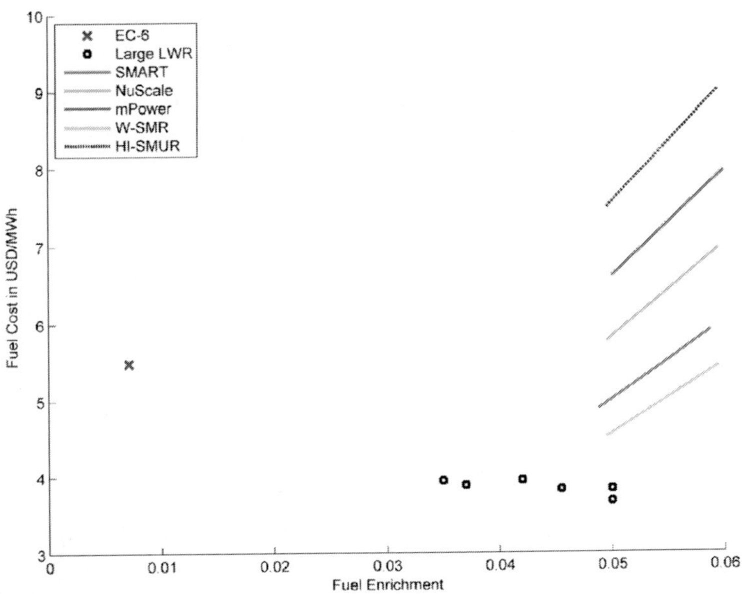

Figure 5. Fuel cost sensitivity to average fuel enrichment for PWR-derived SMRs compared to large reactors.

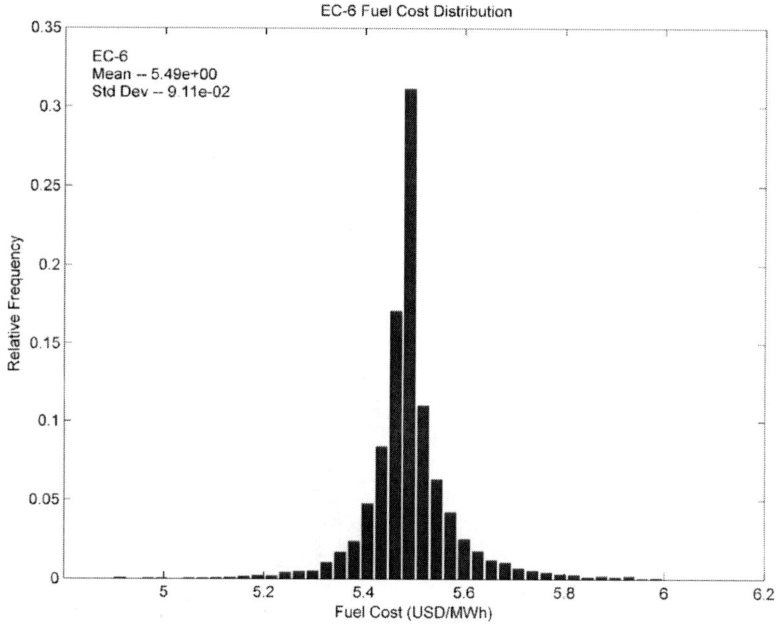

Figure 6. Fuel cost distribution for enhanced CANDU-6.

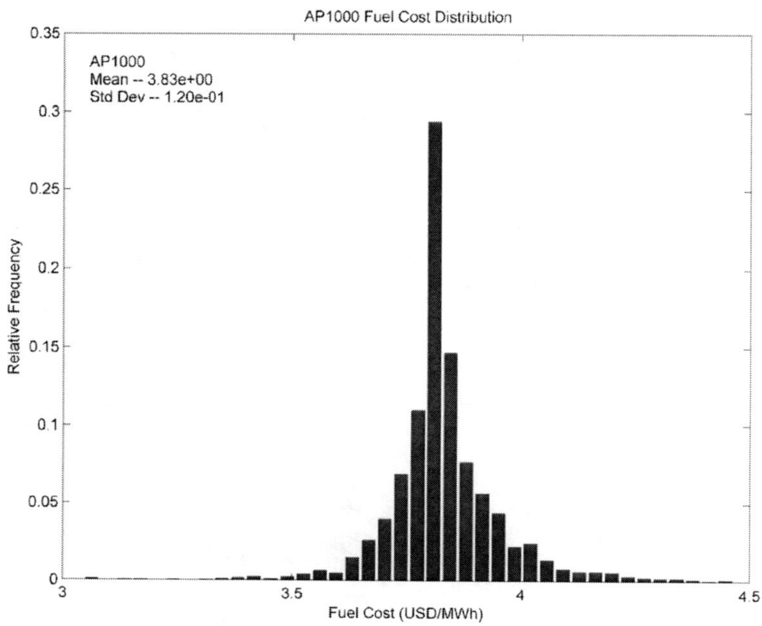

Figure 7. Fuel cost distribution for AP1000. 4.

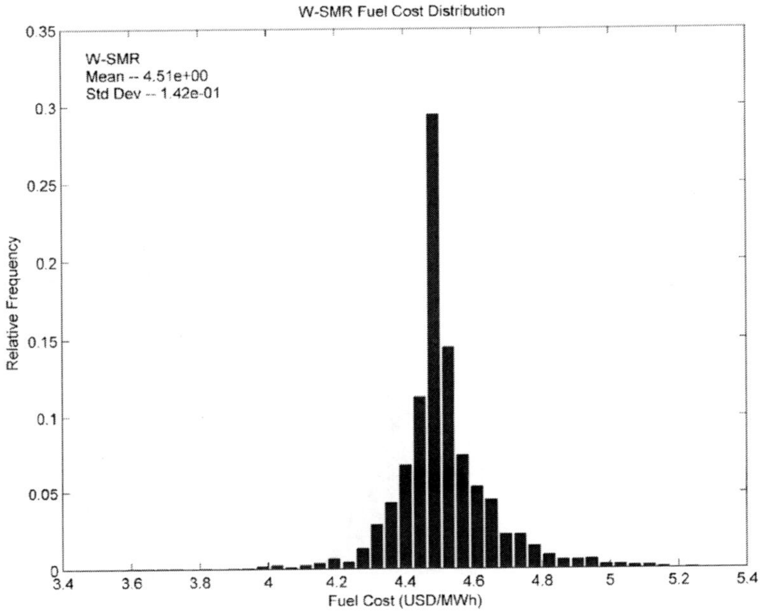

Figure 8. Fuel cost distribution for Westinghouse SMR.

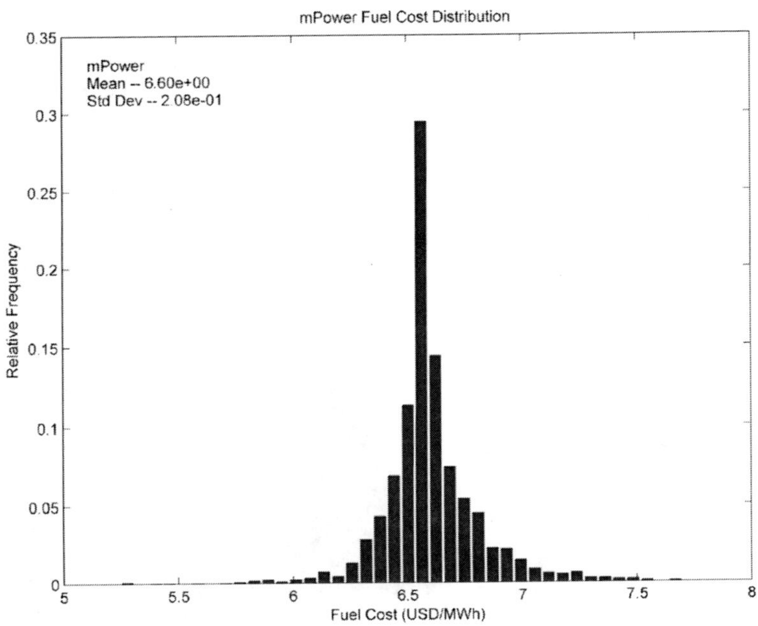

Figure 9. Fuel cost distribution for mPower.

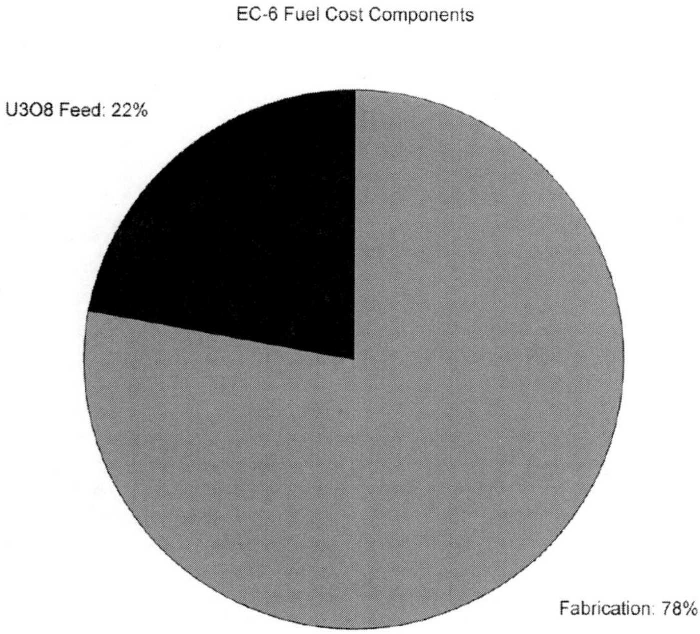

Figure 10. Fuel cost components for enhanced CANDU-6.

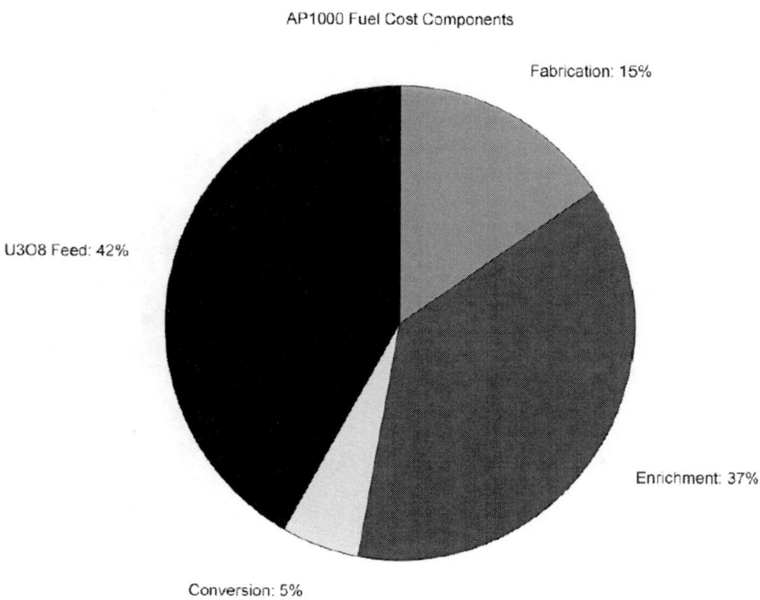

Figure 11. Fuel cost components for AP1000.

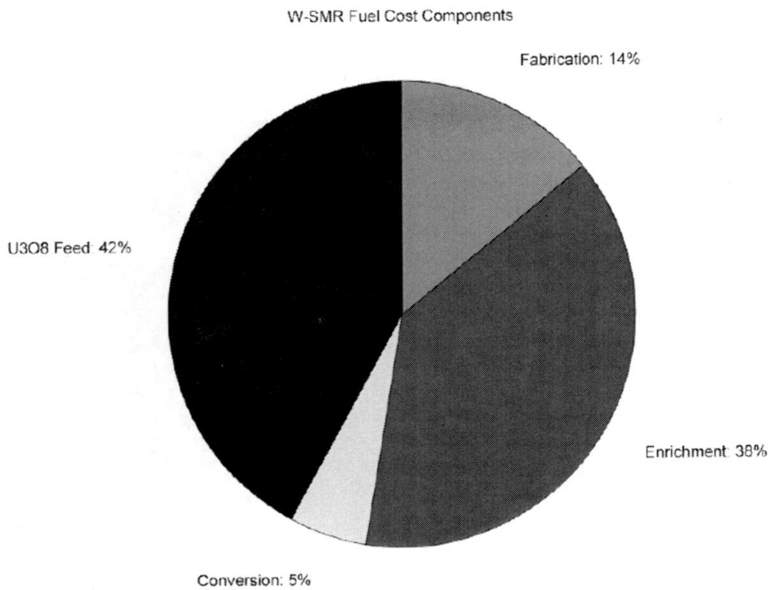

Figure 12. Fuel cost components for Westinghouse SMR.

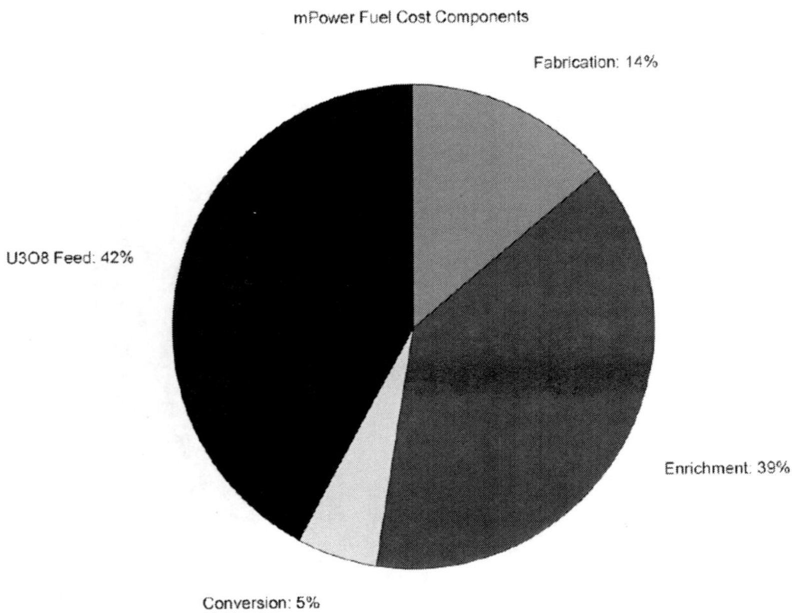

Figure 13. Fuel cost components for mPower.

There are a few caveats to this analysis: The resulting averages presented above should not be used to compare one design to another as the model inputs give a snapshot of reactor design parameters in 2011 and simulated fuel market parameters in 2014. Specific designs may have experienced design changes in enrichment, burnup and thermodynamic efficiency. However, the averages are well-suited to compare one class of reactors to another. Plant thermodynamic efficiency is site-specific (ambient temperature of the cooling water source) and can thus change model inputs. Furthermore, as operating experience leads to efficiency in SMR operation, discharge burn-up and thermal efficiency are expected to rise, driving the fuel cost down. This pattern was experienced with LWRs when discharge burn-ups nearly doubled from 1970 to 2005 [26]. However, technologies that allow for increased fuel burn-up will improve the future fuel economics of both large and small reactors.

ACKNOWLEDGEMENTS

The authors would like to acknowledge Texas A&M University, Department of Nuclear Engineering and the Nuclear Power Institute for providing funding to present the results. Special thanks to Eric Webb of UxC for providing historic uranium, conversion and SWU price data. Anonymous reviewer comments improved the paper.

REFERENCES

1. World Nuclear Association (2013) Plans for New Reactors Worldwide. http://world-nuclear.org/info/Current-and-Future-Generation/Plans-For-New-Reactors-Worldwide/
2. World Nuclear Association (2013) Small Nuclear Power Reactors. http://www.world-nuclear.org/info/Nuclear-Fuel-Cycle/Power-Reactors/Small-Nuclear-Power-Reactors/
3. Ingersoll, D. (2009) Deliberately Small Reactors and the Second Nuclear Era. Presentation at UC Berkeley.
4. Greenspan, E. and Brown, N. (2001) Small Innovative Reactor Designs—Useful Attributes and Status of Technology. Rise University, Houston.
5. Michal, R. (2004) Interview with Argonne's David Wade: On the Development of Small Modular Reactors. https://smr.inl.gov/Document.ashx?path=DOCS% 2fReading+Room %2fArgonneDavidWade2004-8-2.pdf
6. Vujić, J., Bergman, R., Škoda, R. and Miletić, M. (2011) Small Modular Reactors: Simpler, Safer, Cheaper? Proceedings of ECOS 2011, Novi Sad, 4-7 July 2011, 2842-2852.
7. Locatelli, G. (2010) SMR Economics Evaluation. Proceedings of the 4th Annual Asia-Pacific Nuclear Energy Forum on Small and Medium Reactors (SMRs): Benefits and Challenges, Berkeley, 17-19 June 2010.
8. Pannier, C. and Skoda, R. (2011) Assessment of Small Modular Reactor Fuel Cost. Proceedings of ASME 2011 Small Modular Reactors Symposium, Washington DC, 28-30 September 2011.

9. Landrey, B. (2010) Introduction to NuScale Power. Proceedings of the 4th Annual Asia-Pacific Nuclear Energy Forum on Small and Medium Reactors (SMRs): Benefits and Challenges, Berkeley, 17-19 June 2010.
10. Babcock & Wilcox (2010) B&W mPower Reactor Design Overview. http://pbadupws.nrc.gov/docs/ML1015/ML101550512.pdf
11. Lee, W. (2010) The SMART Reactor. Proceedings of the 4th Annual Asia-Pacific Nuclear Energy Forum on Small and Medium Reactors (SMRs): Benefits and Challenges, Berkeley, 17-19 June 2010.
12. Fetterman, R., Harkness, A., Smith, M. and Taylor, C. (2011) An Overview of the Westinghouse Small Modular Reactor. Proceedings of ASME 2011 Small Modular Reactors Symposium. ASME, Washington DC, 28-30 September 2011.
13. Singh, K., Rampall, I. and Rajkumar, J. (2011) On the Thermal-Hydraulic Essentials of the Holtec Inherently Safe Modular Underground Reactor (HI-SMUR) System. Proceedings of ASME 2011 Small Modular Reactors Symposium. ASME, Washington DC, 28-30 September 2011.
14. Elemash (2004) VVER-1000 Nuclear Fuel. http://www.elemash.ru/en/production/Products/NFCP/VVER1000/
15. Westinghouse (2003) Westinghouse AP 1000 Advanced Nuclear Plant: Plant Description. http://www.ne.doe.gov/pdfFiles/AP1000_Plant_Description.pdf
16. Tepco (2010) Sustainability Report 2010. http://www.tepco.co.jp/en/ challenge/environ/pdf-1/10report-e.pdf
17. Areva (2007) UK EPR, Fundamental Safety Overview, Vol. 1: Head Document, Chapter A: EPR Design Description. http://www.if.uidaho.edu/gunner/ME443-543/HW/UK-EPR.pdf
18. Suzuki, S., Ogata, Y., Nishihara, Y. and Fujita, S. (2008) Global Development of Mitsubishi Standard APWR as an Effective Countermeasure Against Global Warming. Mitsubishi Heavy Industries, Ltd., Technical Review, 45, 51-54. http://www.mhi.co.jp/technology/review/pdf/e453/e453051.pdf
19. GE Hitachi Nuclear Energy (2010) ESBWR Design Control Document Tier 2. http://pbadupws.nrc.gov/docs/ML1034/ML103440210.pdf
20. CANDU Energy Inc. (2012) Enhanced CANDU 6 Technical Summary. Mississauga. http://www.candu.com/site/media/Parent/EC6%20Technical%20Summary_2012-04.pdf
21. Cochran, R. and Tsoulfanidis, N. (1999) The Nuclear Fuel Cycle: Analysis and Management. 2nd Edition, American Nuclear Society, La Grange Park.
22. The Ux Consulting Company (2011) UxC Prices. http://uxc.com/review/uxc_Prices.aspx
23. Nuclear Energy Agency (1994) The Economics of the Nuclear Fuel Cycle. OECD, Paris. http://www.oecd-nea.org/ndd/reports/efc/
24. Electric Power Research Institute (2010) Parametric Study of Front-End Nuclear Fuel Cycle Costs Using Reprocessed Uranium. EPRI Report 1020659. http://cybercemetery.unt.edu/archive/brc/20120621082115/http://brc.gov/sites/default/files/documents/1020659.pdf
25. Benedict, M. andPigford, T. (1957) Nuclear Chemical Engineering. McGraw-Hill, New York.
26. International Atomic Energy Agency (2007) Current Trends in Nuclear Fuel for Power Reactors. *Proceedings of IAEA* 2007 *General Conference.* http://www.iaea.org/About/Policy/ GC/GC51/GC51InfDocuments/English/gc51inf-3-att5_en.pdf

CHAPTER 9

Nuclear Power for Poland

Andrzej G. Chmielewski

Institute of Nuclear Chemistry and Technology, Warsaw, Poland

ABSTRACT

The present situation in Poland and Europe, regarding electric power generation by source, is discussed in the paper. The results of the implementation of EU competitive-low-carbon economy policy in some most developed countries in the continent, have given already good experimental data for evaluation of this strategy. Analysis of the reports pro- vided by official sources for Germany, Denmark and Finland is a base for EU energy policy evaluation. The combus- tion technologies will be a main energy sources for many years from now. Therefore effects of fossil fuels and biomass combustion on the environment are presented briefly. Finally, the developments regarding Polish Nuclear Energy Pro- gramme are overviewed.

Keywords: Energy Policy; Europe; Poland; Nuclear Power

1. INTRODUCTION

The energy policy is one of the most important issues for all nations and political structures such as the European Union. The EU is committed to reducing greenhouse gas emissions to 80% - 95% below 1990 levels by 2050 in the context of necessary reductions by developed countries as a group. The Commission analysed the implications of this in its "Roadmap for moving to a competitive low-carbon economy in 2050". The scenarios in this Energy Roadmap 2050 explore routes towards decarbonisation of the energy system. Forecasting the long-term future is not possible. However the experiment regarding this policy implementation has already started in some most economically and technologically developed countries in Europe, like Germany and Denmark. Result of this experiment should be adopted to the roadmap as quick as possible, otherwise Europe may lose its pathway to the bright future. The power sector which provides electricity is a backbone of the contemporary economy. More precisely, the big blocks with capacity bigger than 500 MW_e are the crucial for the development of a civilization whose members will live and work in the big metropolises by the year 2050. At least 80% of the world population will be there. Unfortunately, the energy policy is run by politics and the quick-return economy. Looking at some energy programmes one may consider that thermodynamics is not any longer thought in the schools or that sustainable development is just a slogan which helps to buy an election.

2. POLISH POWER SECTOR

Poland is a growing European economy. However, energy consumption per capita is much lower than in "Old" EU countries. The same concerns electricity, which is well illustrated in **Figure 1** [1]. This is one of the limitations as the country's economic and social prosperity grows. Therefore, the growing demand for energy is expected (see **Table 1**) [2]. Production of primary energy in Poland is based mainly on fossil fuels. The first place belongs, and will most likely belong for a long time, to hard coal and lignite, which cover 56% of demand.

It is important to mention that even Poland is a big producer of hard coal, and imports of this fuel are growing year by year, from 1.05 Mtoe in 1995, up to 8.16 Mtoe in 2010. In 2008, for the first time in history, import exceeded export, and amounted to 10.1 million tones [3]. Crude oil also has a significant share of 25%. The worst situation concerns electricity generation, where almost 90% is based on the combustion of fossil fuels or biomass (see **Table 2** [4]).

Due to the fact that electricity production is based mostly on solid fuels combustion, the emission of pollutants is tremendous, the numbers are given in **Table 3** [5].

Control of pollutants emission makes coal combustion technology very expensive. The new regulations affecting even smaller boilers have to be observed in the EU by the year 2016. In addition, SO_2, NO_x, total fly ash, new pollutants such as non-methane hydrocarbons, mercury and PM 2.5 will have to be treated in the future. The effects of pollutant emissions are harmful to the environment and human beings. Loss of statistical life expectancy, attributed to anthropogenic contributions to PM 2.5, in some regions is equal to several months [6]. The emission of mercury from coal combustion in Europe is equal to 20 tons per year. The metalorganic compounds formed find their way into the human food chain [7]. Finally, nonmethane cancerogenic volatile organic compounds (VOCs), including polyaromatic hydrocarbons (PAH), are emitted as well [8]. In 2020, the country will probably be forced to buy 100% CO_2 certificates, as sequestration will not solve the problem [9].

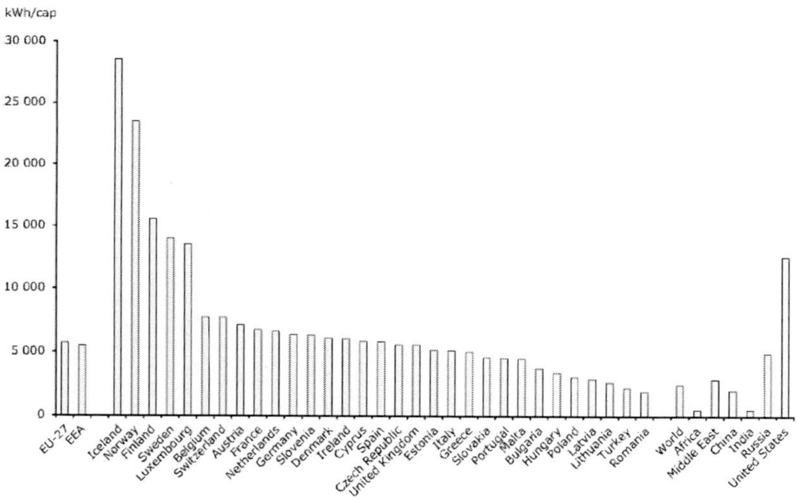

Figure 1. Electricity consumption per capita in different countries of the world [1].

Table 1. Forecast of demand for final energy by sector [Mtoe] [2].

Sector	**2015**	**2020**	**2025**	**2030**
Industry	19.0	20.9	23.0	24.0
Transport	16.5	18.7	21.2	23.3
Agriculture	4.9	5.0	4.5	4.2
Services	7.7	8.8	10.7	12.8
Households	19.1	19.4	19.9	20.1
Total	**67.3**	**72.7**	**79.3**	**84.4**

Table 2. Structure of power installed in electricity produc- tion sector, November 2011 [4].

EPS + EHPS (Hard coal fuel)	56%
EPS (Lignite fuel)	26%
EPS (Gas fuel)	2%
EPS (Hydro)	6%
Renewable (biomass combustion + wind)	7%
Others	3%

Table 3. Emission of pollutants from power sector in 2010 [5].

Pollutant	SO_2	NO_x	$PM_{2.5}$	VOC
[Mg]	509,847	287,324	14,337	18,257

PAH & PCB [kg]	Hg [kg]	Dioxins & Furans [mg i-TEQ]
578	8,771	11,446

3. EUROPE: LESSONS LEARNED

Europe has a dream: carbon-free electricity generation which, unfortunately, is and will remain a dream for this century. The tragedy is that the dreams cost too much. The share of installed electricity capacity by source is given in **Table 4** [10].

Analysis of this table shows that the share of installed capacity, based on combustion processes, is equal to 57.1%. Wood, municipal and industrial waste, biofuels and biogas combustion are counted as "renewable", which is a joke, since a combustion process that is friendly to the environment does not exist. Cherubini and co-authors propose that CO_2 emissions from biomass combustion for bioenergy should no longer be excluded from Life Cycle Assessment studies, or be assumed to have the same global warming potential as anthropogenic CO_2 emission [11].

Therefore, the installed capacity of real clean renewables (hydro, wind, solar and geothermal) is equal to 27.4%, but the real input to the total share gives hydro (16.7%) and wind (8.7%). It is interesting that these tragicomic data seem to be well understood by the authors of the document, since the generation input for "renewables" is given as a total number only. 78.3% of electricity produced in EU-27, in 2010, came from fossil fuel combustion and nuclear power. Unfortunately, 23.6% of electricity production was based on gas combustion which does not agree with sustainable development philosophy. Gas is a raw material for chemical/other industries and household applications, and gas combustion in big power blocks should be considered as a crime against future generations.

Table 4. Installed electricity capacity (*) and generation (**) in EU-27 by source [%].

Fossil fuels	Nuclear	Wood, waste	Liquid biofuels, -gas	Hydro	Wind	Solar	Geotherm.	Other
1	2	3	4	5	6	7	8	9
(*)54.0	15.4	2.4	0.7	16.7	8.7	1.9	0.1	0.1
(**)50.9	27.7		20.6 (so called renewables)					0.8

Two countries in Europe pointed out as the leaders in the renewable energy utilization are Denmark and Germany. In Denmark in 2011, 39.7% of total electricity production was generated by coal. Natural gas accounted for 16.5% of electricity production. Oil and non-renewable waste accounted for 1.3% and 2.2% of the electricity production, respectively. In addition, wind accounted for 28.1%, combustion of straw 2.2%, wood 6.6%, waste 2.7% and biogas 1.0%; solar and hydro 0.0% [12].

In Germany, the production of electricity was based on hard coal, 19%; lignite, 25%; natural gas, 14%; nuclear energy, 18%; heating oil and pumped storage, 5%; wind, 8%; biomass, 5%; water, 3%. photovoltaics, 3%; and biogenic household waste 1%. The annual full-load hours of German power stations in 2010 were; photovoltaics 900 h; pumped storage installations 1100 h; petroleum 1210 h; wind 1380 h; natural gas 3180 h; storage and run-off-river hydroelectric installations 3820 h; hard coal 3870 h, biomass 6400 h; lignite 6600 h; and nuclear energy 7330 h [13]. The German Federal Government plans that renewable energy will provide the main source of electricity by 2030, and projects that 58% of total electricity will be sourced from renewable energy, with wind being the dominant source in 2030, at 30.6%. Electricity from wind is expected to triple over the next years, while electricity from solar will double. Hydro will grow by

35.8% to account for 5.2% of generation, as will the use of biofuels, growing by 39.1% to reach 13.3% of the total. The use of natural gas in electricity generation will also increase over the period to 2030, growing to 22.6% of the total. Coal-fired generation is expected to contract to less than 20% of the total, with nuclear power being completely phased out by 2022 [14].

The numbers from 2010 show that a programme should consider good import-export relations of Germany with neighbours, since the main source of renewable energy will be wind, for which the full load hours equal to 1400 hours annually will not assure continuous electricity supply throughout the year. Imports from a newly constructed nuclear power plant in Kaliningrad, via the Baltic Sea route, may be a solution. The combustion technologies will still cover 56 % (coal 20% + gas 23% + biofuels 13%) of electricity production, which proves that the country will be not able to follow sustainable development policy. One may expect that biomass combustion will be excluded from the list of renewable energy sources and natural gas combustion in the big power blocks has been always considered as a robbery of raw material used by many sectors of economy and household appliances. The example of Denmark, and especially Germany, is proof that Trianer is right; renewables will not solve the greenhouse problem, and will not provide a sufficient electricity supply for the future [15].

The other example of electricity production mix is illustrated, now and in the future, by Finland; one of the most innovative countries in the world. Its economy is highly industrialized, with sizeable high-tech manufacturing, electronics and chemical sectors operating alongside a significant forestry and paper industry. Finland's energy consumption per capita is the highest in the International Energy Agency—Organisation for Economic Co-operation and Development (IEA-OECD) countries. Finland's electricity supply mix is very diversified, with nuclear, hydro and bioenergy accounting for around 31.6%, 16.9% and 15.6% of electricity supply. respectively, and gas, coal and peat also contributing to the energy mix. Furthermore, as part of the Nordic electricity system, Finland is one of the most advanced electricity markets in the world. Yet security of electricity supply remains a high priority concern. Supply concerns are exacerbated by the fact that Finland currently imports up to 2,000 MW of electricity from its neighbours during peak hours, as domestic electricity supply is limited. In 2010, the Parliament adopted decisions-in-principle for two new nuclear power stations, in addition to Olkiluoto 3, which is currently under construction. Finland hopes to achieve relative self-sufficiency, being able to cover peak load situations and possible disturbances of imports, through the development of biomass-fired power, and particularly the construction of additional nuclear power capacity [16].

4. NUCLEAR POWER FOR POLAND

The Ministry of Economy has issued a programme regarding power sector development, "Energy Policy for Poland until 2030" [17], and considers nuclear power as an important component of the electricity supply mix. The Council of Ministers of the Republic of Poland adopted resolution 4/2009, related to the Polish Nuclear Power Program (PNPP) development on 13 January, 2009. The Ordinance of the Council of Ministers of 12 May, 2009, on establishing the Government's Commissioner for Nuclear Energy and Undersecretary of State, Ministry of Economics was the next step in this process. The Resolution of the Council of Ministers of 11 August, 2009, was adopted; "Framework time schedule for nuclear power activities". The 13 May, 2011, amendment to the Atomic Law and other laws entered into force on 1 July, 2011. The additional law of 29 June, 2011, on the preparation and realization of investments in nuclear facilities and accompanying investments entered into force on 1 July, 2011.

An IAEA-led team of international experts that reviewed Poland's programme for introducing nuclear power found that significant progress has been made in the development of the country's nuclear infrastructure. They also noted good practices and made recommendations for further actions. The experts, assembled at Poland's request by the IAEA, conducted an Integrated Nuclear Infrastructure Review (INIR) mission in Poland from 18-22 March, 2013.

PGE (Polska Grupa Energetyczna; Polish Power Group) plans to install around 3000 MW_e of nuclear capacity, with its first unit expected to be online by 2025. In November 2011, the short list of three potential sites was announced to the public: Choczewo (Choczewo commune, poviat of Wejherowo, Pomeranian voivoidship); Gąski (Mielno commune, poviat of Koszalin, West Pomeranian voivoidship), and Żarnowiec (Krokowa commune, poviat of Puck, Pomeranian voivoidship) (**Figure 2**).

PGE signed MOUs regarding collaboration with big companies involved in engineering, construction and operations of NPPs: French EDF (2009), GE Hitachi (2010), and Westinghouse Electric Company LLC (2010). On 7 February, 2013 Polish company PGE EJ 1 sp. z o.o. concluded the contract with a consortium composed of Worley Parsons Nuclear Services JSC, Worley Parsons International Inc. and Worley Parsons Group Inc. The contract pertains to the performance of site characterization works and licensing support required, in order to complete the first Polish NPP building project led by PGE EJ1. The first Polish NPP is to generate approx. 3000 MW_e, as mentioned earlier.

The public attitude to nuclear power plants construction in Poland changed over the years 1987-2011. Just after the Chernobyl accident, only 30% of responding citizens declared support for nuclear energy. This has changed in the past few years. The radical breakthrough was observed in 2009, when 50% of population tested said "yes" to the nuclear option, with only 40% against. The situation changed after the crisis in Japan, reducing public confidence and support for nuclear energy. According to the last polls, only 40% of responders supported nuclear power, and 53% were against. In comparison with previous years, the amount of indecisive answers was small (7%). As in other countries, the operation of radioactive waste disposal facilities arouses even more fear and distrust [18].

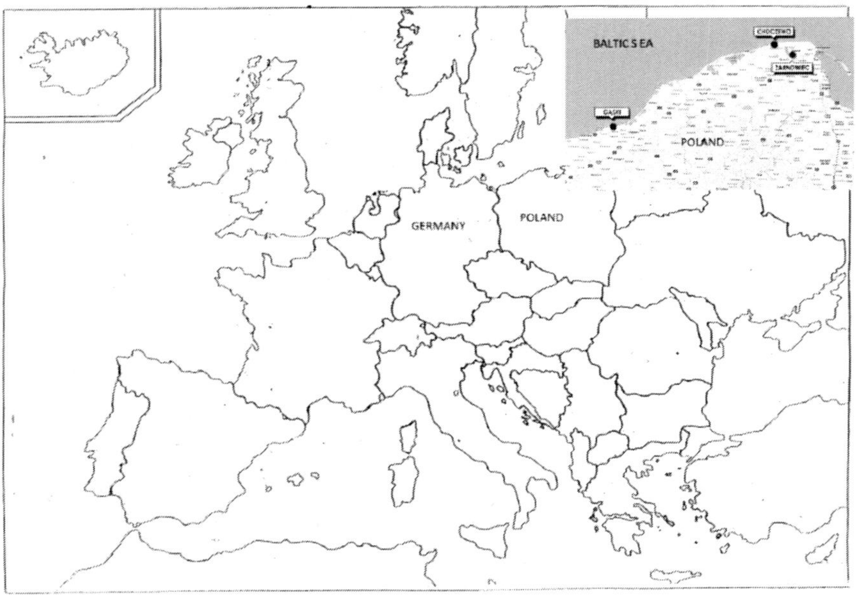

Figure 2. Possible localization for first Polish NPP.

5. R & D AND INDUSTRIAL PROGRAMS SUPPORTING PNPP

Polish Universities and R & D institutes are involved in the implementation of the activities supporting the Polish Nuclear Power Programme. The most active among the academic centres are Warsaw University of Technology, Warsaw University, and the Mining Academy in Cracow, Gdansk University of Technology, and Silesian University of Technology. With regard to R & D institutions, these are the National Centre for Nuclear Research, Swierk; the Institute of Nuclear Chemistry and Technology (INCT), Warsaw; the Central

Laboratory for Radiological Protection (CLRP), Warsaw; and the Institute of Nuclear Physics PAS, Cracow. INCT and CLRP have signed agreements with the National Atomic Energy Agency (NAEA), to support this regulatory body in the fields related to Technical Support Organization (TSO).

The activities of the President of the National Atomic Energy Agency (NAEA), as a central organ of governmental administration, for the issues of nuclear safety and radiological protection, is regulated by the Act of Parliament of 29 November, 2000, Atomic Law (O.J. 2007, No 42, Item 276) and relevant regulations. Further tasks of the NAEA President result from many other legal acts. Since 1 January, 2002, the NAEA President is supervised by the minister competent in environmental issues. The NAEA President executes their tasks through the National Atomic Energy Agency, whose internal organization is established by the statute conferred by the Environment Minister.

The Institutes have signed agreements in collaboration with CEA, France. In 2009, twenty Polish scientists participated in a nuclear "tour de France", visiting for six weeks the most significant manufacturing sites and laboratories in Poland. Specialists in training, research and industry—from mining to waste management, from design to decommissioning—discussed with them technical issues, as well as public relations. During the second phase, at the end of 2010, the Polish scientists received intensive twelve-week training at the Institute of Nuclear Science and Technology in Saclay. The most fundamental aspects of nuclear science were addressed, as well as radiation protection, safety and security, operation and waste management. The trainees had access to central driving simulators. Thirteen educators were hosted by French companies and laboratories for twelve weeks at the end of 2011. They stayed at the CEA in Cadarache, Saclay and Marcoule, at Andra in Chatenay-Malabry, at EDF in Lyon, at Areva in La Défense, and at the CNRS in Orsay [19].

The National Centre for Research and Development established a so-called strategic project, entitled "Technologies Supporting Development of Safe Nuclear Power Engineering" [20]. The Project comprises the following research tasks:

- Development of high temperature reactors for industrial purposes (consortium leader: Academy of Mining and Metallurgy, Cracow);
- Research and development of technology for controlled thermonuclear fusion (consortium leader: Henryk Niewodniczanski Institute of Nuclear Physics, Polish Academy of Sciences);
- Meeting the Polish nuclear power engineering's demand for fuel: fundamental aspects (research network leader: University of Warsaw);

- Development of spent nuclear fuel and radioactive waste management techniques and technologies (contractor: Institute of Nuclear Chemistry and Technology);
- Study of possibilities and criteria for participation of the Polish industry in the worldwide expansion of nuclear power engineering (research network leader: Warsaw University of Technology);
- Development of nuclear safety and radiological protection methods for nuclear power engineering's current and future needs (research network leader: Central Laboratory for Radiological Protection);
- Study of hydrogen generation processes in nuclear reactors under regular operation conditions and in emergency cases, with suggested actions aimed at upgrading nuclear safety (research network leader: Institute of Nuclear Chemistry and Technology);
- Study of processes occurring under regular operation of water circulation systems in nuclear power plants, with suggested actions aimed at upgrading nuclear safety (research network leader: Institute of Nuclear Chemistry and Technology);
- Development of methods and performance of safety analyses in nuclear reactors with heat reception disturbances under critical failure conditions (contractor; Warsaw University of Technology);
- Development of a method and performance of an exemplary systemic analysis of the operation of a nuclear plant, with a water reactor under partial association conditions (contractor: Gdansk University of Technology).

The Polish Institutes represent Poland in different international organizations, such as Euratom Supply Agency (ESA), Nuclear Energy Agency (NEA-OECD), International Framework for Nuclear Energy Cooperation (IFNEC) etc. The universities and institutes participate in several FP 7 and EURATOM projects, such as TALISMAN, ASGARD, IPPA, ADVANCE, MULTI-BIODOSE etc.

The role of EU and its programs on Polish program implementation is very important. In June 2013 the Commission has proposed to amend the 2009 nuclear safety directive. The proposal:

- Introduces new EU-wide safety objectives;
- Sets up a European system of peer reviews of nuclear installations;
- Establishes a mechanism for developing EU-wide harmonised nuclear safety guidelines;
- Strengthens the role and independence of national regulators;

- Increases transparency on nuclear safety matters;
- Includes new provisions for on-site emergency preparedness and response.

The EU needs its own verification mechanism to ensure that common safety objectives are achieved. At least every 6 years nuclear installations would have to undergo specific assessments on one or more nuclear safety issues. The assessments would be submitted for EU-wide peer reviews. Such strategy forms solid foundation for construction and operation of NPPs in Europe, and in Poland.

The abounded former project regarding construction of NPP in Żarnowiec was realized in 70% - 80% by Polish industry. Nowadays, Polish companies are involved in the construction of nuclear power plants in Finland and France. The foreign companies seek similar cooperation with Poland located enterprises. For example, General Electric Hitachi has announced a number of other preliminary project development agreements with various companies and organizations, to support future projects in Poland, including the following [21]:

- Fluor Corp., to serve as GEH's engineering, procurement and construction (EPC) partner (2011)• Energoprojekt Warszawa, S.A. (EW), a Warsaw-based engineering firm (2011)• The Institute of Atomic Energy in Poland (POLATOM); a research institute located in Świerk that advises the government on nuclear energy issues (2011)• Stocznia Gdansk, a leading Polish shipyard, for the potential manufacturing of nuclear components for GEH (2011)• RAFAKO S.A., Europe's leading boiler equipment manufacturer, for the potential manufacturing of nuclear components for GEH (2011)• Gdansk University of Technology, West Pomeranian University of Technology, Szczecin University and Koszalin University of Technology (2011)• SNC-Lavalin Polska, a global engineering services firm (2010)• Warsaw University of Technology (2012).

Similar actions have been taken by the AREVA/EDF team. Besides a large scientific collaboration with Polish universities (mainly Warsaw University of Technology (WUT), with AREVA specialists' lectures at WUT in Warsaw and master's degrees in AREVA research facilities), AREVA has been active in the field of the industrial localization in Poland:

- March 2010: a study on "Polish Industry Participation" in the nuclear project• Since July 2010, sourcing and pre-qualification investigations of the Polish industry landscape, with about 60 days of visits in Polish enterprises, and 29 prequalifications. This action is still running.
- September 2010: first AREVA Suppliers' Day in Warsaw.

- October 2011: second ARVA Suppliers' Day in Warsaw.
- April 2012: first AREVA/EDF Suppliers' day in Gdańsk.
- Starting from December 2012: AFCEN codes seminars (continuing).
- April 2013: specific contacts with Polish heavy industry in Pomerania (shipyards) and major civil work companies.
- Continuation of collaboration with leading Polish companies working on AREVA nuclear power plant construction site in Olkiluoto (Finland).

The role of the technical organizations, such as Polish Nuclear Society (PTN) [22] and Environmentalists for Nuclear Energy (SEREN) [23] in NE promotion and education, cannot be overestimated.

Technology foresight for a vision of the energy sector development in Poland till 2030, was performed at the request of the Ministry of Economy [24]. The aim of the project was to indicate energy and fuel sector development directions in the time horizon up to 2030, and identify key energy technologies of strategic importance. Impact assessments ranks for the 20 Delphi statements published as a result of this study placed on the position number 3 statement: "Without nuclear energy, it will be impossible to assure national energy security and the production of electricity covering total demand". The number of experts who participated in the survey was 750. They represented R & D institutes, academy, industry and other organizations, including NGOs [24].

6. CONCLUSION

The lessons learned in Europe, especially in Germany, have proved that renewable energy sources have limited potential regarding environmentally friendly energy (electricity) production. This thesis needs more studies, however more scientific, thermodynamics based evaluation should be adopted as an approach in its evaluation. Any combustion process, including biomass combustion, cannot be considered as pollution-free technology, ensuring sustainability of resources. Gas combustion, leading to lower pollutant emissions, cannot be considered as a sustainable energy source, since this is a raw material for industry, and an energy carrier for municipal applications. In the case of Poland, as in other countries in Europe, coal will be a main energy source for the next few decades. However, the construction of the nuclear blocks of the size 2 × 1500 MW_e by year 2025 will reduce the expected stress on the environment. The next power plant (3000 MW_e) constructed by the year 2035 together with renewable (wind, biogas) energy sector development will ensure that Poland may meet carbon free EU energy policy strategy till 2050.

ACKNOWLEDGEMENTS

Andrzej G. Chmielewski The studies presented in this paper received funding from The National Centre for Research and Development under the strategic project entitled "Technologies Supporting the Development of Safe Nuclear Power Engineering" research tasks 7 & 8. The author appreciates discussions with: Mr Ziemowit Iwanski, GE Energy, Country Executive, Central and Eastern Europe; Mr Adam Rozwadowski, Director, Areva Poland; and Nathalie Beauzemont, EDF, Senior Nuclear Project Manager. Finally thanks go to the reviewers whose valuable remarks helped to improve this paper.

REFERENCES

1. European Environment Agency, 2008. http://www.eea.europa.eu/data-and-maps/figures/electricity-consumption-per-capita-in-
2. Ministry of Economy, "Forecast of Demand for Fuels and Energy by 2030," 2009. http://www.mg.gov.pl/files/upload/8134/Appendix2.pdf
3. T. Olkuski, "New Trend in Polish Coal Trade (in Polish)," Polityka Energetyczna, Vol. 13, No. 2, 2010, pp. 365-375. http://www.meeri.pl/zaklady/zrynek/ zasoby/10_ 14to_pe_z.pdf
4. H. L. Gabryś, "Electric Power Industry of the Year 2012 in Poland in the Light of Energy Balance for 2011 and Not Only (in Polish)," Energetyka, 2012, pp. 139-141. http://elektroenergetyka.pl/upload/file/2012/3-4/Gabrys.pdf
5. KOBiZE, "Poland's National Inventory Report," Warsaw, 2013. http://www.kobize.pl/materialy/Inwentaryzacje_krajowe/2013/NIR-2013-PL-en.pdf
6. "Loss of Statistical Life Expectancy Attributed to Anthropogenic Contributions to PM2.5, 2000 and 2020," 2000. http://www.eea.europa.eu/data-and-maps/figures/loss-of-statistical-life-expectancy-attributed-to-anthropogenic-contributions-to-pm2-5-2000-and-2020

A. P. Weem, "Reduction of Mercury Emissions from Coal Fired Power Plants," Working Group of Strategies and Review, Geneva Informal Document No. 3 2011. www.unece.org/.../Info.doc.3_Reduction_of_mercury_emissions_from_coal_fired_power_plants.pdf http://www.fossil.energy.gov/programs/powersystems/pollutioncontrols/overview_mercurycontrols.html

7. S. H. Safe, "Hazard and Risk Assessment of Chemical Mixtures Using the Toxic Equivalency, Environ," Health Perspect, Vol. 106, No. S4, 1998, pp. 1051-1058. http://www.ncbi.nlm.nih.gov/pmc/articles/PMC1533329/pdf/envhper00539-0123.pdf
8. European Environment Agency, "CO_2 Emission Trends," 2012. http://www.eea.europa.eu/pressroom/publications/ghg-trends-and-projections-2012/
9. EU Energy in Figures, "Luxembourg: Publications Office of the European Union," 2012. http://ec.europa.eu/energy/publications/doc/2012_energy_figures.pdf

10. F. Cherubini, G. Guest and A. H. Strømman, "Application of Probability Distributions to the Modeling of Biogenic CO_2 Fluxes in Life Cycle Assessment," Bioenergy, Vol. 4, No. 6, 2012, pp. 784-798.
11. Danish Energy Agency, "Energy Statistics, 2011," Copenhagen, Denmark, 2012. http://www.ens.dk/enUS/Info/FactsAndFigures/Energy_statitstics_and_indicators/Annual%20Statistics/Documents /Energy%20Statistics%202011.pdf
12. Federal Ministry of Economics and Technology (BMWi), "Germany's New Energy Policy—Heading towards 2050 with Secure, Affordable and Environmentally Sound Energy," Berlin, 2012. http://www.bmwi.de/English/Redaktion/Pdf/germanys-new-energy-policy
13. OECD/IEA, "Energy Policies of IEA Countries: Germany (2013 Review)," OECD/IEA, Paris, 2013.
14. T. Trainer, "Can Renewables etc. Solve the Greenhouse Problem? The Negative Case," Energy Policy, Vol. 38, No. 8, 2010, pp. 4107-4114. http://www.dieoff.com/_Energy/TrainerRenewables.pdf http://dx.doi.org/10.1016/j.enpol.2010.03.037
15. OECD/IEA, "Energy Policies of IEA Countries: Finland (2013 Review)," OECD/IEA, Paris, 2013.
16. Ministry of Economy, "Energy Policy for Poland until 2030," Warsaw, 2009. http://www.mg.gov.pl/files/upload/8134/Polityka%20energetyczna%20ost_en.pdf
17. G. Zakrzewska and T. K. Anderson, "Transparency and Public Participation in Radioactive Waste Management," Materials Research Society Symposium Proceedings, Vol. 1475, 2012, pp. 51-56.
18. http://www.i2en.fr/en/news/47-poland-uk.html
19. http://www.ncbir.pl/en/strategic-programmes/technologies-supporting-development-of-safe-nuclear-power-engineering/
20. http://www.genewscentre.com/Press-Releases/GE-Hitachi-Nuclear-Energy-Continues-to-Build-Nuclear-Infrastructure-in-Poland-3b8d.aspx
21. http://www.ptn.nuclear.pl/
22. http://www.seren.org.pl/
23. K. Czaplicka-Kolarz, K. Stańczyk and K. Kapusta, "Technology Foresight for a Vision of Energy Sector Development in Poland till 2030. Delphi Survey as an Element of Technology Foresighting," *Technological Forecasting & Social Change*, Vol. 76, No. 3, 2009, pp. 327-338. http://dx.doi.org/10.1016/j.techfore.2008.05.007

CHAPTER 10

Dynamic Interaction Confinement

Gabriel Barceló

Advanced Dynamics S.A., Madrid, Spain

ABSTRACT

The importance of developing new technologies to obtain energy by means of nuclear fusion procedures is beyond question. There are several different and technically possible models for doing this, though to date none of these has been able to attain an industrial reactor with an end performance greater than unity. We still find ourselves at the initial phase, after many years, as a result of having failed as yet to come up with a commercially productive machine. Nuclear fusion research has defined a prototype reactor based on a fluid conductor, isolated materially in a physical container and confined by means of magnetic fields. In this fluid-plasma which interacts with magnetic fields, fusion reactions are caused that release energy, while at the same time a quantity of movement and angular momentum is moved or "rotated" and transported. However, turbulence is caused in these magnetic confinement fusion processes that reduces system efficiency and prevents the obtaining of sufficient net energy from the nuclear reactions. This paper aims to propose new dynamic hypotheses to enhance our understanding of the behaviour of the plasma in the reactor. In doing so, we put forward a profound revision of classical dynamics. After over thirty years studying rotational dynamics, we propose a new theory of dynamic interactions to better interpret nature in rotation. This new theory has been tested experimentally returning positive results, even by third parties. We suggest that these new dynamic hypotheses, which we hold applicable to particle systems accelerated by rotation, be used in the interpretation and design of fusion reactors. We believe that this proposal could, in addition to magnetic confinement, achieve confinement by simultaneous and compatible dynamic interaction. Accordingly, we are of the opinion that it would be possible to get better performance and results in the design of fusion reactors by way of simultaneous magnetic and dynamic interaction confinement.

Keywords: Nuclear Fusion, Rotational Dynamics, Non-Coaxial Spins, Dynamic Interactions, Dynamic Interaction Confinement

1. INTRODUCTION

We are currently in the phase of developing new technologies to obtain energy by means of nuclear fusion procedures. There are several different, technically possible models for doing this, all of which are based on the simulation of fusion processes that we assume and occur in our universe, though to date none of these has been able to attain an industrial reactor with an end performance greater than unity.

It was recently published [1] that scientists from the Lawrence Livermore National Laboratory (LLNL) in the USA have carried out a series of experiments on the thermonuclear ignition of the reactor at the National Ignition Facility (NIF), which have confirmed an improvement in fusion reaction performance, releasing more energy than that absorbed by the fuel used in the process.

For any fusion reaction to occur between two atomic nuclei, these have to overcome the Coulomb repulsion force. This force refers to the inverse-square law between charges. An example of the Sun itself has been taken to achieve the desired result, by way of an attempt to simulate in a reactor the conditions that hold with respect to the Sun. Notwithstanding, the nuclear process that the design of these experimental nuclear fusion reactors is trying to achieve is different to that of the solar one. The aim is to get Helium, a neutron and fusion energy from Deuterium and Tritium in accordance with the following formula:

$$\text{H2(D)+ H3 (T)} \rightarrow \text{He4} + \text{n} + 18\text{ MeV}$$

Plasma confinement is the most difficult obstacle to solve in thermonuclear fusion reactions with the current technological prototypes. Plasma [2] is an aggregate state of matter in which thermal agitation is capable of overcoming the electrical attraction electrons feel for atom nuclei. The

Sun's content is one of the best examples of plasma as it is capable of causing nuclear fusion reactions that produce energy output.

As a result of the extremely high temperatures at which nuclear fusion occurs, hydrogen or helium isotopes that are used as fuel are not in a solid, or liquid or gaseous state, but rather in a "fourth state", which has become known as plasma. Plasma is, therefore, an ionised fluid in which reactions can take place that could not occur, or would not be sufficiently effective, in solids, liquids or gases. Plasma as a reactive medium is the subject of numerous studies at both basic and applied levels. Most of the foundational articles on plasma physics first saw the light of day in the middle and second half of the 20th century [3] , however, since then, theory and laboratory techniques have struck out into diverse branches.

Given that plasma, under its pressure and temperature conditions, can modify the structure of that which contains it, nuclear fusion reactor fuel "containers" cannot be exclusively physical, but rather require confinement procedures to avoid the content damaging the container. Magnetic confinement, and more recently, inertial confinement, techniques have been used to deal with this problem. However, great quantities of energy are required for these technological solutions to work. Indeed, to date, more energy has been required to ignite the reactor than it is capable of producing, and thus the end balance is still negative, given that these nuclear fusion reactors generate energy amounts below that which is required to start up and maintain the fusion process.

Consequently, the current goal is to achieve plasma ignition with a positive energy balance to enable its industrial use. This goal could be achieved, either by means of a new confinement procedure, based on other principles of physics, or by improving the performance of known procedures.

This paper proposes a revision of the grounds of classical dynamics that apply to these reactors, particularly when using fluids accelerated by rotation. Indeed, new specific dynamic hypotheses are put forward.

2. METHODS FOR ACHIEVING FUSION

Confinement enables us to avail of matter at a very high temperature without the containing recipient losing any of its initial structural characteristics or becoming contaminated. As indicated above, the two main technological procedures to isolate the plasma within its container are magnetic and inertial confinement. Other confinement systems have been proposed and studied, albeit without success.

Magnetic confinement consists of containing material in a state of plasma inside an enclosure inside which, in turn, a magnetic field has been created, in accordance with a preconceived design, so that the positive and negative particles that go to make up the plasma do not enter into contact with the inner walls of the enclosure. This procedure is based on the Lorentz force, which acts on any charged particle in motion in a magnetic field, by virtue of which it experiences a force perpendicular to the magnetic field vector and the displacement vector.

Inertial confinement consists of achieving the conditions required to produce nuclear fusion by endowing the fuel particles with an amount of movement such that with the colliding of these the Coulomb barrier is overcome, thus making it possible for the nuclear fusion reaction to occur.

Inertial confinement can be achieved by means of an electromagnetic system made up of laser rays focused on a target, or accelerated heavy ions focused on a small, spherical deuterium-tritium target. Indeed, even a direct or indirect approach can be designed.

In the cosmos we have gravitational confinement [4] , which occurs at the core of stars by means of fusion reactions, owing to the powerful gravitational field. However, it is not technologically possible to imitate this process on Earth.

To sum up, we can define the conditions required for controlled nuclear fusion in a reactor:

- Create a fuel and raise its temperature to values high enough to overcome Coulomb repulsion.
- Design a material and non-material container to confine plasma for long enough for the energy that is generated in the fusion to exceed the losses caused by radiation.
- Design a continuously and repeatedly operating reactor for commercial use at an acceptable maintenance cost and that exercises a feasible environmental impact.

To date, work has been done on developing the prototype of a reactor in Europe based on a conductor fluid interacting with magnetic fields, which is isolated materially in a physical container and confined therein by means of magnetic fields. Fusion reactions are caused in this fluid-plasma that release energy, while at the same time a quantity of movement and angular momentum is moved or "rotated" and transported.

3. MODELS OF EXPERIMENTAL REACTORS

The I. V. Kurchatov Institute of Atomic Energy of the ex-URSS Academy of Sciences developed the first Tokamak fusion reactor in the 1950s, designed by the Russian physicists Igor Tamm and Andrei Sakharov.

Other reactor models were built thereafter, such as the Joint European Torus (JET) (Figure 1) in England, which is also a fusion reactor of the same type. Funded by the European Commission, it has been in operation since 1983 and is the biggest magnetic confinement fusion reactor in the world today.

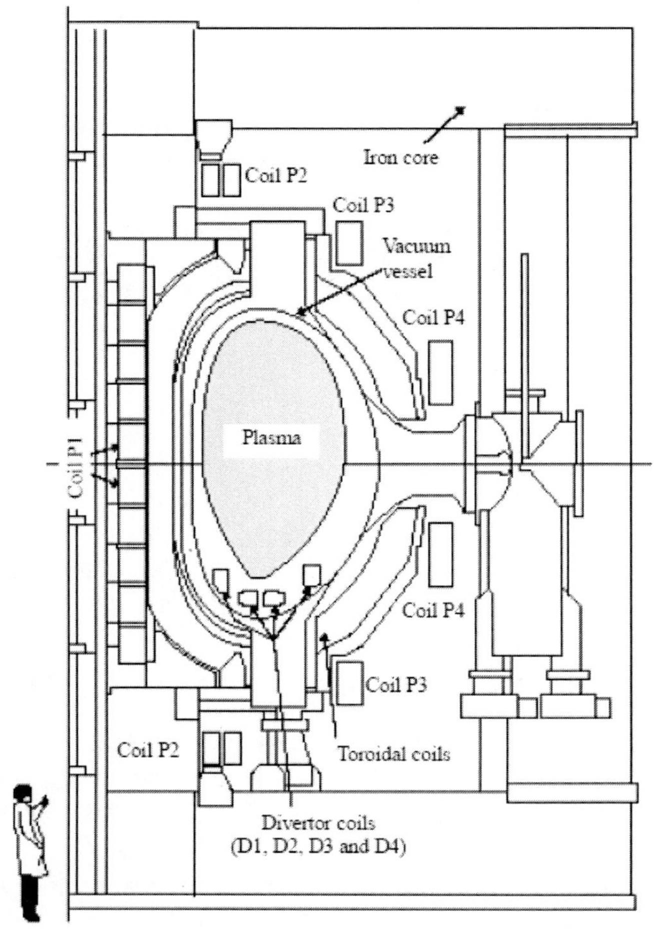

Figure 1. Joint European torus (European fusion development agreement—EFDA).

In 1991 it achieved a performance of 0.7, the coefficient between the system's input energy and the energy obtained from the reactor. Therefore, even though it is an efficient prototype that confirms the physics on which it is based, as well as its technological possibilities, it is still not a commercially viable reactor.

This round furnace is capable of handling a fuel mixture of deuterium and tritium which is supposedly to be used in future commercial fusion energy power plants and which is to be tested beforehand at the International Thermonuclear Experimental Reactor (ITER). ITER is an international experiment designed to show the scientific and technological feasibility of a full-scale fusion power reactor.

ITER is also based on the Tokamak magnetic confinement concept, in which plasma is contained in a toroidal-shaped vacuum chamber. It is, therefore, a cylindrical, ring-shaped fusion reactor, which explains why those of this type are also known as "torus" reactors. The fuel is heated to temperatures in excess of 150 million °C thus forming hot plasma. The strong magnetic fields that are used to keep the plasma away from the walls are created by superconducting coils that surround the container and by an electric current sent through the plasma. The problem lies in the enormous difficulty of compressing the fuel uniformly. In stars, gravity compresses the hydrogen into a perfect sphere thus the gas is heated uniformly and homogeneously. Such uniformity and homogeneity is very difficult to achieve under the reactor's current technical conditions.

4. PLASMA DYNAMICS

The dynamics of a conducting fluid that is interacting with intense magnetic fields created by sets of external coils are complex. Consequently, a lot of research being done into nuclear fusion as an energy source focuses on this issue.

Plasma movement is a complex variable in magnetic confinement that affects the capacity to confine it for long enough to obtain net energy from the fusion reactions. Plasma must be kept in constant circular motion within the reactor. Nevertheless, it is thought that each plasma particle following that circular path is at the same time exercising an orbital movement, as is shown in Figure 2.

The velocity fields that affect each particle generate this complex path, though the behaviour of the plasma in the reactor does not depend on dynamic laws alone, given that its very nature determines the influence of other physical variables. As plasma is a charged particle fluid, electrodynamics also affects its behaviour.

We recommend Enrique Larrea's treatise [6] for an analysis and study of the different theories and meanings of the classical electromagnetic field from the perspective of the theoretical and philosophical models developed, and of classical electrodynamics.

The electrical charges form stable links in matter in the form of atoms, molecules, crystalline networks, etc. The electrodynamic nature of these links is expressed on an atomic scale, in such a way that the matter normally manifests itself as neutral.

Nevertheless, if the temperature of the neutral matter is raised to values comparable to those of the link energies, these links break in such a way that their constituent parts, the electrons and the ions in the case of atoms, are no longer linked to each other, but rather each charged particle moves independently in the electromagnetic fields created by the particles themselves, or by input from without.

Plasmas formed by ionisation from initially neutral constituents tend to contain the same number of positive and negative charges in terms of volume and have a strong tendency to remain quasi-neutral. Consequently, small deviations in neutrality give rise to intense electric fields that are rapidly compensated by the free charges in the plasma.

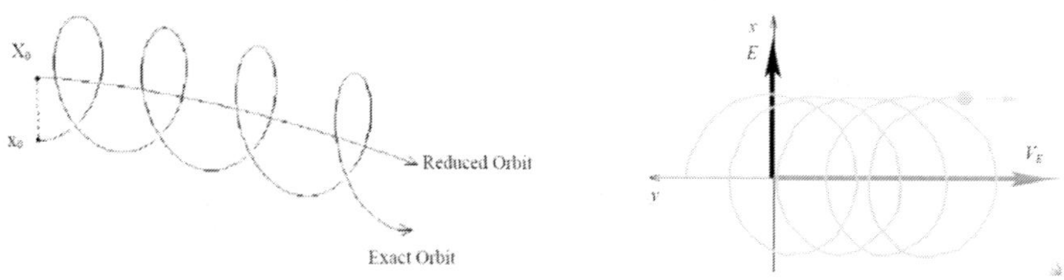

Figure 2. Exact and reduced single-particle orbits in a magnetic field. [5] [7].

Moreover, in the experiments carried out in magnetic confinement fusion reactors, turbulence has been seen to be generated that reduces system efficiency and prevents the obtaining of sufficient net energy from the process.

5. GYROKINETIC THEORY

The study of magnetised plasma dynamics has come to form a new scientific discipline known as gyrokinetic theory. This theory is based on

Maxwell and Fokker-Planck system of equations, consequently the resulting distribution functions and fields are in constant development [7] . Nevertheless, the theory is evolving in parallel with technological developments, while its analytical and computational resolution is proving to be extremely complex [8] . The theory has been gradually developed from its beginnings, with the following representing a historical summary of this evolution:

Brief History of Reduced Plasma Dynamics [9] .

- *Adiabatic motion of charged-particle motion*
  - *Alfven* (1940)
  - *Northrop & Teller* (1960)
  - *Kruskal* (1962)
  - *Northrop* (1963)
  - *Littlejohn* (1979-1983)
  - *Cary & Littlejohn* (1983)
  - *Review: Cary & Brizard* (2009)
- *Oscillation-center dynamics & Ponderomotive Hamiltonian*
  - *Dewar* (1973-1978)
  - *Cary & Kaufman* (1977)
- *Reduced fluid dynamics*
  - *Strauss* (1976)
  - *Hasegawa & Mima* (1977)
  - *Hazeltine* (1983) + *Morrison & Hazeltine* (1983)
- *Linear gyrokinetic theory*
  - *Taylor* (1967)
  - *Rutherford & Frieman* (1968)
  - *Catto* (1978)
  - *Catto, Tang & Baldwin* (1981)
- *Nonlinear gyrokinetic theory*
  - *Frieman & Chen* (1982)
  - *Dubin, Krommes, Oberman & Lee* (1983)
  - *Hahm, Lee & Brizard* (1988)

- *Brizard* (1989)
- *Hahm* (1996)
- *Sugama* (2000) & *Brizard* (2000)
- *Qin & Tang* (2004)
- *Review: Brizard & Hahm* (2007)

In a lot of cases, possible concomitance has been sought with astrophysical plasmas [10] which are supposedly to be found in the universe. The inspiration for such theories is to be found in the studies carried out by H. Alfvén [11].

The rotational nature of astrophysical plasmas has been taken into account in its development [12] , an issue that is reiterated and studied in fusion reactor plasmas. Accordingly, the idea is to have a model constituted by a continuous process, which is what is required for its industrial development. Current developments in our knowledge concerning the magnetic confinement of plasma for fusion defines a toroidal design of magnetic field force lines, which in successive toroidal transits travel over a surface topologically equivalent, in geometrical terms, to a torus, as is shown in Figure 3. Indeed, this is the basis of the JET Tokamak design and that of later reactors.

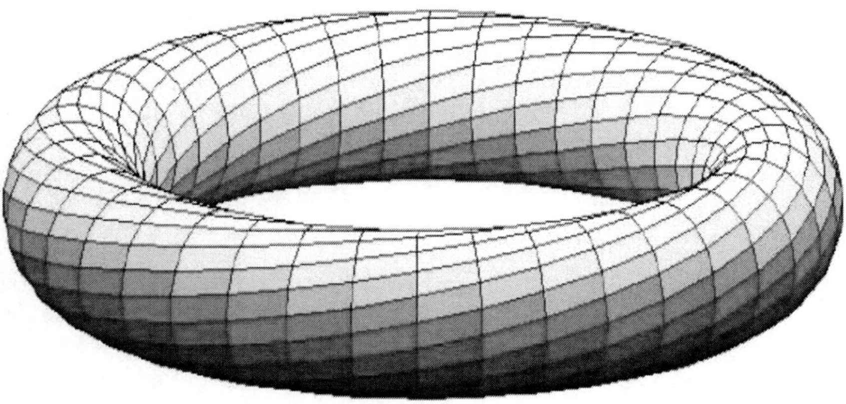

Figure 3. Gyrokinetic particle simulation. Implementation of Mixed-Mode Parallelism© Princeton plasma physics laboratory.

Notwithstanding, it has been confirmed that circulation in the experimental reactors is more turbulent than that which is supposed in stellar dynamic systems and, as we have pointed out above, this reduction of fluid uniformity can cause undesired losses in the system's power of confinement, or even, the possible cavitation of the physical container.

The speed of the plasma is an important physical field in the study of current laboratory systems, and particularly, in the improvement of its material isolation. The rotation profile can exercise considerable influence in the confinement time of the magnetically confined fluid. Profiles with a sharp radial velocity variation can cause a reduction of the turbulent transport and, therefore, could lead to improved confinement.

The study of plasma microturbulence caused by the electromagnetic interactions of its particles is one of the current goals in an attempt to improve the performance of these reactors. Understanding this microturbulence in plasma would enable its control and thus have a direct favourable impact on the building of more efficient fusion reactors.

Simulations conducted for numerous situations, and their comparison, as suggested in Figure 4, can be found in the bibliography. It has been confirmed that the existing gyrokinetic models are sufficiently valid for shorter times than the real time scales of the evolution of the reactor's dynamic profiles, thus it is thought that the effect of turbulence on the electric field and plasma velocity provide incorrect values for the electric field. Consequently more precise mathematical models are needed to represent the real behaviour of plasma at a longer time scale, called the transport scale.

This was in fact just the approach taken by the Spanish aeronautical engineer and physicist, Félix Parra, in his doctoral thesis written at the Massachusetts Institute of Technology (MIT) under the supervision of Dr. Peter J. Catto. This thesis won him the 2011 prize awarded each year by the American Physical Society (APS): *For demonstrating limitations in the gyrokinetic theory of the radial electric field for plasmas in an axisymmetric magnetic field and formulating an alternative procedure—insights that have inspired research around the world* [13]

As we have already pointed out above, plasma rotation is another essential factor in this analysis of the turbulent transport of momentum in axisymmetric systems. In magnetic confinement fusion systems such as the aforementioned ITER [14] , the plasma circulates in the container at a constant movement, which we could define as rotation with respect to its walls. Notwithstanding, it has been shown [15] that the plasma in the reactor can initiate spontaneous circular movement or rotation, without the need for any external dynamic momentum input. The theoretical development of this behaviour is still under study [16] . According to Parra: The origin of the intrinsic rotation is still unclear [17].

There has been some theoretical work in turbulent transport of momentum using gyrokinetic simulations [18] - [23] and different mechanisms have been proposed as candidates to explain intrinsic rotation [24].

The presence of an external magnetic field introduces a source of plasma anisotropy. Variations or fluctuations in the thermodynamic fields are much greater perpendicularly to the field than parallel to it.

According to recent investigations by Dr. Parra: The ion toroidal rotation in a Tokamak consists of an E × B flow due to the radial electric field and a diamagnetic flow due to the radial pressure gradient... the momentum pinch for the rotation generated by the radial pressure gradient is calculated and is compared with the Coriolis pinch. This distinction is important for subsonic flows or the flow in the pedestal where the two types of flows are similar in size and opposite in direction. At the edge, the different pinches due to the opposite rotations can result in intrinsic momentum transport that gives significant rotation peaking [25].

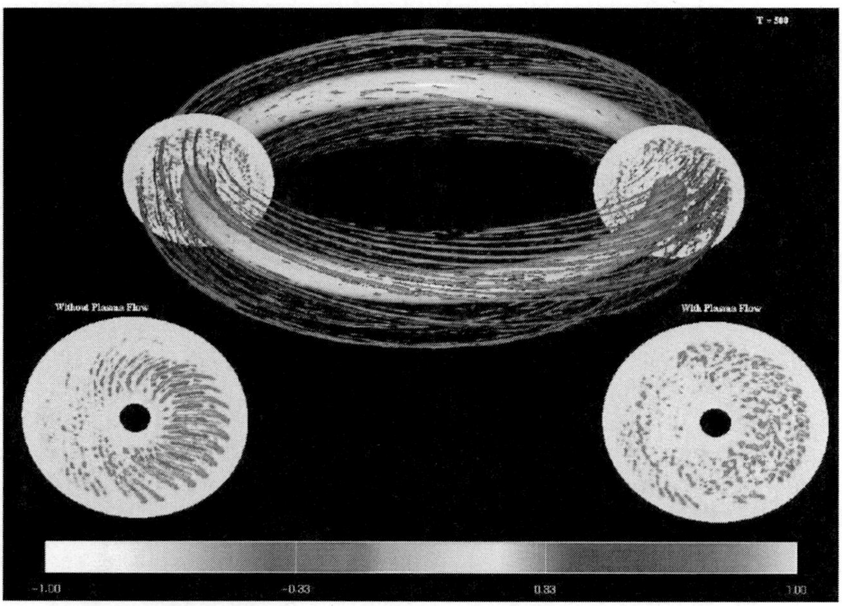

Figure 4. Comparison of nonlinear mode structure: effects of zonal flows. GTC simulations, Gyrokinetic Toroidal Code (GTC)© Princeton plasma physics laboratory.

In recent articles, Dr. Parra also puts forward intrinsic rotation evolution equations in conventional Tokamak reactors [26] [27].

6. THEORY OF DYNAMIC INTERACTIONS

Against this background of study and constant advancement in unravelling the physical behaviour of Tokamak- type nuclear reactors, we suggest a revision of the dynamic criteria being applied, given that we believe there may well be inappropriate interpretations in the principles of classical rotational dynamics at play.

We propose a dynamic model for confined plasma equivalent to the flight of the boomerang, which enables its return to its point of departure and, therefore, a natural dynamic confinement. We advise the study of boomerang dynamics, given that it would afford us a better understanding of the dynamics of confined plasma by dynamic interaction, along with an improved reactor result.

Nevertheless, we believe that the boomerang is nothing more than an illustrative example of the true behaviour of nature when mass is subjected to simultaneous non-coaxial rotations. We hold that its theoretical dynamic principles can be applied in general to bodies or particles with mass, thus we suggest that the classical rotational dynamics theory be rigorously tested and verified.

Over the last thirty years we have been analyzing the behaviour of bodies subjected to accelerations by non- coaxial rotations. We propose a new interpretation of the dynamic behaviour of the boomerang and, in general, of rigid bodies exposed to simultaneous non-coaxial rotations. We have developed a new rotational, non-inertial dynamics hypothesis, which can be applied to understand both, the flight of the boomerang, as well as celestial mechanics. [...]

The Theory of Dynamic Interactions (TDI) claims that the boomerang returns to its origin as a result of being a body that is subject to two simultaneous rotations on different axes (Figure 5). As of the moment it is first released into the air, in addition to the impulse driving it forward, it is made to revolve around an axis that is perpendicular to its plane [28].

In accordance with the Theory of Dynamic Interaction: The anisotropic distribution of the velocities (V_c) generated by the torque shall be compounded with the initial linear velocity (V_T), enabling a curved path... The velocity V_R = (v_x, v_y, v_z) of the new path of its center of mass will be defined by the matrix multiplication of the ro-tating operator by the velocity vector V_0 = (0, V_0, 0) [29]:

$$V_R = \vec{\vec{\Psi}} V_0 \tag{1}$$

This is the equation for movement that we have deduced for bodies with intrinsic rotation and translation when they are subjected to momenta that are non-coaxial with their rotation. In our opinion, this applies to the flight of the boomerang, but we also suggest that it would enable the definition of the path of each plasma particle, as long as the established baseline hypotheses are met.

We have conducted several test experiments on the theory under proposal e.g. moving objects with intrinsic rotation on their main axis semi-submerged in water and subject to a righting torque, consisting of the weight and force of floatation, the relative positions of which do not coincide [30] . This righting torque was not coaxial with the intrinsic rotation. The moving object was, therefore, subject to two non-coaxial rotations, as is the case with the boomerang, but we noticed how the rotation caused by the righting torque did not generate any new moving rotations, but rather was transformed into a change of path of the center of mass [31] . Experiments were also conducted on terra firme [32] , which can also be seen in video [30] [33] [34].

In accordance with the theory being put forward, and as can be appreciated in Figure 6, we suggest that each particle, instead of following a linear path, as would be expected from the classical mechanics equations of Newton-Euler, it would follow a closed path owing to the coupling of the velocities' fields that are caused on every point of each particle.

The issue can be summed up as follows: If we apply this dynamic theory to a Tokamak reactor, every plasma particle with intrinsic rotation will follow a closed path like that of the boomerang when it is subjected to a non-coaxial momentum in addition to its intrinsic angular momentum. Consequently, there are different authors [32] [36] who share these same hypotheses with a view to understanding the dynamic behaviour of bodies or particles in those circumstances in which they have translational velocity, intrinsic rotation and are simultaneously subjected to a non-coaxial momentum with its own rotation [37] . Notwithstanding, in our hypothesis, based on a possible analogy, we are extrapolating the observed behaviour in macroscopic moving bodies to the reactor's plasma particles.

In our opinion, this dynamic model also coincides with the behaviour of hurricanes on the Earth and also resembles atmospheric toroidal vortex rings.

It must be pointed out that in our investigations we have come to a rational deduction to the effect that in such circumstances, the kinetic translational energy can be transformed into kinetic rotational energy, and vice versa, and in general, energy transfers can occur at the core of particles endowed with intrinsic angular momentum: [...] kinetic energy can be transferred, increasing its rotation velocity, its linear velocity or modifying its state of potential [38].

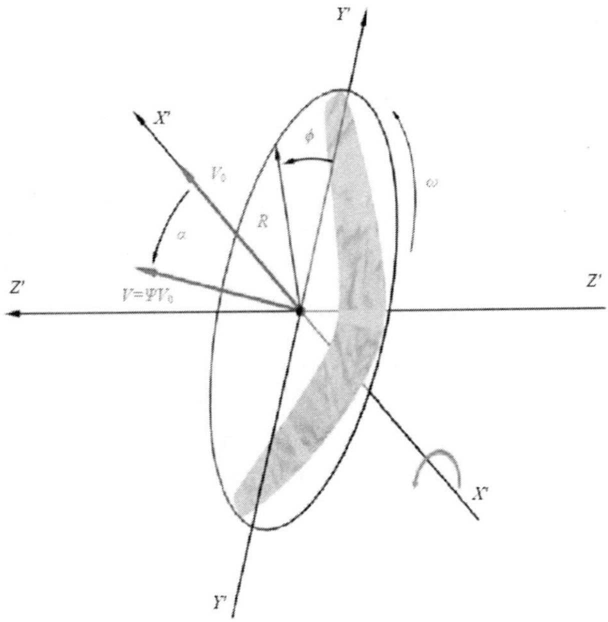

Figure 5. Rotations that generate changes in the boomerang's path as a result of the coupling of the velocity fields that act on its center of mass [28].

At the beginning of this paper we stated that magnetic confinement is based on the Lorentz force. This force manifests itself when an electric charge in movement moves in a magnetic field. We have stated in other works that, in our opinion, the origin of this force is analogous to that of the inertial forces that manifest themselves in TDI circumstances [39] and which oblige particles to follow a closed path. Therefore, and in accordance with our hypothesis, the same dynamic effect and the same effective confinement could be obtained doubly, by the magnetic effect and by the dynamic effect.

Both ideas being proposed should, we feel, be further explored and, where appropriate, confirmed experimentally; on the one hand, that the Lorentz force is analogous to an inertial dynamic force, similar to TDI ones, and on the other hand, that this theory be applied to the reactor's plasma.

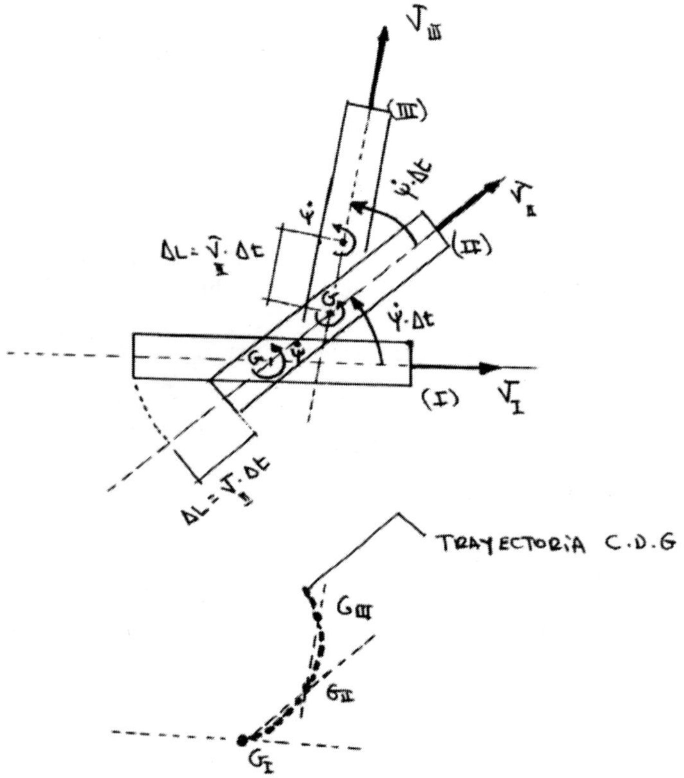

Figure 6. The position of the center of mass varies depending on the coupling between the field of the linear velocity and the velocity field caused by the non- coaxial torque [35] . (Figure Alejandro Alvarez).

7. CONCLUSIONS

Nuclear fusion is a necessary energy alternative into which countries are researching and in which they trust. However, investment is still needed to develop the requisite technology, or even, to conceive the new physical ideas involved in obtaining the energy.

It does not seem likely in the short term that it will be possible to come up with new continuous nuclear fusion procedures and technology without using plasma, which represents another great research challenge in physics.Even though over fifty years of theoretical and technological developments have passed, the road yet to be travelled to develop industrial fusion reactors is still a long one, both in terms of physics, as well as the engineering required.

As far as magnetic confinement reactors are concerned, plasma dynamics is still a work in progress. We are as yet ignorant of the exact dynamic behaviour of plasma and, moreover, when designing these generators it is still necessary to reduce their turbulent tendency and improve their momentum transport capacity.

We propose a profound revision of the principles of classical rotational mechanics, especially as regards particles subjected to accelerations by simultaneous and non-coaxial rotations. After conducting our research, we put forward certain specific, alternative dynamic hypotheses.

In addition to the plasma's material container, we can envisage another simultaneous, non-material container, along with the use of magnetic confinement techniques. We suggest the exploring of a new type of dynamic confinement based on the TDI and one that is compatible with magnetic confinement.

Gyrokinetics is the workhorse for modern research on low-frequency microturbulence in magnetised plasmas. It is a very powerful tool that has been used successfully in many different applications in both fusion and astrophysics:

Analytical theory of drift waves and related modes.

Statistical mechanics of magnetised systems.

Simulations of turbulent transport.

Interpretation of solar-wind data.

Theory of momentum transport.

Simulations of intrinsic rotation [40].

Moreover, in our opinion, the TDI could also be employed as a complementary theoretical instrument that would lead to greater reactor efficiency. Consequently, its dynamic hypotheses ought to be studied in the field of gyrokinetic theory.

The equation for movement that we are proposing in the Theory of Dynamic Interactions for these non-iner- tial situations is very simple (1) [...]. It becomes possible to determine the path of [28] plasma particles in translation, when these particles are endowed with intrinsic rotation and are subjected to rotation actions on another, different axis. [...]

The Theory of Dynamic Interactions on which this proposal is based is fully described in different texts and articles, in which it is possible to see how this theory affects, in a far reaching way, the basic principles of dynamics

[41] , not to mention astrophysics [42] , cosmology [43] , atomic physics and technology. An analysis of this theory reveals its capacity to answer numerous unknowns in dynamics and celestial mechanics and, moreover, explains the behaviour of rotating bodies [31].

These new dynamic hypotheses we are putting forward and which we hold applicable to particle systems accelerated by rotation can, we suggest, be used in the interpretation and design of fusion reactors. Accordingly, we persist in our belief that magnetic confinement in a reactor can be likened to confinement by dynamic interactions based on the Theory of Dynamic Interactions. Applying this criterion we are proposing would enable a twin physical-theoretical principle to isolate plasma and try to minimize its turbulence: each plasma particle should have spin and be subjected to a non-coaxial magnetic momentum from without and another gravimetric one. Accordingly, plasma confinement will be based on two different, albeit analogous, physical principles, generating double intransigence in its path.

The dynamic model based on boomerang dynamics could improve the performance of these energy power plants, minimize turbulence and improve momentum transport. Consequently, we insist on the exploration of the ideas expounded in this paper with respect to the origin of the Lorentz force and the true, dynamic nature of plasma made up of particles with intrinsic rotation when they are subjected to new, non-coaxial momenta.

REFERENCES

1. Hurricane, O.A., et al. (2014) Fuel Gain Exceeding Unity in an Inertially Confined Fusion Implosion. Nature, 506, 343-348. http://dx.doi.org/10.1038/nature13008
2. Bittencourt, J.A. (2004) Fundamentals of Plasma Physics. Springer, Berlin. http://dx.doi.org/10.1007/978-1-4757-4030-1
3. Chen, F.F. (1984) Introduction to Plasma Physics and Controlled Fusion: Plasma Physics. Springer, Berlin. http://dx.doi.org/10.1007/978-1-4757-5595-4
4. Clayton, D.D. (1983) Principles of Stellar Evolution and Nucleosynthesis. McGraw-Hill, New York; Reissued by University of Chicago Press, Chicago.
5. Brizard, A.J. and Hahm, T.S. (2007) Foundations of Nonlinear Gyrokinetic Theory. Reviews of Modern Physics, 79, 421.
6. Larrea Bellod, E. (2010) Los significados del campo electromagnético y sus transformaciones históricas. Bubok. C/ Belén 13, Madrid. Also: Larrea Bellod, E. (2008) Teorías sobre el origen de la luz y el campo electromagnetic. Bubok. C/Belén 13, Madrid.
7. Alonso de Pablo, J.A. and Almeida, J.S. (2013) La rotación y el transporte de momento en plasmas astrofísicos y de laboratorio.
8. Abel, I.G., Plunk, G.G., Wang, E., Barnes, M., Cowley, S.C., Dorland, W. and Schekochihin, A.A. (2013) Multiscale Gyrokinetics for Rotating Tokamak Plasmas: Fluctuations, Transport and Energy Flows. iv: 1209.4782v4 [physics.plasm-ph]
9. Brizard, A.J. (2010) Lectures on Gyrokinetic Theory. Part I: Physical and Mathematical Foundations of Gyrokinetic Theory. Summer School on Self-Organization in Turbulent Plasmas and Fluids. Max-Planck-Institut für Physik Kom- plexer Systeme. Dresden.

10. Howes, G.G., et al. (2006) Astrophysical Gyrokinetics: Basic Equations and Linear Theory. The Astrophysical Journal, 651, 590-614. http://stacks.iop.org/0004-637X/651/i=1/a=590
11. Alfvén, H. (1950) Cosmical Eletrodynamics (The International Series of Monographs on Physics).Clarendon Press. http://ia600703.us.archive.org/23/items/CosmicalElectrodynamics/ Alfven-Cosmical Electrodynamics.pdf
12. Parra, F.I. and Calvo, I. (2011) Phase-Space Lagrangian Derivation of Electrostatic Gyrokinetics in General Geometry. Plasma Physics and Controlled Fusion, 53, Article ID: 045001. http://stacks.iop.org/0741-3335/53/i=4/a=045001
13. Text of the Declaration of the Jury of the American Physical Society (APS) on Awarding Félix Parra Díaz, Aeronauti cal Engineer from the Universidad Politécnica de Madrid, with the 2011 Marshall N. Rosenbluth Award for His Thesis: Extension of Gyrokinetics to Transport Time Scales, Massachusetts Institute of Technology (MIT).
14. Ikeda, K. (2007) Progress in the ITER Physics Basis. Nuclear Fusion, 47, S203. http://stacks.iop.org/0029-5515/47/i=6/a=E01
15. Rice, J.E., Ince-Cushman, A.C., Reinke, M.L., Podpaly, Y., Greenwald, M.J., LaBombard, B. and Marmar, E.S. (2008) Spontaneous Core Toroidal Rotation in Alcator C-Mod L-Mode, H-Mode and ITB Plasmas. Nuclear Fusion, 50, Artcle ID: 124042. http://dx.doi.org/10.1088/0741-3335/50/12/124042
16. Parra, F.I., Barnes, M., Calvo, I. and Catto, P.J. (2012) Intrinsic Rotation with Gyrokinetic Models. Physics of Plasmas, 19, Article ID: 056116. http://link.aip.org/link/?PHP/19/056116/1
17. Parra, F.I., Barnes, M., Peierls, R. and Catto, P.J. Sources of Intrinsic Rotation in the Low Flow Ordering. Centre for Theoretical Physics, University of Oxford, Oxford, Plasma Science and Fusion Center, Massachusetts Institute of Technology.
18. Peeters, A.G., Angioni, C. and Strinzi, D. (2007) Toroidal Momentum Pinch Velocity due to the Coriolis Drift Effect on Small Scale Instabilities in a Toroidal Plasma. Physical Review Letters, 98, Article ID: 265003. http://dx.doi.org/10.1103/PhysRevLett.98.265003
19. Waltz, R.E., Staebler, G.M., Candy, J. and Hinton, F.L. (2007) Gyrokinetic Theory and Simulation of Momentum Transport and Energy Exchange. Physics of Plasmas, 14.
20. Roach, C.M., Abel, I.G., Akers, R.J., Arter, W., Barnes, M., et al. (2009) Gyrokinetic Simulations of Spherical Tokamaks. Plasma Physics and Controlled Fusion, 51, Article ID: 124020. http://dx.doi.org/10.1088/0741-3335/51/12/124020
21. Camenen, Y., Peeters, A.G., Angioni, C., Casson, F.J., Hornsby, W.A., Snodin, A.P. and Strintzi, D. (2009) Transport of Parallel Momentum Induced by Current-Symmetry Breaking in Toroidal Plasmas. Physical Review Letters, 102, Article ID: 125001. http://dx.doi.org/10.1103/PhysRevLett.102.125001
22. Casson, F.J., Peeters, A.G., Camenen, Y., Angioni, C., Hornsby, W.A., Snodin, A.P., Strintzi, D. and Szepesi, G. (2009) Anomalous Parallel Momentum Transport Due to E × B Flow Shear in a Tokamak Plasma. Physics of Plasmas, 16, Article ID: 092303. http://dx.doi.org/10.1063/1.3227650
23. Highcock, E.G., Barnes, M., Schekochihin, A.A., Parra, F.I., Roach, C.M. and Cowley, S.C. (2010) Transport Bifurcation in a Rotating Tokamak Plasma. Physical Review Letters, 105, Article ID: 215003. http://dx.doi.org/10.1103/PhysRevLett.105.215003
24. Wang, L. and Diamond, P.H. (2013) Gyrokinetic Theory of Turbulent Acceleration of Parallel Rotation in Tokamak Plasmas. Physical Review Letters, 110, Article ID: 265006.
25. Lee, J.P., Parra, F.I. and Barnes, M. (2013) Turbulent Momentum Pinch of Diamagnetic Flows in a Tokamak. Nuclear Fusion, 54, Article ID: 022002.
26. Parra, F.I. and Barnes, M. (2014) Intrinsic Rotation in Tokamaks. Theory. arXiv: 1407.1286v1
27. Parra, F.I., Calvo, I., Burby, J.W., Squire, J. and Qin, H. (2014) Equivalence of Two Independent Calculations of the Higher Order Guiding Center Lagrangian. Physics of Plasmas, 21, Article ID: 104506.
28. Barceló, G. (2014) Theory of Dynamic Interactions: The Flight of the Boomerang. Journal of Applied Mathematics and Physics, 2, 569-580. http://dx.doi.org/10.4236/ jamp.2014.27063

29. Barceló, G. (2005) The Flight of the Boomerang. El Vuelo del Bumerán, Marcombo, Barcelona. http://www.dinamicafundacion.com/
30. Theory of Dynamic Interactions. http://www.youtube.com/watch?v=P9hGgoL5ZGk&list= PL3E50CF6AEBEED47B http://www.youtube.com/watch?v=XzTrGEtJGXU&list= PL3E5 0CF6AEBEED47B http://www.youtube.com/watch?v=dtMqGSU9gV4&list= PL3 E50CF6AEBEED47B http://www.youtube.com/watch?v=qK5mW2j2nzU&list= PL3 E50CF 6AEBEED47B
31. Rico-Avello, G.B. Analysis of Dynamic Field Systems Accelerated by Rotation. Dynamics of Non-Inertial Systems. DeMSET-2011 Congress, Miami. http://www.coiim.es/ forocientifico/ FORO%20CIENTFICO/Documentos/DeMSET_2011_GBarcelo.pdf
32. Pérez, L. (2013) New Evidence on Rotational Dynamics. World Journal of Mechanics, 3, 174-177. http://www.scirp.org/journal/wjm http://dx.doi.org/10.4236/wjm. 2013. 33016
33. Pérez, L.A. (2013) Reflecting New Evidence on Rotational Dynamics. http://vimeo. com/68763196
34. Bauluz, E. New Dynamic Hypotheses. This Video Showed the Experimental Tests Carried out by Advanced Dynamics S. A. to Proof and Justify the Theory of Dynamic Interactions. http://www.advanceddynamics.net/
35. Barceló, G. (2013) Technological Applications of the New Theory of Dynamic Interactions. Global Journal of Researches in Engineering: Mechanical and Mechanics Engineering-G, 13, Version 1.0 https://globaljournals.org/GJRE_Volume13/E-Journal_GJRE_(G)_Vol_13_ Issue_5.pdf
36. González, M.D. (2013) Dinámica de sistemas con spin: Un nuevo enfoque. Fundamentos y aplicaciones. ADI Ser, Ma- drid.
37. González, M.D. Equation of Motion of Systems with Internal Angular Momentum-II. arXiv:physics/0603207.
38. Barceló, G. (2013) Theory of Dynamic Interactions: Laws of Motion. World Journal of Mechanics, 3, 328-338. http://dx.doi.org/10.4236/wjm.2013.39036
39. Barceló, G. (2008) A Rotating World (Un mundo en rotación). Editorial Marcombo: Barcelona. http://www.dinamicafundacion.com/
40. Krommes, J.A. (2011) The History and Present Status of Gyrokinetic Theory. International Sherwood Fusion Theory Conference, Austin, 4 May 2011.
41. Barceló, G. (2010) On the Equivalence Principle. 61st International Astronautical Congress, American Institute of Ae- ronautics and Astronautics, Prague. http://www. coiim.es/forocientifico/FORO%20CIENTFICO/Documentos/ON_THE_EQUIVALENCE_PRI NCIPLE.pdf
42. Barceló, G. (2013) Proposal of New Criteria for Celestial Mechanics. International Journal of Astronomy and Astrophysics, 3, 385-391. http://dx.doi/org/10.4236/ijaa.2013.34044
43. Barceló, G. (2013) Imago Universi: A Story of the Human Conception of the Cosmos. Arpegio, Barcelona. http://www.editorialarpegio.com/ http://imagouniversi.com/

CHAPTER 11

SVBR-100 Nuclear Technology as a Possible Option for Developing Countries

Vladimir Petrochenko[1], Georgy Toshinsky[1,2*], Oleg Komlev[1]

[1]JSC "AKME-Engineering", Moscow, Russia

[2]FSUE SSC RF-IPPE, Obninsk, Russia

ABSTRACT

Many developing countries need ecologically clean power sources (PS). The nuclear power plants are such sources. However, a great number of the developing countries do not possess developed large capacity power systems. Moreover, currently in the developing countries, there are no highly skilled personnel to provide construction and reliable and safe operation of the nuclear plants, which are complex and potentially hazardous systems. In some countries, the level of terroristic threat is extremely high. For that reason, there are specific requirements to the nuclear PSs intended for use in the developing countries. In the presented report, the specific requirements which must be met by the NPT proposed for use in developing countries are formulated, basic statements of the SVBR-100 concept are presented, design and principal scheme of the reactor facility are described, major characteristics of SVBR-100 are summarized.

Keywords: SVBR-100, Reactor, Nuclear Power Technology, Nuclear Power Plant, Inherent Self-Protection, Passive Safety

1. INTRODUCTION

Now the population in the world is growing mainly due to developing countries where electricity is in deficit. The most expedient way to reduce the deficit in electricity is implementation of nuclear PSs which can provide power safety of the country, reduce dependence on external suppliers of power and power resources, political and cost conjuncture at the market of fuel resources and transport services, assure socially acceptable tariffs on power production for the population and industry, and as a result to heighten a living standard of the population.

Use of the NPT will also enable to prevent burning of large amounts of fossil fuel that will considerably reduce exhaust of harmful substances in the environment and result in ecological enhancement.

Along with these, when choosing the NPT, some specific features in developing countries should be taken into account. In particular, the most important features are as follows:

- Lack of the skilled personnel for reliable and safe operation of the reactor facility (RF), lack of the backgrounds for personnel's training.
- Lack of builders and assembly workers for providing high quality of assembly works at the NPP site.
- Lack of large electrical grid systems for transmitting electricity to long distances.
- A wide variety of demands both in PS power and purpose (generation of electricity, generation of heat water for the population and industrial enterprises, sea water desalination).
- Lack of adequate political stability in some developing countries.
- A high level of terroristic threat in some developing countries.
- A high risk of unauthorized proliferation of nuclear fissile materials (NFM).
- Lack of infrastructure required for NPT functioning.
- Limited investment opportunities.

- Moreover, when choosing the NPT, one should take into account the general conditions typical for both developed and developing countries. Those conditions are uppermost the following:
- The population's radiophobia that increased much after the disaster happened at nuclear power plant (NPP) Fukushima 1.
- Making use of an anti-nuclear state of public opinion for achieving the political goals.
- In both developed and developing countries the available resources of natural uranium for large-scale development of nuclear power (NP) are economically limited.
- The necessity to assure competitiveness of nuclear PSs with alternative PSs at the sites of their location.
- Postponed problems of handling the spent nuclear fuel (SNF) and long-lived radioactive waste (RAW).

Below the requirements to the NPTs purposed for use in developing countries are formulated, the basic conceptual statements of the innovative NPT based on SVBR-100 reactors for meeting the proposed requirements are summarized and the description of specific features of the NPT based on RF SVBR 100 is presented.

In the presented manuscript the specific requirements, which must be met by the NPT proposed for use in developing countries, are reviewed. The basic statements of the SVBR-100 concept are presented. The design and principal scheme of the reactor facility are described and major characteristics of SVBR-100 are summarized.

2. REQUIREMENTS TO THE NPT IN DEVELOPING COUNTRIES

NP development in developing countries puts forward a number of requirements to the type of the NPT proposed for use, which have been followed from the provisions listed above.

1) There must be assured elimination of the necessity of population evacuation and protection measures beyond the NPP site in events of any failures of the equipment, personnel's errors and their combinations as well as in events of ill-intentioned actions. This is especially viable when the PS is used in a mode of co-generation with production of thermal power, when it is necessary to install nuclear heat power plants (NHPP) not far from the population centers.

2) With due account of the large potential market for small and medium-sized power units in developing countries the RF must be export-available not only from the standpoint of assuring safety (including that in conditions of terroristic attacks), but from the standpoint of NFM non-proliferation.

3) Commercial scales of NP development in developing countries means changeover of nuclear power-units to the category of widespread industrial objects from unique ones. From that standpoint operating the RF must be most simplified concerning the requirements to the personnel's qualification that must be also provided by high level of RF inherent self-protection and passive safety that makes possible simplification of the scheme and design of the nuclear steam supplying system (NSSS) due to elimination of lots of safety systems required for traditional type RFs.

4) Because in developing countries the demands in PS power differ much (100 - 600 MWe), the modular structure of the NSSS is more preferable as it can provide any power divisible by the power of a single module and therefore eliminate many-type designs of RFs.

5) Due to a great number of the required power-units, it is necessary to organize quantity production of the RFs, and their fabrication must eliminate use of the unique equipment to the most possible extent. That is, it must be available at lots of engineering factories and does not hinder development of large-scale nuclear power based on light water reactors (LWR), for which the high-pressure vessels are required.

6) In the future wide-scale implementation of nuclear PSs will require large quantities of nuclear fuel. In conditions of limited resources of natural uranium that calls forth the necessity for development of that power on the basis of fast reactors which can operate in the closed nuclear fuel cycle (NFC) in a mode of fuel self-providing without consumption of natural uranium.

7) The indispensable condition of nuclear PS implementation is their competitiveness with thermal power plants (TPP) on fossil fuel. To meet that requirement it would be important to reduce maximally the schedule terms of plant construction that can be provided firstly by increasing the scopes of factory production of the RFs and transportation of the ready units to the construction site by railway or other kinds of transport with a minimal scopes of works on the site, as well as serial (conveyer) production.

8) Commercialization of the NPT in developing countries must be preceded by construction and operation of the entirely referent experimental-industrial power-unit for the purpose to verify safety, technical and economical characteristics and operating abilities of the

RF. That will make possible minimization of the technical and investment risks, provide the conditions for attraction of private investments and realize the concept "built, own, operate (or lease)" at the point that all residual risks are taken by the Supplier that will assure the most attractable conditions for developing countries.

3. SPECIFIC FEATURES OF THE NPT BASED ON RFSVBR-100

3.1. High Level of Inherent Self-Protection and Passive Safety

- During a historically short period of NP development, use of nuclear power was followed by a number of low probability accidents of various extent of severity, which caused strong exhausts of radioactivity into the environment and/or considerable economical losses:
- Three Mile Island Unit 2 (TMI-2) accident in 1979 (USA). The accident in PWR type reactor resulted in core meltdown due to loss of primary circuit coolant caused by technical failure (the safety valve remained open) and inadequate personnel's actions. Today, the power-unit cannot be restored, radioactivity release has been localized within the containment.
- Chernobyl disaster occurred on 26 April 1986. An explosion of steam in reactor RBMK at Unit 4 resulted in catastrophic exhaust of large quantities of radioactivity into the atmosphere. The explosion was caused by prompt neutrons runaway because of violation of the operating regulations and defects in the reactor design.
- In 1995 fire occurred at fast sodium reactor "Monju" (Japan). It happened as a result of non-radioactive sodium leak in the intermediate circuit pipeline and inadequate personnel's actions. It had taken 15 years to perform repair-and-renewal works.
- In 2011 the disaster happened at NPP Fukushima 1 (Japan) because of the earthquake and following tsunami. Long blackout of BWR type reactors led to termination of removal of heat decay from the reactor cores and pools-storages of SNF. It was followed by meltdown of fuel and exhaust of radioactivity caused by steam discharge from the primary circuit and explosions of hydrogen-air mixture generated due to intensive steam- zirconium reaction.

- The initial events for these accidents are very unlike. These are such as personnel's errors, technical failures, defects in the reactor designs and extremal external impacts. However, there is one common cause of the severe consequences of all these accidents which are the result of releasing the various types of potential energy accumulated in different materials, first of all, in the RF coolant:
- Compression energy of water coolant.
- Chemical energy of interaction of water steam with zirconium.
- Chemical energy of interaction of hydrogen produced by steam-zirconium reaction with air oxygen.
- Energy of chemical interaction of sodium with air oxygen.

Due to these accidents, in some countries the population's trust in the NP safety was decreased and in order to enhance the confidence to the NP the necessity to make corrections in the strategy of long-term development of NP arose.

First, the necessity for development of the reactors with safety level eliminating the severe accidents, which require to take protection measures or even population evacuation, was highlighted in paper [1] in 1985 before the Chernobyl disaster occurred. In that paper those class reactors were named "inherently safe reactors". The general principles for construction of those reactors were justified in paper [2] in 1990 after the Chernobyl accident happened.

In this paper it is highlighted that for the population the possibility of the catastrophic consequences of the nuclear accident is much more important than the low probability of its realization. From the standpoint of the nuclear community and educated part of the population that perception of the NP is not reasonable. It may seem strange as in accordance with the reliably stated statistical data, the risks caused by operation of industrial enterprises and their fuel-power infrastructure are many orders greater than the corresponding risks for NP while the existing NPTs are used [3].

Nevertheless, that population acceptability of the NP should be taken into consideration as an objective factor and the high safety level of the NPP should be validated by "transparent" arguments without use of probabilistic analysis methods that is possible in case of eliminating the very reasons of those accidents.

In the system approach, safety and hazard are considered as interconnected concepts. Therefore, in order to understand better how to achieve a high safety level of the nuclear power facility, the nature of hazard peculiar to nuclear power systems should be analyzed.

The hazard from the RF is determined by three factors:

1) Radiation potential accumulated, i.e., total radioactivity (more exactly, radiotoxicity) contained in the reactor facility.
2) Residual heat release of radioactive fission products decay.
3) Amount of radiotoxicity-released products into the environment for different initial events.

The first and second factors do not depend strongly on the RF type, because total radioactivity contained in the RF and determined mainly by the amount of fission products is associated primarily with thermal power of the reactor and duration of its operation at this power level, i.e., by energy production. These factors are peculiar to each NPT and cannot be reduced.

The third factor depends strongly on the RF type and is determined by reactivity margin, reactivity feedbacks, design features, and potential energy accumulated in the RF materials (nuclear energy, internal thermal energy, coolant compression energy, chemical energy), which in an event of its release can cause exhaust of radiotoxicity into the environment. Therefore, the hazard associated with the RF (for identical power levels and operation time) will be determined by the third factor that can be reduced at the design stage of the project.

The nuclear fission energy, which can be released under conditions of reactivity accidents, can and must be minimized as early as at the reactor design stage by limiting the reactivity margin, use of negative reactivity feedbacks, and by various engineering solutions in the control and protection system (CPS), which exclude a possibility of prompt neutrons runaway of the reactor[1].

At the same time the potential (non-nuclear) energy stored in coolant used in the RF primary circuit coolant is an inherent property of coolant material and cannot be changed by engineering solutions.

Upgrade of safety of the NPPs based on traditional type RFs (with high amount of potential energy stored in coolant) requires increase of the number of safety systems and defense-in-depth barriers, which decrease the probability of arising the severe accidents (but do not eliminate deterministically the reasons for their happening) and weight of their consequences. While assessing that probability, the failures of the basic equipment, safety systems, protection barriers, and personnel's errors are considered as random events.

However, because of high diversity and complication of developing processes and lack of the certain initial data required for calculation, there are many uncertainties in the results of safety substantiation by probabilistic safety analysis (PSA) methods as applicable to severe accidents, their probability being very low (~10^{-6} per reactor-year and less). Therefore, these results do not possess the credibility value as necessary.

Moreover, use of probabilistic analysis methods makes no sense if we consider such initial events, (for example, over-normal external environmental effects or man-caused impacts, ill-intended actions) when all active safety systems (or non-active safety systems with movable parts), which are in a standby mode, and the protection barriers can be disabled, and radioactivity release can achieve a disastrous level. A well-known fact is explosions performed by terrorists at NPP "Lemoniz" (Spain) in 1978 and 1979 [4]. The importance to consider such scenarios is verified by the fact that they have also been addressed by the IAEA [5].

Along with that, the PSA methods are useful and necessary to assess the frequency of the accidental events including such severe ones as core damage in case it is not followed by release of potential energy that can cause destruction of protection barriers and consequent exhaust of radioactivity. Such approach to the requirements to safety of Generation IV reactors (the concept of about zero-point radiology risk for the population in an event of any external initiating impacts coupled with a probabilistic approach to the frequency of core damaging by internal initiating impacts) is considered in [6].

For the risk of radioactivity release from different type RFs being at a similar, socially acceptable level, the number of safety systems and defense-in-depth barriers, which strongly determine the NPP technical and economical characteristics, can be reduced in case of decreasing the potential energy accumulated in the RF, mainly in coolant. At this point, it is important that the high safety level at a low value of potential energy stored in coolant can be achieved virtually "free of charge", mainly, due to elimination of the very causes of severe accidents, i.e., deterministically.

The effect of potential energy stored in coolant on safety and economic parameters is considered in paper [7]. Water coolant accumulates all kinds of stored potential energy which can cause damage of the RF; coolant compression energy, energy released in events of steam-zirconium reaction and explosion of hydrogen and air mixture 21 MJ/m^3. There is no pressure in case of sodium coolant, but there is chemical energy 9.4 MJ/m^3. Under certain conditions its release can cause fire and explosion.

These factors of hazard can be compensated by the certain economic losses. So, there is a conflict between economic and safety requirements. The highlighted conflict is eliminated in case of using heavy liquid metal coolants (HLMC) such as eutectic lead-bismuth alloy and pure lead, as well as molten salts. In those reactors there is no stored potential energy of coolant compression and chemical energy, which under certain conditions can cause damage of fuel and primary circuit, loss of coolant and destruction of protection barriers.

Therefore, the SVBR-100 type reactors assure deterministical elimination of the severest accidents which cause radioactivity exhaust and require population evacuation. That makes possible with a minimal risk use such reactors in developing countries.

The main effect in providing a high safety level of the SVBR-100 RF is achieved due to use of the fast neutron reactor, HLMC and integral design of the reactor with entire elimination of pipelines and valves with radioactive coolant beyond the reactor monoblock (RMB) vessel (see Figure 1). This has been verified by performed computations and development works [8].

The reactor possesses a negative void reactivity effect and negative feedbacks, the efficiency of the strongest absorbing rod does not exceed 1$—the efficient delayed neutrons fraction. And those coupled with technical realization of the CPS eliminate prompt neutron runaway of the reactor.

The high boiling point of coolant heightens reliability of heat removal from the core, and safety due to lack of the heat transfer crisis. Also, being coupled with a provided safeguard casing of the monoblock vessel, that eliminates loss of coolant accidents (LOCA) and high pressure radioactive exhausts.

The low pressure in the primary circuit reduces the risk of tightness failure. It makes possible to decrease the thickness of reactor vessel walls and diminishing the limitations imposed on the rate of temperature change according to the thermal-cycling strength conditions.

The RF components do not contain materials releasing hydrogen as a result of thermal and radiation effects and chemical reactions with coolant, water and air. Therefore, in an event of tightness failure in the primary circuit the likelihood of chemical explosions and fires is virtually eliminated.

The RF inherent self protection and passive safety properties make it possible to couple realization of much of safety functions and normal operating functions. At this point, the safety systems do not contain elements, in which actuation can be blocked in an event of failure or impact of any human factors:

- Removal of residual heat decay is provided passively by natural circulation of LBC in the primary circuit. This is realized by transferring heat via four independent channels into the SG to the secondary circuit water and then to water of the passive heat removal system (PHRS) tank, with removal of produced steam to the atmosphere. For that purpose condenser is mounted in the PHRS tank under the water level. In the process of heat decay removal the steam from the separator is transferred to the condenser and condensate is discharged into the separator via the steam pressure control valve. The valve is closed when the steam pressure is reduced in order to eliminate overcooling of the condensate (see Figure 2). This represents a grace period of about three days long without exceeding the allowed temperature limits.

- In an event of large leak in several SG tubes, localization of SG leak and RMB protection are also provided passively when the steam pressure in the gas system is increased over 0.5 MPa. This is provided by actuation of the bursting membranes and discharge of steam into the bubbling device. (Operating experience has revealed that in an event of small leak in the SG, the RF does not need to be shut down at once).
- When LBC temperature is increased over a dangerous value, the rods of the additional emergency protection system, which are mounted in "dry" channels and do not have drivers on the reactor lid, actuate passively by gravity due to fusible locks, which are made of the alloy with a corresponding melting temperature and hold the rods in the upper position at normal temperature modes.

Moreover, in an event of postulated failure of all four channels of the PHRS, it is provided to flood the reactor well with water from the bubbling devices mounted above and transfer heat via the mono block vessel, air gap and protective casing to the water with further removing the produced steam to the atmosphere.

To reveal safety potential of the SVBR-100 RF, there were performed computations of radiational consequences caused by damage of the power-unit containment, damage of the reinforced-concrete overlapping over the reactor, tightness failure in the primary circuit gas system with direct contact between the coolant surface and atmospheric air, damage of all four channels of the SG PHRS and simultaneous total "blackout" of the NPP [9]. The superposition of such initial events is much beyond the normative safety regulatory documentation requirements including IAEA requirements [10] and can occur under the following circumstances: military actions, NPP capture by terrorists, extreme nature and man-caused actions. Therefore, today it is not considered while making justification of NPP safety.

The results of considerations and computations have revealed [9] that the integrity of the reactor monoblock vessel installed in the reinforced-concrete well below the ground level is maintained, the core will not be damaged in case makeup of the reactor well with water will be organized not later than twenty four hours after all channels of the SG PHRS have been damaged (transfer of necessary makeup water is provided by five ten-ton fire-engines per a day), and reactivity exhaust into the environment has not achieved the values requiring population evacuation.

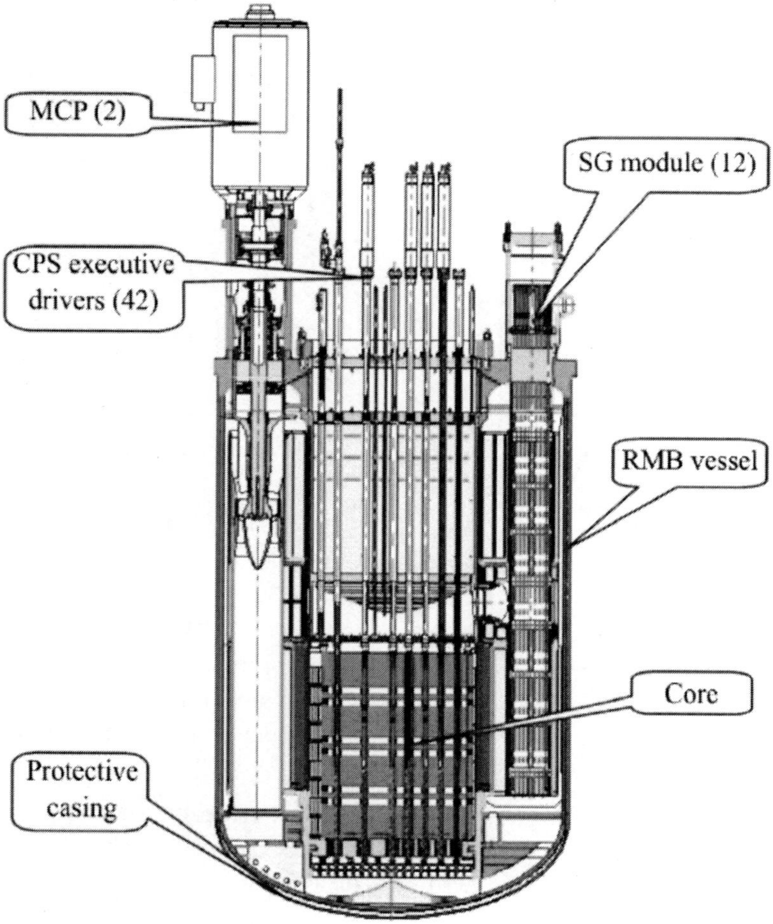

Figure 1. Reactor monoblock.

These enable speaking not only about RF SVBR-100 tolerance to the equipment failures and personnel's errors and their multiple superpositions but also about tolerance to malevolent actions when all special systems can be intentionally disabled.

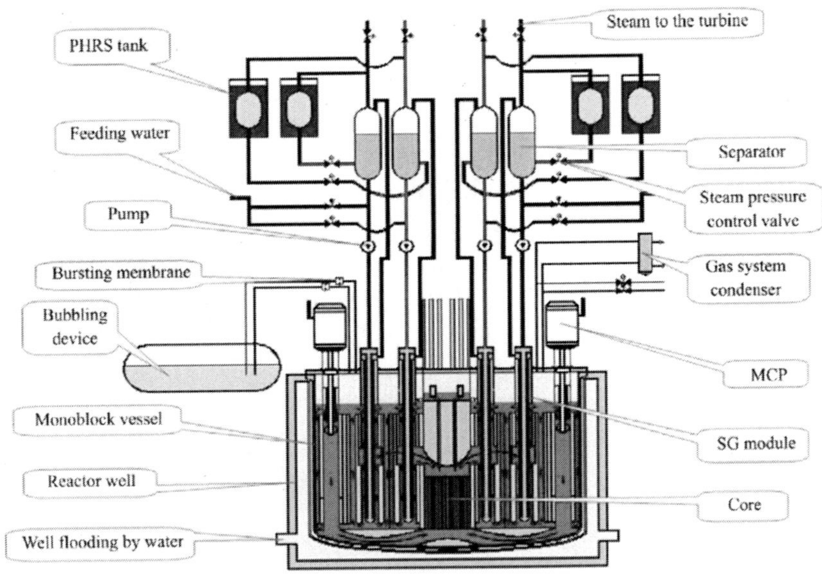

Figure 2. Hydraulic diagram of RF SVBR-100.

3.2. Brief Description of RF SVBR-100

These enable speaking not only about RF SVBR-100 tolerance to the equipment failures and personnel's errors and their multiple superpositions but also about tolerance to malevolent actions when all special systems can be intentionally disabled.

The considered NPT [11] is grounded on critical review of unique experience of development and operation of LBC cooled reactors at Russian NSs (15 reactors, 80 reactor-years) [12] and experience of operating the fast reactors with sodium coolant.

Characteristics of the SVBR-100 RF are as follows:

1) A fast neutron reactor with chemically inert LBC, i.e. eutectic lead-bismuth alloy in the primary circuit. Boiling point of LBC is 1670°C, melting point of LBC is 124.5°C.

2) An integral design of the reactor, when the whole primary circuit equipment is mounted in a single strong RMB vessel. Valves and LBC pipelines are completely eliminated (see Figure 1).

3) The reactor monoblock with a safeguard casing is installed in the reinforced-concrete well.

4) The basic components of the reactor monoblock and reactor facility are designed as separate modules. At this point, an opportunity of their replacement and repair is ensured.

5) On ending the core lifetime, fuel unloading will be performed at once, cassette-by-cassette, and fresh fuel will be loaded as a single cartridge (new core). After the fuel sub-assemblies (FSA) have been unloaded from the reactor, they are put in steel capsules with liquid lead and then to the storage with natural circulation of atmospheric air.

6) Repair of the primary circuit equipment and refueling are performed without LBC draining under maintaining it in liquid state due to residual heat release from the core or operation of the heating system. At this point, experience of operating the NS RF has been accounted: when those works were performed, air penetrated into the primary circuit that caused accumulation of slags and blocked the coolant flow rate.

7) A two-circuit scheme of heat removal and a steam generator (SG) with multiple compulsory circulation (MCC) over the secondary circuit are used (see Figure 2).

8) Natural circulation (NC) of coolants in heat-removal circuits of the reactor monoblock is sufficient to ensure passive removal of heat from the reactor without dangerous over-heating of the core.

9) The circulation scheme of LBC provides elimination of water/steam ingress into the core in an event of SG leak due to effective gravitational separation of steam on a free LBC level in the monoblock.

3.3. Reasoning the Option for Reactor Power Level

The option for the reactor power level to be 100 MWe or 280 MW and, consequently, the option for the reactor dimensions has been reasoned by the following:

1) The circulation scheme of LBC provides elimination of water/steam ingress into the core in an event of SG leak due to effective gravitational separation of steam on a free LBC level in the monoblock.

2) On the other hand, this is the maximal power at which the dimensions of the reactor monoblock assure an opportunity of its transportation in complete factory readiness by railway. Therefore, the opportunities of choosing the sites for NPP construction will be considerably extended, the labor inputs and the schedule of NPP construction will be essentially reduced.

3) The selected power level assures the favorable conditions for passive removal of residual heat decay via the reactor monoblock vessel without dangerous increase of fuel elements' temperature and these principally simplify the design and safety systems of the RF.

4) The solution to the problem of seismic resistance of the RF is simplified due to comparatively low weight of the monoblock.

In accordance with computations, the selected power level assures the following [13]:

- Lifetime duration is ~50,000 full power hours provided the mastered oxide uranium fuel is used (CBR = 0.84).
- CBR ≥ 1 provided the MOX fuel is used, and the reactor operates in the closed fuel cycle in a mode of fuel self-providing when lifetime duration is 76,000 full power hours.
- Lifetime duration is ~76,000 full power hours provided uranium nitride fuel is used (CBR = 0.91) and burn-up reactivity margin is less than 1$ or lifetime duration is ~150,000 full power hours.
- CBR ≥ 1 provided the mixed nitride fuel is used, and the reactor operates in a mode of fuel self-providing at burn-up reactivity margin being less than 1$ and lifetime duration is 76000 full power hours or the reactor operates in a mode of extended breeding with CBR = 1.13 at a doubling time of plutonium being ~45 years and lifetime duration being ~200,000 full power hours.

For the experimental-industrial power-unit with RF SVBR-100 the lifetime duration of 50000 full power hours has been validated for oxide uranium fuel. For other kinds of fuel the extended lifetime needs experimental verification.

The basic parameters of RF SVBR-100 are summarized in Table 1 (reference variant).

3.4. Decreasing the NFM Proliferation Risk

The solution to the problem of non-proliferation can be only achieved by coupling both technological and political measures. The relationship of those measures will be different for nuclear and non-nuclear countries. During the recent decades all nuclear countries, which are the members of the "Nuclear Club" community and legally possess the nuclear weapons, have solved this problem successfully, by use of the measures of physical protection, accounting, and safeguard. For that reason, the additional measures on technological maintenance of non- proliferation will be justified if they do not reduce the NPT competitiveness.

Table 1. Basic parameters of RF SVBR-100.

Parameter	Value
RF thermal power, MW	280
RF electric power, MWe	100
Steam-producing rate, t/h	580
Produced steam parameters: pressure, MPa/temperature, °C	6.7/278
LBC temperature: inlet/outlet, °C	340/490
Average power density of the core, kW/dm^3	160
Average linear load of the fuel element, kW/m	26
Fuel: type U loading, kg average enrichment in U-235, %	UO_2 ~9200 ~16.7
Core lifetime, thousands of full power hours	50000
Time interval between refueling, years	~7 - 8
RMB dimensions: diameter × height, m	4.53 × 7.55
RMB weight without core and coolant, t	360

When NPPs are used in developing countries, the additional measures on technological maintenance of non-proliferation should be implemented, together with political measures and international control [14].

The duration of fuel lifetime makes possible perform refueling not often than once a 7 - 8 year period. At this point, the heavy refueling equipment is not delivered to the User-Country and refueling is performed by the Supplier-Country under the IAEA aegis. Fresh fuel is delivered to the NPP site directly before loading in the reactor. The unloaded SNF will be stored in the protected storage at the NPP site five years or less in order to reduce residual heat release to the values, which assure its safe transportation in special containers to the Supplier-Country. All transport and technological operations must be performed under the aegis of IAEA.

When uranium fuel is used, uranium enriched in less than 20% will be used. At the stage of closing fuel cycle and SNF reprocessing, 2% of fission products built-up in the SNF and all minor actinides (MA) will remain in the re-fabricated fuel, with an exception of curium that is separated and stored up to decay in plutonium with its returning back to the fuel cycle. Handling that fuel needs the special technological equipment. Use of that fuel makes easier account of fuel movements. Breeding zones, in which plutonium for weapons can be built-up, are also absent in the reactor.

When NPPs are used in developing countries, delivery of fresh fuel and return of SNF for storage and reprocessing in the Supplier-Country must be realized on a nondiscrimination basis by International Nuclear Fuel Cycle Centers in accordance with initiatives of Russia and USA Presidents.

3.5. Modular Structure of Power-Unit NSSS

Due to a modular structure of the NSSS of the power-unit it is possible [15] :

1) To provide higher reliability (tolerance to failures of the power-unit being a system composed of separate RFs) and safety (to reduce the potential radiation risk) as compared with a power-unit based on a large capacity single reactor.

2) Not to organize a large capacity reserve PS in the areas of decentralized power supply.

3) To provide the loading factor (LF) to be not less than 90% due to long reactor operation without refueling. The LF will be only determined by reliability factors of the turbine installation. When each RF is shut down for refueling or maintenance, the power-unit's capacity will be reduced much less as compared with that of the power-unit based on a single reactor of large unit capacity.

4) To provide organization of large-scale (conveyer) production of reactor monoblocks (tens of monoblocks annually) and continuous work load of engineering factories. Thus, the manufacturing costs will be considerably reduced. As fabrication of the reactor monoblock of the RF does not require the unique engineering equipment as it is for high pressure vessels of thermal reactors, there is an opportunity to form a competitive market of producers.

5) To use the standard designs for different capacity power-units and factory production-line methods for their assembly and construction. Along with high quantity production of RFs, these assure reduction in the schedule and cost of power-units construction.

6) To locate the modular small and medium-sized NPPs in the centers of power consumption. Therefore, the expenditures for construction of high-voltage lines will be eliminated.

7) To provide an opportunity of phase implementation of the power-unit in operation with gradual raising of power capacity as the assembly and starting-up and adjustment works for the scheduled module have been completed. Thus, the pay-back term of the capital investments can be reduced due to earlier output of production and earlier beginning of credit repayment as compared with a power-unit based on the reactor of large unit capacity.

These all heighten much the consumer benefits of RF SVBR-100.

The NSSS modular structure and shipment of factory-ready modules assure reduction of the investment cycle of NPP construction that is extremely viable for approaching the NPP technical and economic parameters to those of modern TPPs with short investment cycles. Therefore, the financial risks can be considerably reduced.

As there are only two states of RF functioning, namely, operating and shut down, control of the modular NSSS is carried out by one operator using the common power master unit. If there is any fault in a single RF, it is automatically shut down and is cooled down autonomously, away from the turbine installation systems.

On expiring the RF lifetime (50 - 60 years) and unloading the spent nuclear fuel and LBC, the basic RF element-reactor monoblock—will be decommissioned and placed in a storage of solid radioactive waste. A new reactor monoblock will be installed instead. The other elements of the RF and power-unit can be decommissioned and replaced as well, i.e. the renovation can be performed. At this point, the lifetime of the modular NPP will be only limited by that of reinforced-concrete construction structures and can be expanded up to 100 - 120 years while the costs being much less as compared to those required for construction of the new power-unit. When the power-unit has been completely decommissioned, practically no radioactive materials will remain in the NSSS building after the reactor monoblocks have been dismantled. Thus, the cost of decommissioning will be considerably reduced.

3.6. High Potential for Improvement

The innovative Project of the NPP with reactor facilities SVBR-100 is in fact the First Generation Design based on a conservative approach. It has predetermined a high potential for further improvement of the Project that will be realized as the corresponding R & D have been accomplished and operating experience has been gained.

Especially:

- Increasing the LBC temperature at the reactor outlet while increasing the value of maximal temperature of the fuel element's cladding from 620°C to 650°C (there are all the necessary backgrounds) will provide (as the computations have revealed) growth of the reactor thermal power by 10% without considerable change of the reactor design.
- Use of the once-through SG generating super-heated steam assures that the efficiency of the thermodynamic cycle will be heightened by about 10%, capital costs will be lowered, RF design will be simplified.
- Use of nitride fuel can provide twice increase of the reactor lifetime (the operability of fuel elements should be verified) and correspondingly reduce the consumption of fuel.

3.7. Commercialization Concept

Despite the maximal possible use of experience of operating the LBC cooled reactors at the NSs, the conditions of operating the equipment of NS RFs and NPP RFs are much different. The operating mode of NS RFs is characterized by operation mainly at low power levels under lowered LBC temperatures, whereas the NPP RFs operate mainly at nominal power. Moreover, the requirements to the lifetime of the NPP RFs equipment are much higher than those to the lifetime of the NS RFs equipment. The technical and economical parameters also need direct verification. These all require construction of the experimental-industrial power-unit with RF SVBR-100.

The RF inherent self-protection and passive safety properties can be demonstrated in controlled conditions at the experimental-industrial RF equipped with additional sensors and devices, while simulating all possible combinations of the equipment failures, personnel's errors and malevolent actions.

After performing the certified tests of the experimental-industrial power-unit and verifying the design characteristics the SVBR-100 RF will be ready for commercialization and wide use as a part of the NPP power-units of different capacity and purpose.

4. CONCLUSIONS

- The most expedient way to upgrade the NPP safety and at the same time improve the economic characteristics is use of RFs, in which the value of stored potential energy is the lowest and inherent self-protection and passive safety properties can be realized to the maximal extent.
- Those RFs cannot amplify the external impacts, therefore, the scale of damages will be only determined by the energy of the external impact, the exhaust of radioactivity being localized. Such type RFs will possess the robustness properties ensuring their enhanced stability not only in events of single failures of the equipment and personnel's errors, but also in events of malevolent actions. And that is especially important for development of nuclear power in the developing countries with a high threat of terrorism.
- The innovative nuclear power technology based on multi-purposed standardizated modular fast reactors with HLMC, i.e. SVBR-100, which possess developed inherent self-protection and passive safety

properties, assures a high level of social acceptability for those reactors and provide the conditions for their use in developing countries. Due to low amount of potential energy stored in coolant, specific design of the RF, the opportunity for severe accident such as those at NPPs TMI-2, Chernobyl, Fukushima 1 are deterministically eliminated (on the assumption the initial events are identical).

- The modular structure of the power-unit's NSSS provides an opportunity to changeover to the advanced technologies of the standard design for different capacity power-units on the basis of series factory-manu- facture of standardizated reactor modules and changeover to production-line methods for their assemblage. This will make possible considerably reduce the schedule period required for NPP construction as well as provide technical maintenance of the reactor modules on a servicing base. Thus, the number of the operating personnel and corresponding expenditures will be reduced.
- The conservative approach adopted in the RF design has predetermined a high potential to further improvement of the RF. Realization of the designated measures that requires performing the corresponding R & D will make it possible to bring closer the specific capital cost of constructing the modular NPP and the construction schedule to the values, which are typical for modern TPPs. This will heighten the NPP competitiveness at the investment market and provide the conditions for attraction of non-governmental investments.
- The considered NPT possesses the properties making possible realization of the concept: "build, own, operate (or lease), deliver fresh fuel, take back the spent fuel, take the risks associated with potential nuclear detriment during the lifetime" that is most attractive for developing countries and though that concept is more burdensome for the Supplier-Country. Of course, if the User-Country wants to develop the own nuclear technology and not only use its products, the different concepts of the life cycle with extended participation of the User-Country are possible.
- Specific features of the proposed NPT make expedient its use in developing countries as a source of electric and heat energy that is located at a short distance from the consumer and for water desalination
- Development of experimental-industrial power-unit with the SVBR-100 type reactor facility is provided in Federal Target Program "New Generation Nuclear Power Technologies for 2010-2015 Years and Future Trends up to 2020". There will be only one-time expenditures for demonstration of the present NPT as on the basis of the tested "standard" reactor module the nuclear power-units of different

capacities and purposes can be constructed without performing the additional R&D. The project will be realized within the frameworks of a state-private partnership, for which a joint venture consisting of State Atomic Energy Corporation "Rosatom" and private company JSC "Irkutskenergo" is organized, its name is JSC "AKME-engineering". It is expected that the presented NPT will be ready for commercialization in the developing countries in 2022- 2025.

ACKNOWLEDGEMENTS

The authors would like to thank SSC RF-IPPE employees S.V. Budarina, K.G. Melnikov, I.V. Tormyshev for the assistance in preparation of the present Paper.

REFERENCES

1. 1. Spievak, I. and Weinberg, A.M. (1985) Inherently Safe Reactors. Annual Review of Energy, 10, 431-462. http://dx.doi.org/10.1146/annurev.eg.10.110185.002243
2. 2. Forsberg, C.W. and Weinberg, A.M. (1990) Advanced Reactors, Passive Safety, and Acceptance of Nuclear Energy. Annual Review of Energy, 15, 133-152. http://dx.doi.org/10.1146/annurev.eg.15.110190.001025
3. 3. Arutyunyan, R.V. (2013) Have We Got Safety Culture? Atomnaya Strategiya, 83, 9-11. http://www.proatom.ru/
4. 4. Nuclear Terrorism. http://www.gochs.info/p669.htm
5. 5. IAEA (2006) Advanced Nuclear Plant Design Options to Cope with External Events. IAEA-TECDOC-1487. http://www-pub.iaea.org/MTCD/publications/PDF/ te_1487_web.pdf
6. 6. Ammirabile, L. European Project SARGEN_IV: Safety Approach and Assessment of GEN IV Reactors. http://www.iaea.org/NuclearPower/Downloadable/Meetings/2013/2013-02-26-02-27-TM-SFR/day-1/4-L.Ammirabile-EC.pdf
7. 7. Toshinsky, G.I., Komlev, O.G., Tormyshev, I.V. and Petrochenko, V.V. (2013) Effect of Potential Energy Stored in Reactor Facility Coolant on NPP Safety and Economic Parameters. World Journal of Nuclear Science and Technology, 3, 59-64. http://www.scirp.org/journal/wjnst
8. 8. Toshinsky, G.I., Komlev, O.G., Stepanov, V.S., et al. (2007) Principles of Providing Inherent Self-Protection and Passive Safety Characteristics of the SVBR-75/100 Type Modular Reactor Installation for Nuclear Power Plants of Different Capacity and Purpose. Paper No. 175598. International Conference Advanced Nuclear Fuel Cycles and Systems (GLOBAL'07), Boise, 9-13 September 2007.
9. 9. Bolhovitinov, V.D., Pankratov, D.V., Toshinsky, G.I., et al. (2003) Assessment of Radiation Consequences Caused by Large Tightness Failure in the Primary Circuit Gas System of RF SVBR-75/100 with Simultaneous Coolant Heating up to 600°C. Paper No. 2203 (CD-ROM), 3rd Conference Heavy-Liquid Metal Coolants in Nuclear Technologies (HLMC-2003) within the Frameworks of Russian Scientific and Technical Forum Nuclear Fast Neutron Reactors, Obninsk, 4-12 December 2003.
10. 10. IAEA (2012) Safety of Nuclar Power Plants: Design, Specific Safety Requirements. IAEA Safety Standards Series No. SSR-2/1.

11. 11. Zrodnikov, A.V., Toshinsky, G.I., Stepanov, V.S., et al. (2008) Innovative Nuclear Technology Based on Modular Multi-Purpose Lead-Bismuth Cooled Fast Reactors. Progress in Nuclear Energy, 50, 170-178. http://dx.doi.org/10.1016/j. pnucene. 2007.10.025
12. 12. Zrodnikov, A.V., Toshinsky, G.I., Stepanov, V.S., et al. (2004) Lead-Bismuth Reactor Technology Conversion: From Nuclear Submarine Reactors to Power Reactors and Ways to Increase the Investment Attractiveness of Nuclear Power Based on Fast-Neutron Reactors. Proceedings of IAEA International Conference "Fifty Years of Nuclear Power—The Next Fifty Years", Report IAEA-CN-A3, Obninsk.
13. 13. Novikova, N.N., Komlev, O.G. and Toshinsky, G.I. (2006) Neutronic and Physical Characteristics of Reactor SVBR-75/100 with Different Types of Fuel. Proceedings of International Congress on Advances in Nuclear Power Plants, CD-ROM, Paper No. 6355, Reno, 4-8 June 2006.
14. 14. Gromov, B.F., Toshinsky, G.I., Zrodnikov, A.V., Dragunov, Yu.G., Stepanov, V.S., et al. (2001) Nuclear Power Complex Based on SVBR-75/100 Small Reactors Cooled by Lead-Bismuth Liquid Metal Coolant. Competitiveness, Simplified Life Cycle, Safety, Non-Proliferation. Proceedings of IAEA International Seminar on Status and Prospects for Small and Medium Sized Reactors, Cairo, 27-31 May 2001.
15. 15. Zrodnikov, A.V., Toshinsky, G.I., Komlev, O.G., et al. (2006) Nuclear Power Development in Market Conditions with Use of Multi-Purpose Modular Fast Reactors SVBR-75/100. Nuclear Engineering and Design, 236, 1490-1502.http://dx.doi.org/ 10.1016/j.nucengdes.2006.04.005

CHAPTER 12

Technological Perspectives for Propulsion on Nuclear Attack Submarines

Luciano Ondir Freire, Delvonei Alves de Andrade

Nuclear and Energy Research Institute, Instituto de Pesquisas Energeticas e Nucleares, Sao Paulo, Brazil

ABSTRACT

This work aimed at proposing a new combination of technologies to improve military performances and reduce costs of nuclear attack submarines, without overlooking safety constraints. The last generation of nuclear attack submarines increased size to meet safety and operational requirements, imposing huge burden on costs side, reducing fleet size. The limitations of current Technologies employed were qualitatively discussed, explaining their limitations. There are new technologies (plate and shell heat exchangers) and architectural choices, like passive safety, and segregation of safety and normal systems, which may lead to reduction of costs and size of submarines. A qualitative analysis was provided on this combination of technologies, stressing their commercial nature and maturity, which reduced risks. The qualitative analysis showed the strong and weak points of this proposal, which adopted the concept of strength in numbers. Concluding, new Technologies enabled the existence of 3800 t nuclear attack submarines with powerful propulsion systems and good acoustic discretion.

Keywords: Plate and Shell Heat Exchangers, Nuclear Attack Submarines, PWR

1. INTRODUCTION

State of art nuclear attack submarines (Virginia, Astute and Suffren classes) have nuclear reactors that depend on electric energy and active engineered safety features to keep a safe state in case of design basis accidents. It is important to note that even after shutdown, the reactor core keeps generating residual heat, which may melt the core internals if emergency core cooling system does not work. Therefore, design needs to add redundancies on emergency long term electrical generation (diesel generators) to render acceptable the risk of total loss of electric power following a design basis accident on reactor. Such redundancies, along increase in payload, imposed greater displacements on the last generation of nuclear attack submarines if compared with previous generation. The surface displacements reached the range 7000 - 8000 tons for submarines using loop reactors and about 5000 tons for the submarine using an integral architecture reactor [1].

Evidently, this displacement growth had drawbacks in propulsion performance, due to growth of wetted surface, causing augmentation of drag, besides costs growth. Each nuclear class component, produced under nuclear quality standards, has an acquisition cost about one order of magnitude greater than its normal equivalent [2], and requires constant inspections during service life.

In the field of land commercial nuclear power plants, vendors started to propose simpler solutions with less components and passive safety, which is able to cool the core after a design basis accident without electrical energy supply or active components. Such designs were commercially successful and many units of AP1000 are under construction, some soon entering in operation [3]. An example of passive safety is the nuclear containment of International Reactor Innovative and Secure, IRIS, which assures core cooling by gravity in case of loss of coolant accidents and reduces containment footprint [4]. In IRIS, all components are integrated in a single pressure vessel, eliminating pipes and reducing frequency of loss of coolant accidents, simplifying the nuclear plant and reducing its volume, what is interesting for naval applications. In case of accidents, the residual heat is removed by containment walls to external fluid, eliminating the need of pumps or any active device in the long term, which reduces the need of electric energy.

In the end of eighties' decade, advances in fabrication technology, like laser welding, allowed the fabrication of shell and plate heat exchangers, which have similar capabilities as shell and tube heat exchangers, but are more compact and reliable [5]. Shell and plate heat exchangers work with pressures temperatures almost as elevated as shell and tube heat exchangers, reaching pressures of 150 bar and temperatures of 600°C [5]. Shell and plate heat exchangers have suitable ranges of temperature and pressure for application on pressurized water reactors and have much smaller volume, as seen in Figure 1 [5].

Figure 1. Comparison between shell and plate heat exchanger (top) and shell and tube heat exchanger (below) with same thermal load

The nuclear reactor core generates high energy ionizing radiation which requires a radiological shield of about 1 meter thick [6]. The reactor shielding, also called primary shielding, blocks gamma rays and high energy neutrons from fission reactions inside. the active nuclear reactor core. In loop reactors, the gamma decay of the short lived ^{16}N to ^{16}O in pipes and steam generator imposes the adoption of another shielding to protect the crew [6], which implies in great addition of weight to the ship. The adoption of integrated designs greatly reduces the weight and volume inside a ship because there is no need of a secondary shielding to protect the crew from ionizing radiation from short lived isotopes generated in the coolant within the core.

In integral designs, the same shielding blocks the fission radiation and the gamma radiation from ^{16}N isotopes. In other words, integral designs eliminate the need of secondary shielding, roughly reducing plant weight by half. The total plant volume is also reduced due to elimination of space for primary pipes, steam generators, seating and space for visual inspection during service life.

French integral nuclear reactors use one shell and tube heat exchanger on top of reactor vessel to generate steam. This solution has limitations to maximum power once tube sheets cannot have diameters beyond a certain value due to technological constraints. Once submarine diameter limits also the height of the steam generator, this volume limitation reduces the maximum power that can be extracted from the nuclear reactor, reducing the maximum speed of the submarine.

On the other hand, the shell and plate heat exchangers allow construction of integrated reactors with any shape, allowing greater powers, achieving performances like those of loop reactors in a reduced footprint. There is already a patent that shows an illustration of such concept [7].

2. ASSUMPTIONS

It is assumed the budget for defense is constant and independent of technology adopted. This assumption is justified by the fact most countries use a fraction of gross domestic product to define military expenditures. Therefore, it is possible either to have many cheap submarines or a few expensive ones. Another assumption is that vertical launchers are not required, which is justified by the fact most nuclear attack submarines does not have vertical launchers. It is assumed that a nuclear attack submarine must be able to realize blockade operations against enemy merchant ships and be able to outrun the fastest ships, which run at around 25 knots nowadays. This assumption is justified by previous use of submarines in Second World War. It is assumed that only proven technologies should be employed. On the other side, the technologies must be far from becoming obsolete once a submarine program has a long life and requires spare parts during its life. This assumption is justified because a delays or downtimes in such expensive systems as nuclear attack submarines represent enormous financial damages. Last, but not least, the machinery of a submarine needs to be silent at low speed but not at high speed. This assumption is justified by the fact that at high speed, hydrodynamic noise is preponderant.

3. METHOD AND DEVELOPMENT

Only a qualitative analysis of weights and volumes and power of propulsion systems is performed, comparing past solutions and examples. Only orders of magnitude are taken into account, without precise calculations. For example, the six tenths law is used to estimate costs in function of performances. Another example is the noise level in function of the shaft speed of a rotating machine. This work shows an option without providing calculations which are not interesting for pre conceptual studies.

The intrinsic characteristics of each nuclear reactor architecture (integral and loops) are discussed in terms of integration constraints of weight and volume. Second step is to demonstrate the possibility of using plate and shell heat exchangers to further reduce nuclear reactor. The analysis goes on by reviewing the types of propulsion systems and how to reduce their volume keeping performance, reliability and acoustic discretion.

A propulsion system should be light and compact. The best proven architecture to achieve this is the integral architecture because it does not need space for primary pipes and secondary shielding. In order to achieve high power, the steam generation must be done by plate and shell heat exchangers. A possible physical arrangement could be a set of plate and shell heat exchangers in vertical position organized in the form of annulus, as shown in Figure 2 and Figure 3. The base heat exchanger is the XPS50 of GESMEX. The material chosen is steel AISI 316 L with 1.25 mm thickness. Such vertical position for this kind of heat exchanger was already employed successfully in industry. Both feed water piping and steam piping are in the top, easily accessed when the reactor vessel is opened. If a heat exchanger suffers some leakages, it is easy replace it, removing it by the top. In case of inclination of the ship, it is possible to have liquid water exiting some heat exchanger due to gravity effects. The heat exchanger in the lowest position would have bigger flow because of hydrostatic pressure, creating the risk of partial evaporation, which in turn reduces charge losses, increasing still more the liquid water content at outlet. In order to prevent a huge liquid water fraction entering the turbine and causing erosion, the following measures are necessary:

- To add flow restrictions at inlet of secondary side of each heat exchanger in order to reduce flow unbalance;
- To have super heating margin on the outlet steam (or margin on the heat transfer area), so even if there is unbalance, it is not sufficient to overcome margins;
- To stop operation to when inclination angle surpasses design limits, once it is not possible to design a reactor which is able to operate at any inclination. Besides, if the reactor is shutdown, the turbine receives no steam and is in a safe condition.

It was found that at angles of 45 degrees from vertical, three passes heat exchangers with 21% margin in heat transfer area should not have problems with liquid water at outlet.

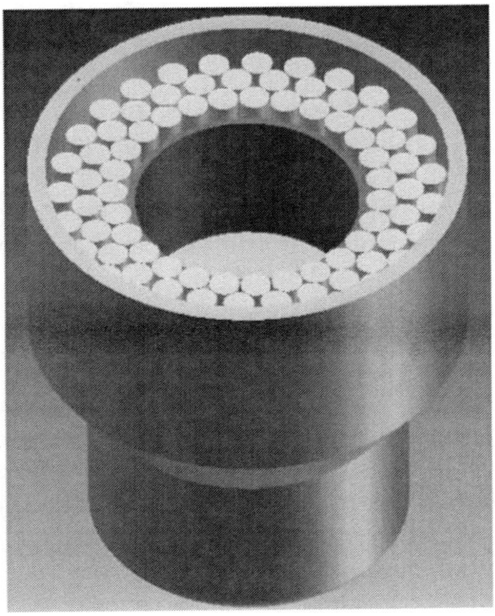

Figure 2. Arrangement in form of annulus.

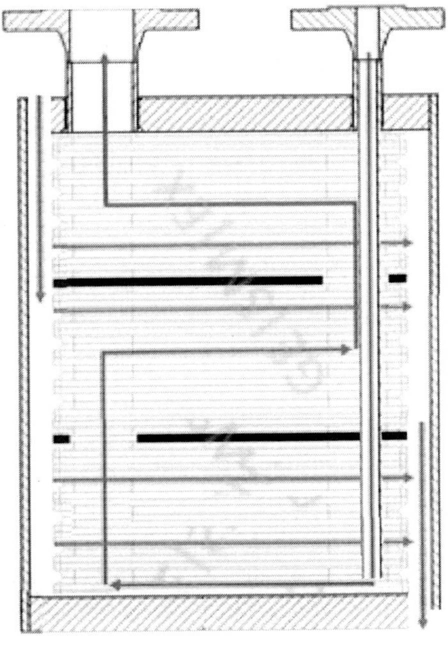

Figure 3. Plate and shell heat exchanger in vertical position.

Typically in PWR, steam generators have a double role: they generate steam to be used in power conversion system and, in accidental scenario, they contribute to perform the emergency secondary heat removal function [8]. This solution imposes constraints in design and operation, like redundancy on steam generators, periodic inspections, classification of part of steam generators in class 1 and part in class 2, segregation and independence.

If the safety function of emergency secondary heat removal is performed by other heat exchangers not used in normal operation, the heat exchanger used for power conversion does not need to be redundant and classified. Of course, failure of non-nuclear safety equipment must not impair safety functions. In current designs, the steam generators tubes are part of reactor coolant pressure barrier and therefore need to be classified as safety class 1 [8]. It is important to note that operation has wear effects over the tubes, which frequently have leakages and need to be plugged.

On the other hand, if plate and shell heat exchangers are used instead of shell an tube, it is possible to place the plate packs inside the pressure vessel. In this case, leakages or ruptures in the plate pack do not constitute a breach in the reactor coolant pressure barrier, as shown in Figure 4.

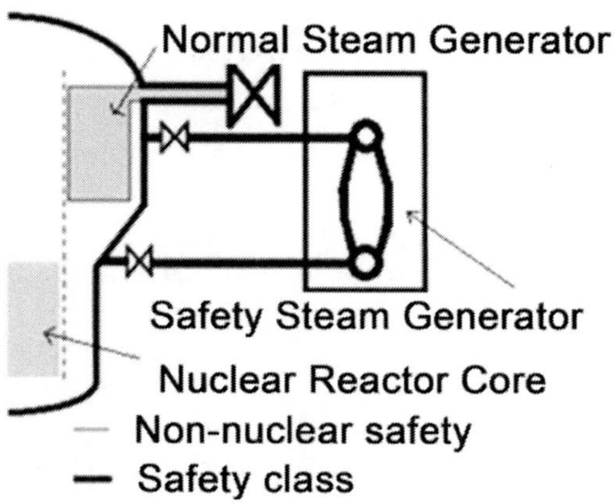

Figure 4. Normal steam generator within reactor vessel and external safety steam generator.

Of course, steam lines and feed water lines need to be class 1 up to isolation valves because the plate packs, being non-nuclear safety, do not receive credit in design basis scenarios. Once plate packs are located inside pressure barrier, a leakage can only cause a limited loss of coolant, which will stop as soon as isolation valves are closed, affecting only plant availability, but not plant safety.

Safety steam generators may be shell and tube type heat exchangers located in an ambient pressure tank, with natural circulation drive for primary coolant, in similar fashion as IRIS [4]. Those heat exchangers must only be pressurized for tests and accidental scenarios.

Once the plate packs work always under compression because the external pressure (primary coolant, around 150 bar) is higher than internal pressure (secondary fluid, around 64 bar), catastrophic ruptures are unlikely to occur. Of course, small leakages may still happen with frequency higher than design basis accidents defined in nuclear standard [8]. In practice, it may be found that the association of one big internal steam generator for normal operation and two or three small steam generators dedicated to design basis accidents may have better availability and safety than current designs. The reason for better safety is that safety heat exchangers do not suffer wear in normal operation and the reason for better availability is smaller frequency of leakages in pipes and plate packs working under compression.

In the power conversion system, there are three main types of propulsion architectures: turbo mechanical (the most common), turbo electric (the most silent) and hybrid (the newest). The turbo mechanic propulsion is the most compact and allows reaching higher speeds because it has less energy conversion stages. It is more powerful but it also is noisy. Figure 5 shows a simplified example.

Turbo electric propulsion systems (Figure 6) use fixed speed power generators, operating at optimum speed, having superior efficiency when operating at lower speeds if compared with turbo mechanical systems. The operation at fixed speed allows machines to be designed to be more silent at this shaft speed, which results in good acoustic discretion properties. On the other hand, electric power generators, rectifiers, drivers, electric motors and associated seating are heavy and occupy a lot of space. Once submarines are limited in terms of internal volume, historically, submarines with this architecture have maximum speeds between 25% and 40% smaller than submarines with turbo mechanical architectures. Examples of submarines with turbo electric architecture are SSN Tulibee and Ruby class submarines.

Finally, the hybrid propulsion systems have the advantages of both previously described architectures. It is compact and allows turbines operation at constant speed, which gives good acoustic discretion properties. Figure 7 presents a variant that is the hybrid parallel architecture. At low power and speed, the clutch allows the power generators operate at high speed, recharging the battery and providing auxiliary power to the ship, while a tiny propulsion motor provides propulsion. To reach high speeds, the clutch connects the turbine to the propulsion shaft line gearbox. The propulsion motor can still enhance the power at the shaft line for a limited time by drawing energy from the battery.

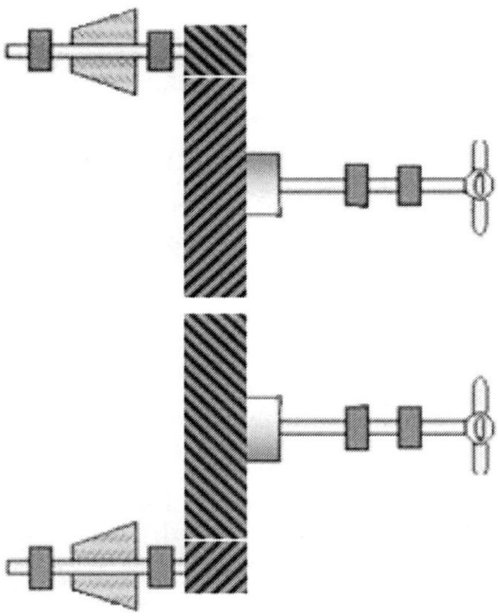

Figure 5. Turbo mechanical propulsion.

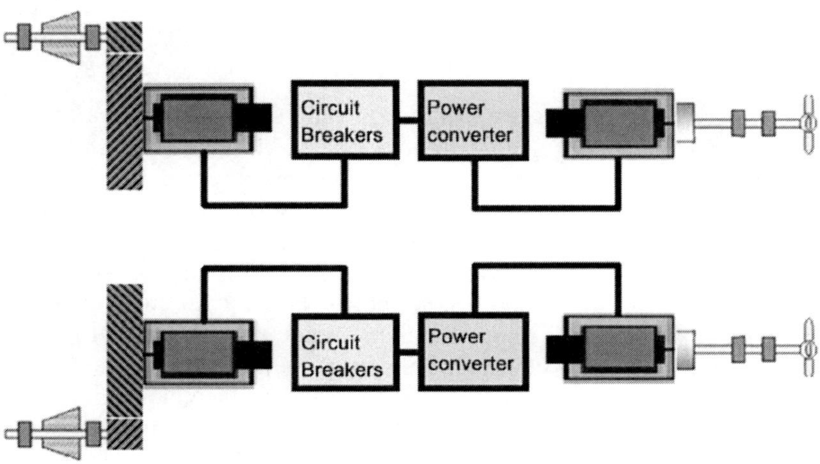

Figure 6. Turbo electrical propulsion.

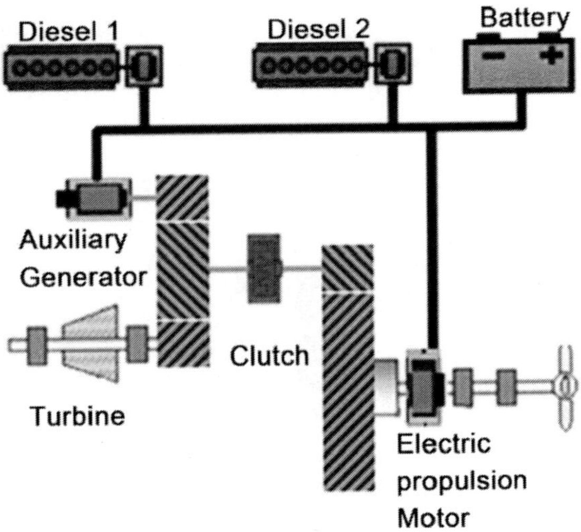

Figure 7. Hybrid parallel propulsion.

Of course, when the clutch connects the turbine to the shaft line and becomes the prime mover, the propulsion system is as noisy as a turbo mechanical one. But, at high speeds, the hydro dynamic noise is higher than machinery noise, which means that there is no gain in being silent at high speeds when the turbulent flow around the hull produces a lot of noise. It is important to note that a submarine stays most of the time at low speed, being the use of high speed a small fraction of time. In the hybrid parallel architecture shown, both the tiny electric propulsion motor and the generator may work as motor or generator. This gives flexibility, providing redundant downgraded modes of operation, keeping a minimum propulsion capability after the failure of any component.

In the power conversion system part, typically submarines have adopted two segregated sides of turbines to achieve reliability goals. Such solution has drawbacks of being expensive and large in terms of weight and volume if compared with a single turbine. On grounds of the six tenths law [9], it may be stated that two turbines are 32% more expensive than a single turbine with same power. Besides, two turbines require almost twice more space than a single turbine with same power because all auxiliary system need to be duplicated. It is important to note that in peace time, after losing one side of propulsion system due to equipment failure, the submarine needs to come back to repairs. Once the two sides of propulsion system are always in operation, the failure rate is doubled, reducing the mission reliability.

A modern solution is the hybrid parallel architecture (Figure 7), employed in merchant ships and tugs. Such solution allows keeping a minimal propulsion capability in case of a single failure, being as reliable as the redundant solution to return to home, but cheaper and more powerful. The mission reliability is superior because the quantity of components in operation is reduced.

4. RESULTS AND DISCUSSION

Adopting a propulsion system with hybrid parallel architecture, an integral reactor with shell and plate steam generators, it is estimated that is possible to reach a thermal power of 64 MW within a hull of 8.8 m of diameter. Once the nuclear safety is assured by passive elements, diesel and batteries do not need to be nuclear safety classified, having smaller redundancies and cost as they will only be required to power the submarine during return home.

This submarine could have a surface displacement similar to the Permit class (3800 t), shaft line power of 11 MW and maximum speed of 28 knots. The hybrid propulsion allows having acoustic discretion similar to SSN Tulibee at low speeds and maximum speed similar to Permit class. The limitation is that both properties are not simultaneous. If the clutch is disconnected, the propulsion system is silent but slow. If it is connected, high speeds may be reached at expense of acoustic discretion.

This submarine acoustic discretion is a central issue. The noisiest components in a nuclear attack submarine, in reducing order, are: turbines, primary pumps and shaft line reducing gears. All those machines are rotating machinery, which have many resonance frequencies. Whenever their shaft speed matches resonance frequencies, high amplitude oscillations are produced, generating noise.

In order to improve acoustic discretion, rotating machinery must have their masses reduced, its rotation speed must be elevated, a single rotation speed must be adopted and their carcass must be decoupled of the hull by flexible seating or cradles. In order to reduce mass, one solution is to employ components dedicated to operation at low speeds of the submarine, when flows are small. In this case, part of primary coolant pumps, of seawater condenser pumps, of condensate extraction pumps and feed water pumps must be sized to low speeds, when the clutch is disconnected.

In a submarine, turbines should work at their nominal rotation due to three reasons: first, to increase energy conversion efficiency; second, to reduce noise; and third, to reduce propagation of its noise in sea water, which is reduced at higher frequencies. Figure 8 shows a 130 Hz nominal rotation shaft vibration in function of its frequency. It is easy to see that a turbo mechanical propulsion turbine at low speeds may be near resonance frequency, and even if it not the case, the turbine noise would be around 3 orders of magnitude greater than at nominal speed. On the other hand, a hybrid parallel architecture uses its turbine at nominal speed when the submarine is at low speeds, fact that reduces vibration, which in turn allows stating that hybrid parallel propulsion has potential to be silent.

About the primary pumps, the use of integral design may allow to use the reactor vessel as a cradle, decoupling it from the hull by the use of flexible seating. Another advantage is that integral designs facilitate natural circulation, allowing the operation at low power without active pumping. However, natural circulation alone is very unstable and unsuitable to submarine operation. A better solution is to have a very small pump with capacity only to stabilize natural circulation. This small pump could have other functions, like driving coolant into purification system. This pump may be very silent if it is fixed on the reactor vessel, which may be decoupled from hull.

Finally, about the shaft line reduction gear, in hybrid parallel architecture, it does not work at low speeds. At low speeds, the prime mover is a tiny silent electric propulsion motor whose frequency of rotation is below detection range of modern passive sonars.

About the use of plate and shell heat exchangers, its fabrication feasibility is assured by the fact plate and shell heat exchangers are a commercial of the shelf solution, already proven in industry. The water chemistry requirements are not set by the heat exchanger, but by turbine. Once the steam is completely vaporized and superheated, any solid impurities in feed water are dragged along the steam and may cause corrosion on the turbine. According to EPRI [10], once through steam generators must work with the lowest possible levels of solid impurities to be adequate with secondary system design. This means that the water chemistry is not a problem to this type of heat exchanger and the integration with a secondary system is feasible. The access for inspection is easy because all pipes are in superior part and each heat exchanger may be easily replaced.

The modular architecture allows easy identification of the heat exchanger with leakages when the pressure vessel is open, once it is enough to pressurize the secondary side with air and see where the bubbles appear. The same way of IRIS, the elimination of pipes reduces the number of possible failures, reduces plant footprint and the size of nuclear containment. This means greater payload for a submarine. Additionally, the integral reactor design eases radiological shielding design and reduces crew exposition.

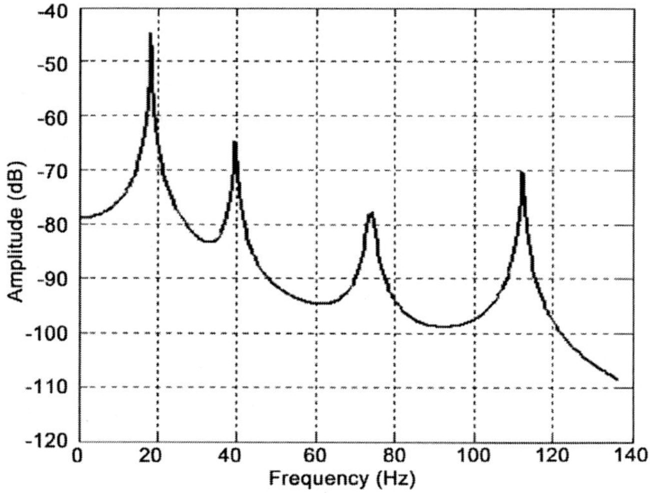

Figure 8. Example of rotor vibration in function of frequency.

The functional segregation of the normal and safety functions has the advantages of use of industry standards for the major part of heat exchangers (large area for power conversion) and makes the emergency secondary heat removal more reliable. Once emergency secondary heat removal heat exchangers are not used continuously, they do not suffer wear.

Although non-nuclear safety class, the power conversion heat exchangers are part of the first level of defense in depth and must avoid frequent incidents. The frequency of problems in plate and shell heat exchangers is estimated to be lower than current designs where tubes work in traction. Because the plate pack is under compression, there should not be propagation of small defects or creep. Of course, small leakages may occur, but they should not cause major plant downtimes.

5. CONCLUSIONS

Adoption of industry proven technologies can allow power densities improvements in nuclear attack submarines propulsion. Besides, the adoption of passive safety can reduce size and cost of nuclear reactors, allowing use of diesel generators without nuclear safety classification. Instead of two batteries, a single battery and much less pumps and pipes could be employed, and the submarine still would be safer than current designs. The use of integral designs can contribute to acoustic discretion, weight reduction, volume reduction and radiation doses on crew members.

With such solutions, it is possible to build nuclear attack submarines with half of surface displacement of state of art submarines Virginia class and Astute class, with a cost less than half of those submarines. The reduced cost allows formation of numerous fleets, whose cost can be even further reduced due to learning effects. The speed should allow outrunning the fastest merchant ships, making this submarine a formidable blockade weapon. And acoustic discretion could be yet superior to state of art submarines.

This new submarine can the built using only mature and proven technologies, tested for about 20 years, which means new developments at level of components are not required. It is just about a new form of integration of existing technologies.

REFERENCES

1. Saunders, S.C. (2015) Jane's Fighting Ships 2014-2015: Yearbook. 117th Edition, IHS Inc., London.
2. US Nuclear Regulatory Commission (2003) Risk-Informed Categorization and Treatment of Structures, Systems and Components for Nuclear Power Reactors, 10 CFR 50.69, Regulatory Analysis.
3. Reuters (2014) China Seen Buying Westinghouse Reactors for $24 Billion Nuclear Energy Projects. http://www.reuters.com/article/2014/04/21/china-nuclear-idUSL3N0ND1GS
4. Carelli, M.D., et al (2004) The Design and Safety Features of the IRIS Reactor. Nuclear Engineering and Design, 230, 151-167. http://dx.doi.org/10.1016/j.nucengdes.2003.11.022
5. Klahm, T. (2009) When Two Worlds Merge—A Compact and Efficient Plate Heat Exchanger Combined with a Pressure and Temperature Resistant Shell. http://www.gesmex.com/fachbericht-e303.html
6. Theodore Rockwell III (1950) Reactor Shielding Manual. Naval Reactors Branch, Division Of Reactor Development, United States Atomic Energy Commission, First Edition, USA.
7. Charles Fribourg (2012) Nuclear Reactor Having Plate or Micro-Channel Heat Exchangers integrated in the Vessel. WIPO Patent WO 2013/153076 A1.
8. American Nuclear Society (1988) Nuclear Safety Criteria for the Design of Stationary Pressurized Water Reactor Plants: ANSI/ANS-51.1-1983(R1988): Technical Standard. USA. 92 p.
9. Williams Jr., R. (1947) Six-Tenths Factor Aids in Approximating Costs. Chemical Engineering, 54, 124-125.
10. EPRI (2000) PWR Secondary Water Chemistry Guidelines—Revision 5. EPRI, Palo Alto, CA, TR-102134-R5.

CHAPTER 13

A domain-specific analysis system for examining nuclear reactor simulation data for light-water and sodium-cooled fast reactors

Jay Jay Billings*, Jordan H. Deyton, S. Forest Hull III 1, Eric J. Lingerfelt, Anna Wojtowicz

**Oak Ridge National Laboratory, PO Box 2008, MS6173, Oak Ridge, TN 37831, USA*

[1] Florida Institute of Technology, 150 W. University Blvd., Melbourne, FL 32901, USA.

ABSTRACT

Building a new generation of fission reactors in the United States presents many technical and regulatory challenges. One important challenge is the need to share and present results from new high-fidelity,high-performance simulations in an easily usable way. Since modern multiscale, multi-physics simulation scan generate petabytes of data, they will require the development of new techniques and methods to reduce the data to familiar quantities of interest (e.g., pin powers, temperatures) with a more reasonable resolution and size. Furthermore, some of the results from these simulations may be new quantities for which visualization and analysis techniques are not immediately available in the community and need to be developed.

This paper describes a new system for managing high-performance simulation results in adomain-specific way that naturally exposes quantities of interest for light water and sodium-cooled fast reactors. It describes requirements to build such a system and the technical challenges faced in its development at all levels (simulation, user interface, etc.). An example comparing results from two different simulation suites for a single assembly in a light-water reactor is presented, along with a detailed discussion of the system's requirements and design.

Keywords: Nuclear reactors LWRSFR Simulation Modeling High-performance computing

1. INTRODUCTION

Modeling and simulation has always played a vital role in nuclear engineering and its applications. It is becoming even more relevant as the community starts examining and constructing new types of reactors. Although existing computational models of the conventional fleet perform very well within the operational experience base of today's power plants, they do not necessarily provide the predictive capability needed to enable the deployment of completely new designs or to enable the use of today's designs for very different modes of operation. New simulation codes are under development in several efforts sponsored by the US Department of Energy's Office of Nuclear Energy, including those from the Nuclear Energy Advanced Modeling and Simulation (NEAMS) program (2011) and the Consortium for Advanced Simulation of Light Water Reactors (CASL) (2011).

These new codes are much different from the existing codes. Where as tried-and-true codes employ sophisticated engineering models calibrated based on experimental observation to describe system behavior, the new code suites leverage high-performance computing (HPC) platforms to provide truly predictive simulators. The new codes can examine the physics of nuclear reactors with an unprecedented resolution at spatial and temporal scales ranging from the microstructure of the fuel all the way to the plant itself, and they are commonly capable of coupling neutronics, fuel mechanics, structural mechanics, and the rmohydraulics (Gastonet al., 2009; Siegal et al., 2007; Turner et al., 2013). While they are very powerful, they inevitably generate many terabytes and even petabytes of data that greatly exceed what any single person can absorb and interpret alone. Even results that are refined through post-processing can still be gigabytes in size.

Additional challenges arise when the user actually examines and reviews the results. Large tables of data, even if well-organized, are not easily consumed by humans because of their size and general lack of sufficient context to eliminate ambiguity. Furthermore, in this particular area, there are often very large numbers (fluxes, _1012) associated with very small numbers (displacements, _105) that make intuitive interpretation of the results difficult. More often than not, users are forced to implement their own codes on top of the post-processing routines to collect ,assimilate, rescale, normalize, and plot the very small amount of data that they actually need, which is a very costly chore in terms of both time and resources. Comparing the results from new simulations to those from old simulations or experiments also requires custom code, thereby multiplying the cost of adoption.

The authors assert that it is not sufficient to leave these challenges unaddressed and "pitch it over the fence" to users and analysts. Instead, just as new codes are under development to address the physics questions, new technologies need to be developed to address data analysis. The system presented in Section 2 seeks to address this issue in three ways. First, it provides specialized data structures and input/output (I/O) libraries designed specifically for storing quantities of interest from nuclear simulations. Second, it provides a user interface that complements the I/O libraries to provide highly tailored views that put data in the proper context and reduce ambiguity. Finally, it provides an extension interface for adding custom analysis routines that can be easily coupled to routines for data mining and tailored analysis. An example of the utility of the system provided in Section 3 demonstrates the ease with which interesting information between two codes, one new and one seasoned, can be extracted from the system.

2. ARCHITECTURE

A high-level overview of the system's logical architecture is presented in Fig. 1. The system is split into three parts, one each for users who examine data, developers who store data, and those interested in implementing custom analysis routines to examine the data. In each case, care was taken to make sure that the workflow of the respective actor was easy and intuitive for that use case.

Five high-level functional and nonfunctional requirements were used to guide the detailed design of the system based on feedback from interviews and experiments with prototypes, all of which were derived from the assumption that the simulation results must be shared with one or more people.

1) The data should be organized according to the natural layout of a nuclear reactor and not on the discretized geometry of the simulation code.

2) A data point has a position in three dimensions plus time with a value, uncertainty, and units.

3) A reactor is composed of Parts, which are represented by custom data structures, and data can be stored for every Part (literally on every Part) available.

4) The information is stored in a cross-platform, self-describing, binary compatible, open format.

5) The data must be easily and equally accessible from languages used for user interfaces and from languages used to author simulation codes.

The need to compare data from two simulations led directly to requirement 1. Storing output data on the grid or mesh used by the simulation would allow only immediate and direct comparisons with other results from the same code unless grid-point mapping or mesh-to-mesh transfers (Tautges and Caceres, 2013) were performed. However, storing data based on the layout of a reactor—on cores, assemblies, pins, etc.—allows for immediate comparison of results, albeit at the cost of pulling that information from the grid or mesh during the simulation. One drawback of this design is that some data could be lost when the grid or mesh is removed. However, this is not strictly the case, since data can be mapped to any Part in the reactor and given any position.

It is extremely important that the data be stored in a binary compatible open format available on multiple platforms and readable in multiple languages (requirements 4 and 5) for three reasons. First, the subject matter experts will examine the data on workstations, not clusters or supercomputers. Second, in many situations, it must be stored in such a way that it can be reviewed if required, possibly at a much later date. Finally, someone may need to view the data who does not have the expertise to write the code to read the database but is otherwise qualified to evaluate it via a user interface.

The definition of a reactor, requirement 3, loosely follows that of real light water reactors (LWRs) and sodium-cooled fast reactors (SFRs). Reactors are hierarchically composed of a set of Parts, where a Part describes both the large-scale structures of a reactor, such as assemblies, and smaller Parts like pellets or "material blocks". Data is stored for these Parts according to the description in requirement 2, including time, space, uncertainty and dimensional information, in addition to the value of interest.

Exact descriptions of the Parts of the nuclear reactors supported by this project are provided in Section 3. The Parts are exactly the same in both supported languages, the user interface, and the I/O libraries. The design is general and based on publicly available information from Wikipedia, the Nuclear Regulatory Commission's public website, and interviews with subject-matter experts. No proprietary designs are considered, and no proprietary information is used or provided, although it is certainly possible to add such Parts to the system with minimal effort.

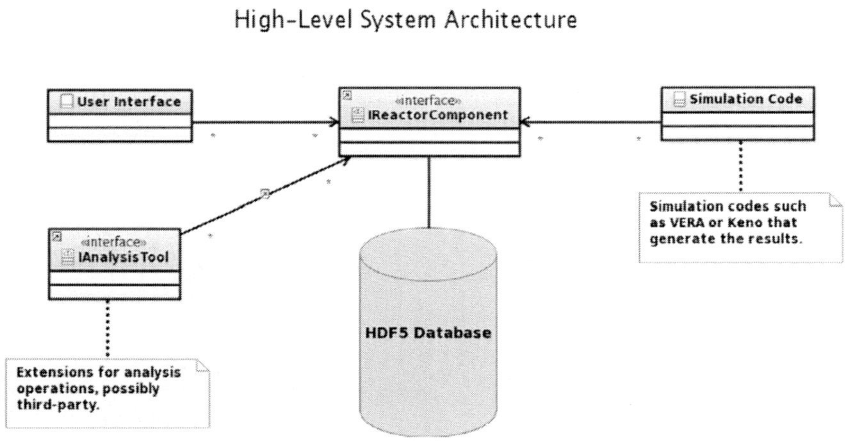

Figure. 1. A high-level Unified Modeling Language class diagram of the system's logical architecture, highlighting its focus on interacting with different Parts, or IReactor Components, of the reactor data.

The Parts are organized in a class hierarchy of data structures that uses object-oriented design principles to take advantage of both the natural hierarchy of reactors and abstractions for similar Parts (inheritance). The two different types of reactors that are currently supported—pressurized water reactors (PWRs) and SFRs— are available in their own modules but are accessed in the same way.

2.1. Input–output libraries

The system provides its own libraries for managing input and output to relieve software developers from the task of writing their own file parsers and emitters that conform to the shared specification. They remove the burden of writing to older data formats for developers of new codes and remove the complication of writing to newer formats, which may require significantly specialized skills, for maintainers of existing codes. Instead, developers instantiate and fill the Parts with data and then store those parts using the provided I/O routines. The workflow is shown, at a high level, in Fig. 2. In this figure, the "NiCE I/O API" represents the I/O library, and it acts as an intermediary between the simulation code and the user interface. The process is relatively simple: the developer loads the data structures needed and tells the library to write them. A user, on the other side, picks a file and the user interface loads that file.

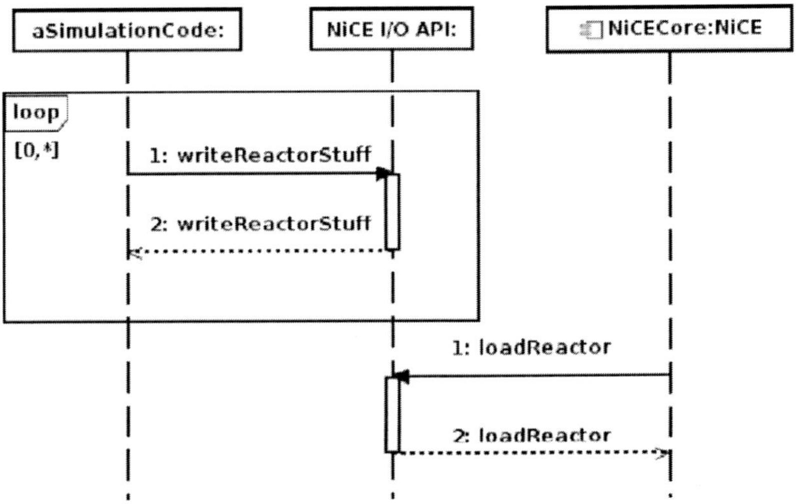

Figure. 2. A Unified Modeling Language sequence diagram that shows the basic process by which the results computed by a simulation code are loaded in the user interface ("NiCE") via the input–output interface, ("NiCE I/O API").

Developers may store whatever data they need on any Part. Data is stored associatively so that sets of data are stored against tags and any tag can be used. It is trivial to store data tagged with "Axial Power" right next to data tagged as "Cross sections" or "Velocity" on the same Part. Writing the data is then very simple: the developer calls a single write operation on the "highest" Part of the reactor, which is normally the reactor core, that contains all of the other Parts the developer has. The simulation code may write multiple reactors to disk, (thus the loop in Fig. 2). Any client can read this file once it is written, such as the user interface or standalone analysis codes, using the inverse read operation with the desired data file passed as an input argument.

The file layout, which is different from the format, of the I/O libraries matches the class hierarchy of the data structures with a few exceptions. For example, it is much more memory-efficient to store the units of values in individual tables and refer to the name of the units with an integer in the file because, in practice, the integer is much smaller than a fixed-length string for representing an arbitrary system of units. A much more important optimization is the reuse of large structures like assemblies and pins, where possible. The layout of a physical reactor will have only a small number of different pin types (or assemblies) relative to the total number of pins in the reactor. Exploiting this fact by reusing the geometry and material information both in the data structures and on disk greatly reduces the amount of memory required. This detailed analysis of the layout of the data has made it possible to read and write data for "full" cores with 50 axial levels, hundreds of assemblies, and 50,000 pins in seconds.

2.2. User interface

The ultimate goal of this system is to put simulation data into the hands of human users, and humans are much more efficient at examining visual images than raw data. The system includes a graphical user interface that presents information in multiple ways, including custom views, color maps, plots, and raw data. The user interface is designed so that users feel like they are manipulating a reactor instead of reading a paper.

The most important design requirement for the user interface is that it enables users to perform comparative analysis. That is, users must be allowed to compare results among multiple reactors side-by-side, with at least one of the reactors accepted as a 'gold-standard" reference. For that matter, users must be allowed to examine the same part of a reactor with multiple views of its own data—geometry, color maps, plots, and so on.

Each Part has at least two associated graphical views, one each for geometric information, including assigned materials, and state point data, which is defined as the values of the quantities of interest at a given time step. Views can be switched via a simple toolbar or by right-clicking on the canvas. It is possible to create plots for state point data for some Parts and, using specialized analysis routines, compute direct numerical comparisons (and create plots of those comparisons) between the state point data for two or more Parts. Each view can be saved as a PNG image file with a handy button on the toolbar.

Three-dimensional core geometries are supported by the system, but the user interface breaks this into two *xy* and z views. The *xy* view shows a cross section of the current Part at the specified axial level. The z view is next to the *xy* view and shows the axial geometry.

The user interface also exposes any external analysis routines that are available to users in a menu, as well as any options that can be configured for those routines. Results from the analyses are captured and presented to the user in a simple list. The file names on the local hard drive and their associate timestamps are also presented so that users can easily move the data to another system if needed. One external analysis routine is provided by default—to compute percentage differences among pins—and it is treated as a somewhat special case in that its results are automatically plotted, which is not the case for other analyses.

It is possible to view the data stored in the system using other user interfaces and techniques, c.f. Section 2.4.

2.3. Integration with external analysis routines

The system can be extended to use external analysis routines for teams or developers who want to manipulate the data in a specific way. This mechanism is provided because it is impossible for the core development team to guess all of the possible analyses, much less implement and support the capability in the long term. The routines are referred to as "external" simply because they are plug-and-play.

External analysis routines are exposed to users through the user interface. The external routines can be anything from simple math operations to large wrappers that ship the work off to other libraries. The development team has tested three such extensions to date: the differencing tool for pin powers mentioned previously, a k-means clustering routine to find clusters in the data, and a set of "routines" that wrap a very large visualization toolkit, VisIt (2011), to create different plots from the default set. The differencing routine is the only external routine available by default, but more detailed information on findings with the k-means clustering routines can be found in Pokhriyal et al. (2013). k-means is a clustering algorithm used to find groups in data that can later be used for classification. Ideally, a set of external analysis routines would be available to perform classification and anomaly detection on pins and assemblies as a form of automatic data triage to locate troublesome, erroneous, or simply interesting areas of the core.

Developers can add routines by implementing the "IAnalysisTool" interface shown in Fig. 1, as well as a few other interfaces associated with it, and declaring the analysis tool as a service to the framework. Analysis routines are written in Java and are dynamically consumed by the framework at runtime. Currently, only assemblies are passed to analysis tools, not reactor cores. This capability is not available in the C++ implementation, because that implementation is primarily focused on I/O for the simulation codes.

Much more detailed descriptions for all of these interfaces are presented in the project documentation at the NiCE Project Page (2011–2014).

2.4. Technologies

The system is built with mostly off-the-shelf components and assembled as part of the NEAMS Integrated Computational Environment (NiCE), (NiCE Project Page, 2011–2014; Billings et al., 2013; Billings et al., 2014). The code for developers is written in C++ so that it can be easily used with C/C++ and Fortran codes. Version five of the Hierarchical Data Format (HDF5) (The HDF Group, 1997–2014) is used to read and write the data structures to and from disk. This greatly reduces the amount of work required to satisfy the cross-platform, multi-language requirements, as HDF5 is natively available in many languages and is completely open. HDF5 also provides multiple ways to access the data in "hdf" files in Ascii, including a command line utility—h5dump— that will dump the contents to disk and a graphical utility—hdfvi ew—for viewing the contents of any HDF5 file.

The graphical user interface is implemented in the same system as NiCE, which uses the Eclipse Rich Client Platform (RCP) (McAffer et al., 2010), and is written in Java. RCP is the platform on which the Eclipse Integrated Development Environment is built, is very flexible, and is cross-platform. It has many tools and utilities for working with graphics and connecting those graphics to data.

The k-means algorithm discussed in Section 2.3 was implemented in-house for NiCE, based on open literature, and is available in the source code. It is not enabled by default because the development team is exploring the possibility of using third-party capabilities to replace the home-grown version, namely Apache's Mahout (2011–2014), and enable more data mining capabilities.

2.4.1. Relationship with other tools

Other tools exist that are capable of visualizing this kind of output, including VisIt and ParaView (2015), both of which are capable of reading the HDF5 files generated by this system. The primary differences between more mainstream visualization systems and this system is the strong focus on the nuclear energy domain and the addition of interactivity, specifically picking and selection. Most general purpose tools do not look at data from a strictly domain perspective.2 In fact, most general purpose tools only consider the data from the very general standpoint of the mesh, which is generally different across different types of physics in multiphysics simulations.

Interactivity is very important because of the needs of the user to quickly find the data that interests them, which extends beyond simply showing something in 3D and allowing for rotations, etc. A general purpose tool might allow users to single-out individual pins through some process of slicing or bounding the domain, but that is not the same as actually picking or selecting an individual pin and only examining that pin. "Diving deeper" with general purpose tools may also lose any particular domain context, whereas picking Parts in our system maintains the entire domain view of the data.

3. PARTS

The system is composed of many Parts across the different levels of the reactor and is capable of storing data for PWRs and SFRs. The following discussion presents each Part in the system, the types of data that can be assigned to the Parts, and the visualizations available in the user interface.

The data presented for PWRs is taken from real simulations performed with the Virtual Environment for Reactor Analysis (VERA) (Turner et al., 2013) and KENO (Rearden) for Problem 3a of the VERA Benchmarks (Godfrey, 2013). This problem represents a single 17 by 17 PWR fuel assembly at the beginning of its life and at constant temperature. The assembly is refined with 49 axial levels. VERA was modified for our purposes to write the results directly to file using the data structures and I/O libraries from our system. We obtained the KENO results from the author of Godfrey (2013) and converted them into our system using a stand-alone program. Additional assemblies were added to the geometric configuration of Problem 3a to show the system's ability to work with assemblies other than fuel assemblies, such as in-core instruments.

The results shown in the pictures for SFRs are generated and do not represent a physical system. The SFR pictures are presented solely to instruct readers about the capabilities of the data structures and the user interface.

All of the graphical views below exploit object-oriented design principles to minimize the amount of custom code needed to draw the different shapes. In many places, the same code is reused to draw Parts from PWRs and SFRs. The simplicity of using this system is also noteworthy, as the estimated time for a new user to go from importing data to exporting a plot like those shown below is merely minutes and does not require any knowledge of file formats, data layouts, or scripting languages.

3.1. Part properties

Each Part has a set of properties assigned to it that represent the most common properties of the given piece, such as material, pitch, diameter, and so on. These properties can be manipulated programmatically and are viewable within the user interface in a "Properties View", as shown in Fig. 3.

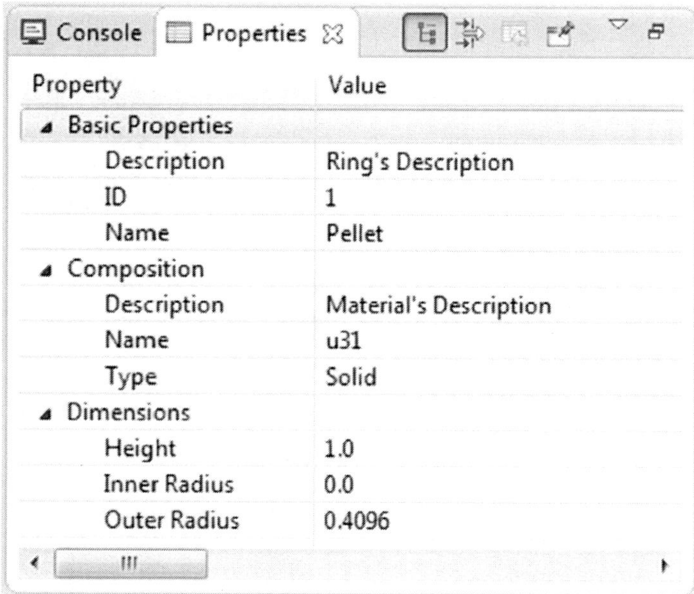

Figure. 3. A view that shows the properties of the selected part, including its composition and geometry.

Spatial data about a Part is stored by setting a *data provider* that includes *data* elements as described by requirement 2 in the Architecture section. Data providers are simply containers and are capable of managing arrays of data at different times. Data providers can be stored on any Part (as required by the IReactor Component interface). However, in practice, this can destroy performance, so some Parts manage the data providers for other Parts.

3.2. Cores

Reactor cores are the highest Parts in the hierarchy at present. The PressurizedWaterReactor class represents PWRs. This class extends a generic LWR base class that defines some properties and operations common across different types of LWRs, including both PWRs and boiling water reactors.[3] Assemblies of any type are stored in their own grids, so that one type of assembly can share the same position as another type, which allows for control rod ("spider") assemblies to be added. Control banks, fuel assemblies, in-core instruments, and "rod cluster" assemblies can be added to PWRs. Additionally, the fuel assembly pitch and the grid labels can be configured on the PressurizedWaterReactor class.

Fig. 4 shows the geometric view of a PWR core in the user interface for the 3a problem. The grid labels are read as configured on the PressurizedWaterReactor class and displayed along the top horizontal and left vertical axes. Each assembly is represented by a single square on the grid. The single green assembly in the middle is a 17 by 17 fuel assembly, and the yellow assemblies are control banks. The type of assembly shown can be configured using the "Assembly Type" dropdown button.

The SFReactor class represents SFRs. Its interface is nearly identical to that of the PressurizedWaterReactor class, but the allowed assembly types and properties are different. Fuel, control, reflector, shield, and test assemblies can be added to SFRs. As for PWRs, size can be obtained for SFRs; but instead of only an assembly pitch, SFRs offer both a lattice pitch and a flat-to-flat distance. Flat-to-flat distance is stored on the core for convenience, and the system assumes that all assemblies have the same flat-to-flat distance.

The graphical representation of an SFR with fuel assemblies is shown in Fig. 5. The view shows the hexagonal layout of the core and functions exactly like the view of the PWR core.

3.3. Assemblies

Multiple assembly types can be configured for both PWRs and SFRs. Each assembly is composed of a collection of rods, and each rod is mapped to a location on a grid. "Rods" in this case are not necessarily fuel rods. They could be control rods, for example. The data providers that contain state point data for the rods, of any type, are stored on the assemblies at a specified grid location instead of directly on the rods. This is done to optimize data access times and storage (although they can still be stored on the rods themselves if needed).

PWR assemblies can store their sizes and rod/pin pitches. SFR assemblies can store their size duct thicknesses in addition to their sizes.

SFRs make a distinction between a "pin" and a "rod", which is discussed in detail in Section 3.4. PWRs make no such distinction and refer to both simply as "rods".

PWRs support.

- **Fuel assemblies** of burnable fuel that is used in the reactor.
- **Control banks** used to regulate the power within the reactor.
- **Incore instruments** that represent assemblies configured with detectors or sensors within a core.
- **Rod cluster assemblies** of (typically neutron absorbing) rods placed in and moved between fuel assemblies during refueling outages.
- SFRs support
- **Fuel assemblies** of burnable (inner fuel, outer fuel) and blanket (optional) assemblies.
- **Control assemblies** that represent primary and secondary (shutdown) assemblies.
- **Reflector assemblies** that are configured to reflect neutrons.
- **Shield assemblies** for shielding against radiation.
- **Test assemblies** that represent assemblies used for testing materials and fuels

The geometry of one quarter of the PWR fuel assembly from Problem 3a is shown in Fig. 6. Rods are represented by circles, with control rods represented by blue circles and fuel rods represented by red circles. The blue spaces between rods represent the coolant. The rods are arranged according to the pitch and size configured on the assembly. The "main" view on the left represents the axial level (1 in the figure) selected using the slider, spinner, or clickable axial view on the right. Like the core view, the grid labels are taken on, configured on, and read from the assembly and displayed on the top horizontal and left vertical axes.

The tool bar can be used to switch this view from the geometric configuration to a view of the simulation data, as depicted in Fig. 7. The axial pin powers for Problem 3a for the entire fuel assembly are shown in this view. Each square represents a rod, and the color mapping is bluer for lower values and redder for higher values. (The circles from the geometric view have been replaced with squares to make it easier to show the data values.) The color mapping in this figure was normalized for the selected axial level (level 28), but it can also be set relative to the whole assembly or all assemblies.

Figs. 8 and 9 show the same views for SFRs. Both views have different grids from their PWR counterparts, but the same principles apply. Each circle in the geometric view represents a fuel pin in the SFR assembly and each hexagon in the data view represents the same fuel pin.

The type of data displayed in the data view, for either reactor type, can be switched with the "Data Feature" button. For the PWR example in Problem 3a, both axial and total pin power data.

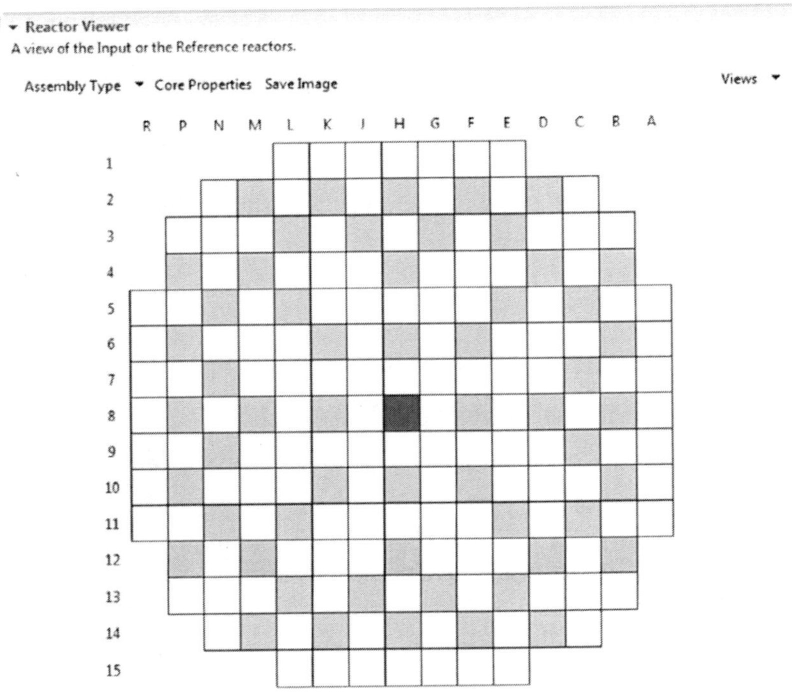

Figure. 4. A view of a pressurized water reactor core with one fuel assembly (green) and many control banks (yellow). (For interpretation of the references to color in this figure legend, the reader is referred to the web version of this article.)

The type of data displayed in the data view, for either reactor type, can be switched with the "Data Feature" button. For the PWR example in Problem 3a, both axial and total pin power data are available and can be selected. Any data stored on the assembly in a data provider will appear in the list and can be selected.

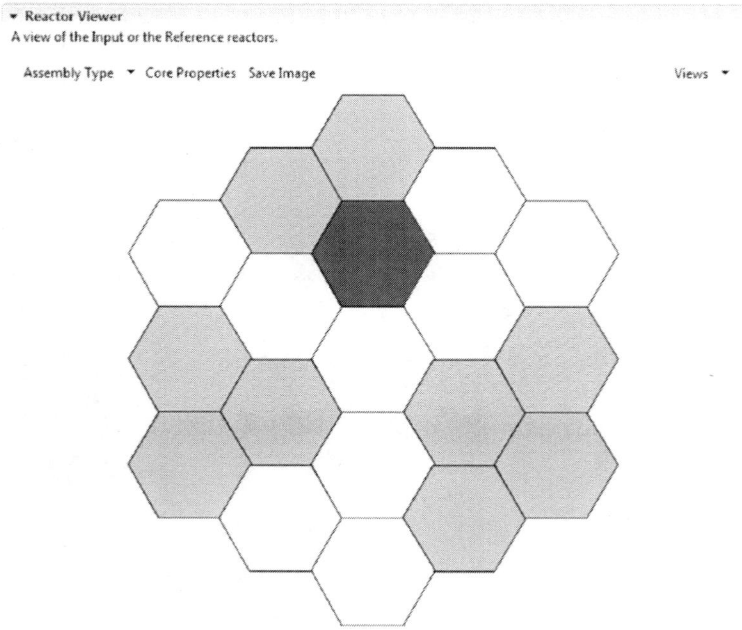

Figure. 5. A view of a small sodium-cooled fast reactor core with fuel assemblies in yellow and green (selected). (For interpretation of the references to color in this figure legend, the reader is referred to the web version of this article.)

3.4. Pins and rods

PWR assemblies are filled with rods made of either fuel or poisons or, in some cases, simply empty. The structure of a rod is defined by a collection of "material blocks" that describe the materials in a rod between two points along its axis. Material blocks are axisymmetric and are composed of concentric rings of materials. For example, a material block in a PWR fuel rod has a fuel ring, a fill gas ring, and a cladding ring. The clad and fill gas are typically defined separately from the material blocks for the fuels. Each rod can also store a pressure.

Each ring is defined by an inner and outer radius, a height, and a material type. Materials can be gases, liquids, or solids and can be labeled to match the materials in the simulation.

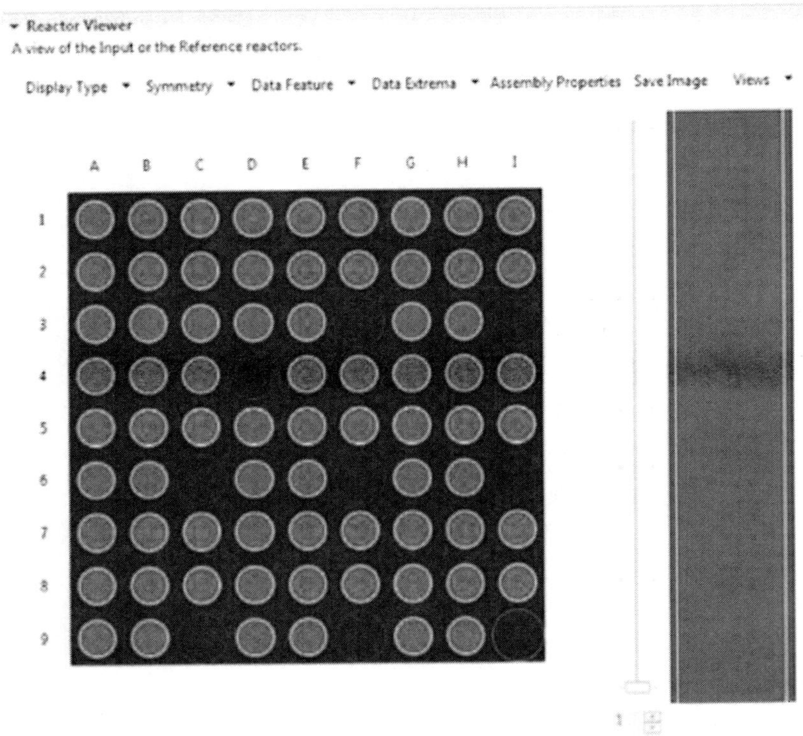

Figure. 6. A geometric representation of one quarter of a pressurized water reactor fuel assembly with control rods in blue and fuel pins in red. (For interpretation of the references to color in this figure legend, the reader is referred to the web version of this article.)

The SFR capability makes a further distinction between "pins" and "rods". In SFRs, the pin is the basic unit of fuel assemblies and control assemblies. Rods are used only in reflector assemblies. Pins are identical to rods in PWRs.

Fig. 10 shows the geometry of a fuel rod in Problem 3a. The rings of material are represented by the concentric circles of red, yellow, and green areas for the fuel, fill gas, and clad, respectively. Similar to the view for assemblies, the axial level can be adjusted using the slider, the spinner, or the clickable axial view. The data view for this rod showing the pin power at the 42nd axial level is shown in Fig. 11. The fuel is homogeneous in Problem 3a, so only one material block and ring are shown.

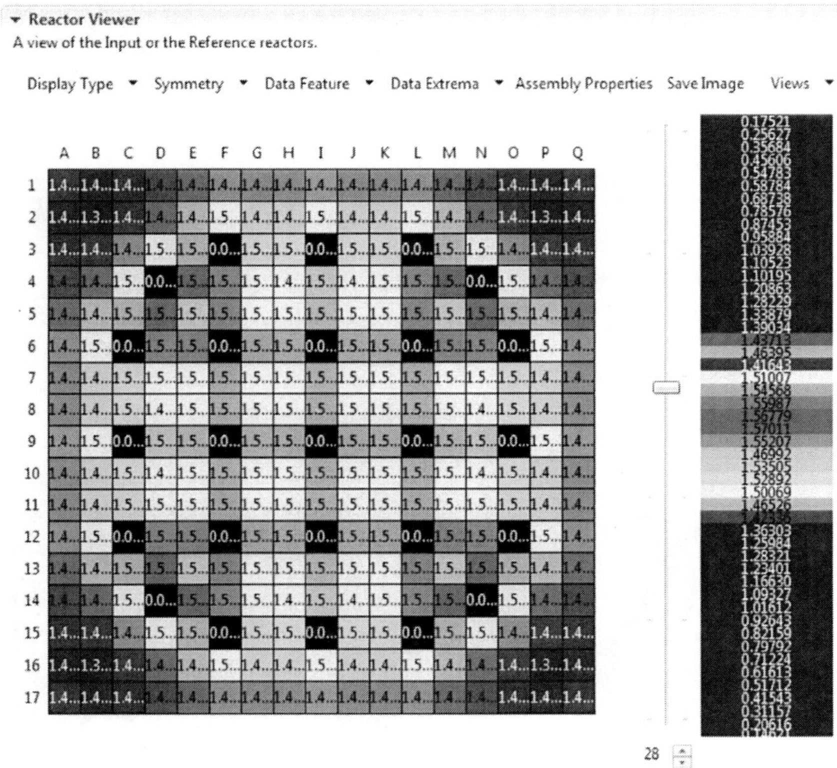

Figure. 7. Axial power mapped onto an assembly from a pressurized water reactor with areas of higher power in red and lower power in blue. (For interpretation of the references to color in this figure legend, the reader is referred to the web version of this article.)

The views for SFR pins are not shown because they are, for the most part, identical to the views for PWRs.

3.5. Plots and comparisons

The user interface provides a limited plotting capability for information stored on assemblies and pins/rods. This allows users to make quantitative comparisons quickly and easily. Plots of the axial pin power for select fuel pins in Problem 3a are shown in Figs. 12 and 13. The height from the bottom of the pin is shown on the horizontal axis, and the axial pin power on the vertical axis.

A plot showing the normalized percentage differences between the axial powers from VERA and KENO for Problem 3a is shown in Fig. 14. This plot was created using the external analysis routine described in 2.3 and shows very close agreement between the two codes.

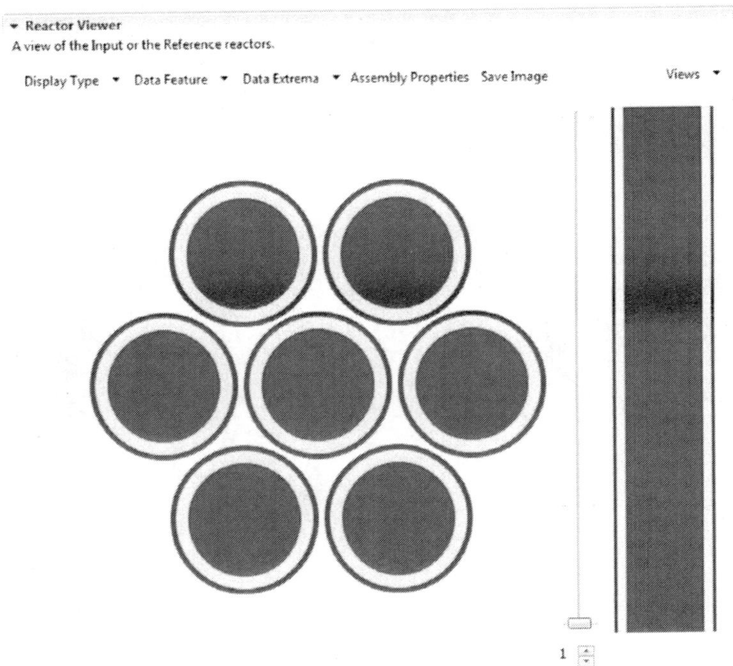

Figure. 8. A geometric representation of a seven-pin sodium-cooled fast reactor assembly completely composed of fuel rods.

4. AVAILABILITY AND ONGOING WORK

Source code and fully compiled binaries of this system are available as part of NiCE at https://github.com/eclipse/ice and http://niceproject.sourceforge.net, respectively. The system is referred to as the "Reactor Analyzer" in NiCE, and several detailed tutorials are available at the same website. Detailed source code documentation is also available. Future versions of the system will be available in NiCE, but in the near future that project will be "converted" into a new project at the Eclipse Foundation called the "Eclipse Integrated Computational Environment". The system will then be available at http://www.eclipse.org. The current version is available under the Eclipse Public License and will soon also be available under the Modified BSD License.

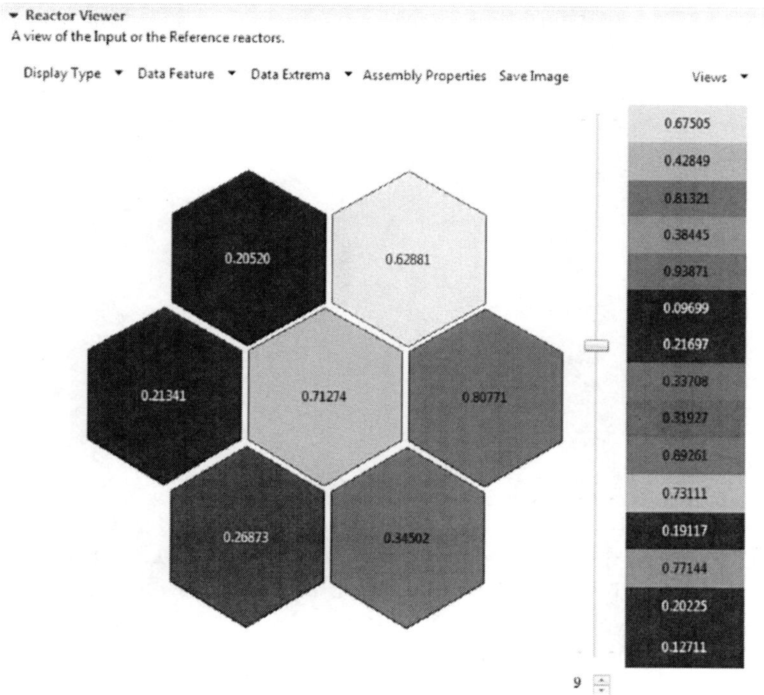

Figure. 9. A view of random data mapped to the seven-pin sodium-cooled fast reactor assembly to show the color mapping capability. (For interpretation of the references to color in this figure legend, the reader is referred to the web version of this article.)

Readers are encouraged to watch the recorded demonstrations published on YouTube.com at http://www.youtube.com/ jayjaybillings.

Detailed Unified Modeling Language models were developed as part of this work; they are also available for download, although the format is proprietary. The authors will gladly export the Unified Modeling Language models into a more friendly format upon request.

The authors are open to feedback and contributions from readers. Those interested in using or contributing to the work through testing, design improvements, or source code development should contact the corresponding author.

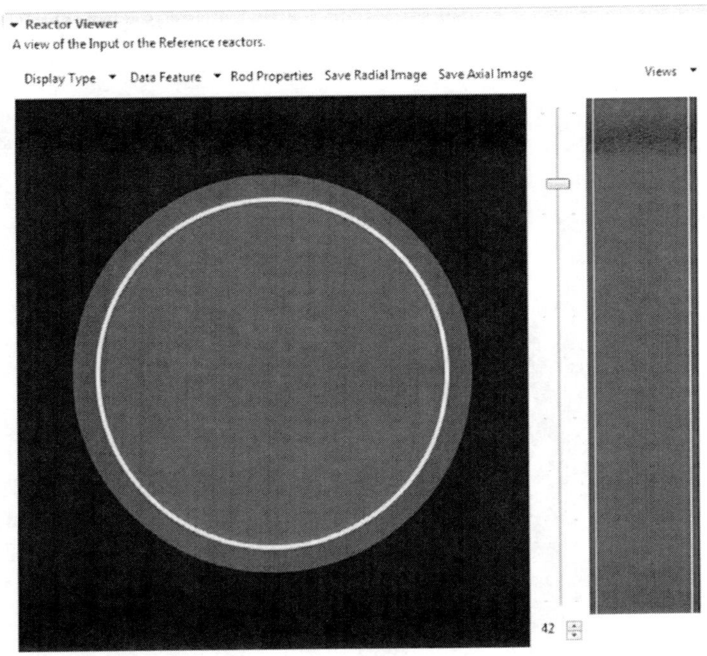

Figure. 10. A geometric representation of a single fuel pin in a pressurized water reactor with red, yellow, and green rings for the fuel, fill gas, and cladding, respectively. (For interpretation of the references to color in this figure legend, the reader is referred to the web version of this article.)

4.1. Integration and coupling

There are two primary modes of coupling and integrating this system with existing workflows for simulators that are written in C/C++ or Fortran. The first it to directly link the code into the target executable and call the application programming interface directly during the simulation. This is the preferred option. The second option is to write a translation code that acts as a post-processor. The post-processor would read the output of the simulation and then make the necessary translation and insertion into the virtual reactor.

A good example of the first case is available in the "examples" directory of the repository. The examples can easily be modified to read in data and write it out to support the second case.

Coupling to the system in Java is extremely straightforward and is detailed in many test cases, including the DenovoReactorFactory. java file in the org.eclipse.ice.client.widgets.reactoreditor.tests bundle.

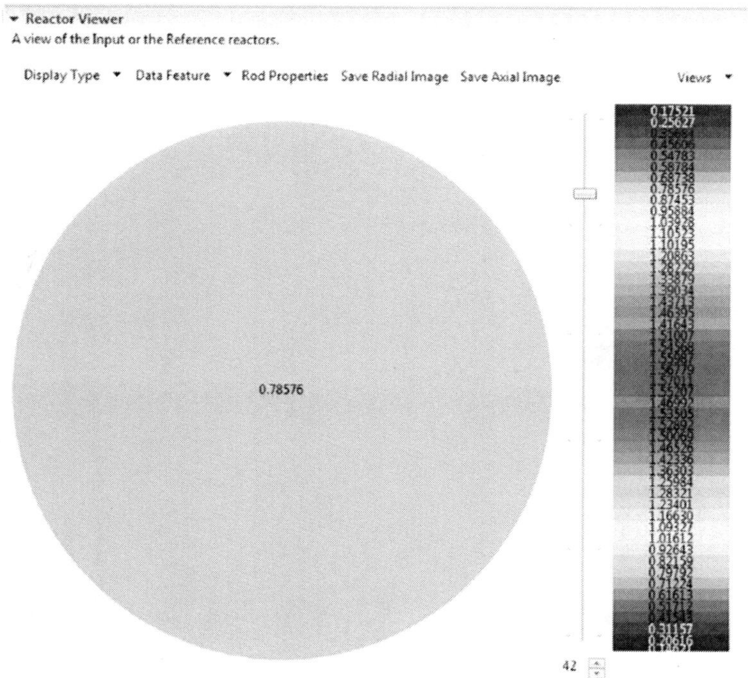

Figure. 11. A view of the axial power for this pin at the selected axial level.

4.2. Limitations

This system is relatively new and has several limitations. It remains highly useful considering these limitations and will only improve with time as they are addressed and other improvements added.

Performance has not been tested with very large amounts of data and many time steps. Scaling to a full core based on the 3a problem has performed well, after work to optimize the code; but even at the size of a full core, Problem 3a is not very big. It is only resolved at 49 axial levels with two sets of state point data and five distinct rod types.

The system has not been tested with real data from SFR simulations. Although the authors do not expect that this will present any issues, that remains a possibility.

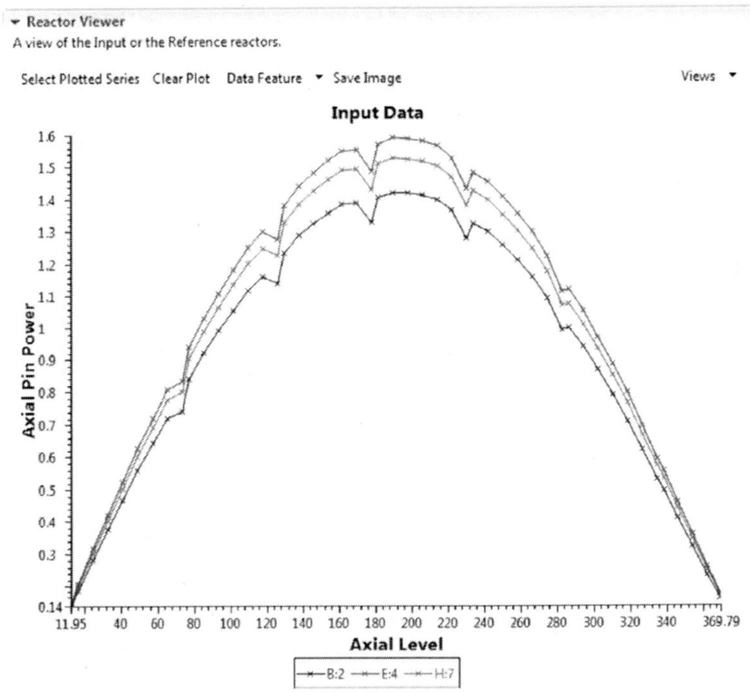

Figure. 12. A graph of the axial pin power across the B2, E4, and H7 pins in the assembly for the "input" simulation performed with VERA.

The system needs to be modified to include more of the regular properties of Parts. For example, the heights of rods and pins are determined by the total heights of their material blocks, but it needs to be possible to retrieve this value more easily. Plenum gas has not been considered.

4.3. Planned future work

There are many possible refinements to this system and a large amount of upcoming work. The most immediate refinement will be the addition of components to represent the pieces of a nuclear plant, such as generator, pipes, and other "plant-level elements". As previously mentioned, it is also important to test the system for much larger amounts of data and with the parallel I/O capabilities natively available in the HDF5 library.

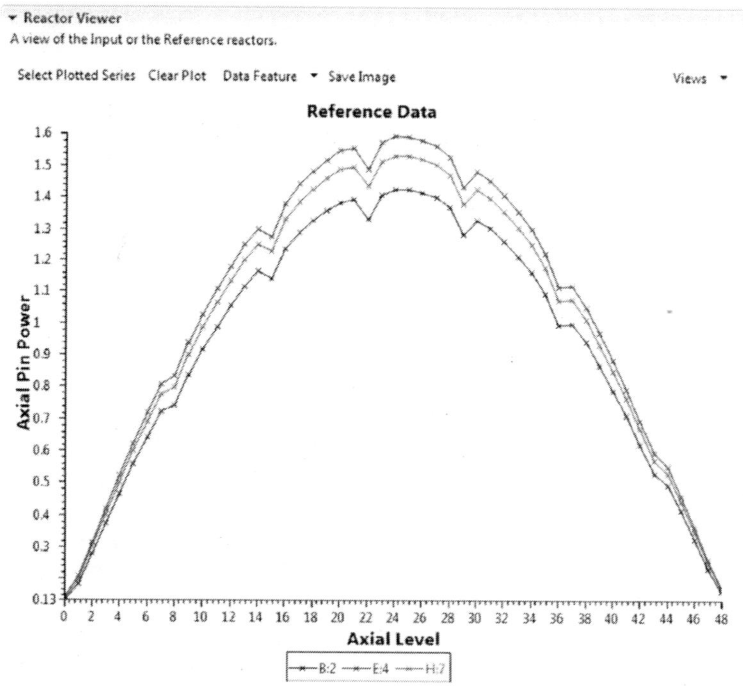

Figure. 13. A graph of the axial pin power across the B2, E4, and H7 pins in the assembly for the"reference" simulation performed with KENO.

Extensions to more languages are planned, including bindings for C, Fortran, and Python. Each of these will be a wrapper around the C++ version, not a stand-alone implementation.

The authors are currently working with members of the SHARP team (Siegal et al., 2007) to test the system with real data from simulations of SFRs.

The model for pins and rods will be improved in the near future after tests with BISON (Williamson et al., 2012) later this year.

Extending the properties available in Parts is straightforward, and the authors are working with collaborators to extend the set of properties for each Part.

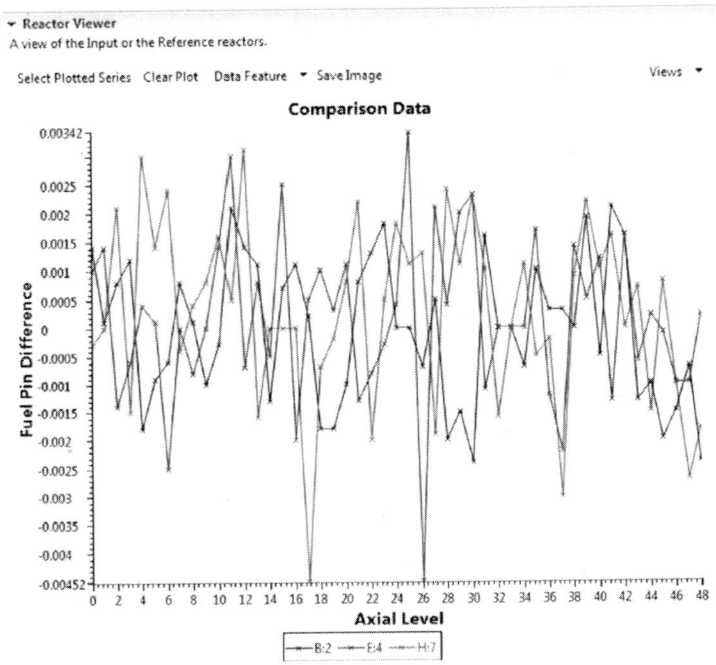

Figure. 14. A graph showing the percentage difference in the axial pin powers for the B2, E4, and H7 pins in the assembly.

5. CONCLUSIONS

Future simulations of new nuclear reactor designs will require new ways to examine the results because of the high fidelity and resolution inherent in state-of-the-art simulation codes. The system presented herein is capable of reducing the analysis burden on both users and developers by organizing the results in an intuitive ,domain-specific way and providing easy-to-use I/O capabilities and a user interface. Its application to a real-world problem with a 17 by 17 PWR fuel assembly from a VERA benchmark problem was shown by generating plots of the percentage difference between axial powers from VERA and KENO. Work remains to cover all of the different parts of LWRs and SFRs in sufficient detail for widespread use, as well as to optimize the system for truly large amounts of data in time and space.

Addressing the "data problem" will be critical to the success of the new modeling and simulation capabilities in development. Certainly, it is necessary to provide at least some streamlined capability to examine the large amounts of data coming from these simulations and, ideally, to make it possible to discover interesting new physics in the results through data mining and machine learning.

ACKNOWLEDGMENTS

The authors are grateful to members of the NEAMS and CASL communities, others whom we have interviewed, and our program managers. The authors are grateful for the assistance of Greg Davidson, Jess Gehin, Andrew Godfrey, Ugur Mertyurek, and John Turner from Oak Ridge National Laboratory (ORNL) and Justin Thomas from Argonne National Laboratory. The authors are also grateful for the financial support provided by the NEAMS, Advanced Reactor Concepts, and CASL programs.

The authors are especially thankful to David Pointer of ORNL for his assistance, guidance, and continued support during the performance of this work and the development of the paper.

This work has been supported by the US Department of Energy, Office of Nuclear Energy, and by the ORNL Postgraduate Research Participation Program, which is sponsored by ORNL and administered jointly by ORNL and the Oak Ridge Institute for Science and Education (ORISE). ORNL is managed by UT-Battelle, LLC, for the US Department of Energy under contract No. DE-AC05-00OR22725. ORISE is managed by Oak Ridge Associated Universities for the US Department of Energy under contract No.DE-AC05-00OR22750.

REFERENCES

1. Apache Software Foundation, Apache Mahout, 2011–2014. https://mahout.apache.org (2011–2014).
2. Billings, J.J., Belt, A.P., Guidry, M., Hull, S.F., Lingerfelt, E.J., Lyon, G.M., McCaskey, A.J.,Pokhriyal, N., Godfrey, A., Mertyurek, U., 2013. Simulate this! – The eclipse platform as an integrated computational environment. In: EclipseCon 2013.
3. Billings, J.J., Deyton, J., Hull, S.F., McCaskey, A.J., Patterson, T., Wojtowicz, A., 2014. Prevent three-eyed fish: analyze your nuclear reactor with eclipse. In: EclipseCon 2014.
4. Consortium for the Advanced Simulation of Light-water reactors web page, 2011. https://www.casl.gov (September 2011).
5. Gaston, D., Newman, C., Hansen, G., Lebrun-Grandie, D., 2009. Moose: a parallel computational framework for coupled systems of nonlinear equations. Nucl. Eng. Des. 239.
6. Godfrey, A., 2013. Vera core physics benchmark progression problem specifications, Tech. Rep. CASL-U-2014-0006-002, Oak Ridge National Laboratory, Oak Ridge, TN, USA (March 2013).
7. McAffer, J., Lemieux, J.-M., Aniszczyk, C., 2010. Eclipse Rich Client Platform, second ed. Addison-Wesley, Boston.
8. NiCE Project Page, 2011–2014. https://niceproject.sourceforge.net (April 2011– 2014).

9. Nuclear Energy Advanced Modeling and Simulation web page, 2011. http://www.ne.doe.gov/AdvModelingSimulation/program.html (September 2011).
10. ParaView Web Page, 2015. http://www.paraview.org (June 2015).
11. Pokhriyal, N., Billings, J.J., Mertyurek, U., Godfrey, A.T., 2013. Anomaly detection for high fidelity core simulations. In: Proc. 2013 American Nuclear Society Annual Meeting.
12. Rearden, B., 2005. SCALE: A Modular Code System for Performing Standardized Computer Analyses for Licensing Evaluation, ORNL/TM-2005/39 I-III, Oak Ridge National Laboratory.
13. Siegal, A., Taugtes, T., Caceres, A., Kaushik, D., Fischer, P., 2007. Software design ofsharp. In: Joint International Topical Meeting on Mathematics and Computation and Supercomputing in Nuclear Applications.
14. Tautges, T.J., Caceres, A., 2009. Scalable parallel solution coupling for multiphysics reactor simulation. In: Proc. SciDAC 2009 Journal of Physics: Conference Series.
15. The HDF Group, 1997–2014. Hierarchical Data Format, version 5, http://www.hdfgroup.org/HDF5/ (1997–2014).
16. Turner, J., Summers, R., Palmtag, S., 2013. Virtual environment for reactor applications (vera), Tech. Rep. CASL-U-2014-0006-000, Oak Ridge National Laboratory, Oak Ridge, TN, USA (March 2013).
17. VisIt Web Page, 2011. https://wci.llnl.gov/codes/visit/home.html (September 2011).
18. Williamson, R.L., Hales, J.D., Novascone, S.R., Tonks, M.R., Gaston, D.R., Permann, C.J., Andrs, D., Martineau, R.C., 2012. Multidimensional multiphysics simulation of nuclear fuel behavior. J. Nucl. Mater. 423

CHAPTER 14

Effect Analysis of Volume Fraction of Nanofluid Al_2O_3-Water on Natural Convection Heat Transfer Coefficient in Small Modular Reactor

Anwar Ilmar Ramadhan[1]*, Ery Diniardi[1], Rasma[2]

[1]*Mechanical Engineering Department, Faculty of Engineering, Universitas Muhammadiyah Jakarta, Jakarta, Indonesia*

[2]*Automotive and Heavy Equipment Department, Faculty of Engineering, Universitas Muhammadiyah Jakarta, Jakarta, Indonesia*

ABSTRACT

Development and use of nuclear energy is currently growing very rapidly, in order to achieve increasingly advanced technology, both in terms of design, economic factors and safety factors. Thermal-hydraulics aspects of nuclear reactors should be done with calculation and near-perfect condition. Including today began development of a nuclear reactor with low power below 300 MW, or commonly called the Small Modular Reactor (SMR). One is CAREM-25 developed by Argentina with a power of 25 MW, where in CAREM already using natural circulation system and the use of nanofluid as coolant fluid. In this research, analytic modeling of thermal-hydraulics nuclear reactor SMR CAREM-25, when the nanofluid Al_2O_3-Water used as cooling fluid in the cooling system of a nuclear reactor.

Further to this analytic modeling will be done on CFD. Analytic modeling with CFD to determine the flow phenomena and distribution as well as the effect of nano-particles of Al_2O_3- Water based on the volume fraction (1% and 3%) of the coefficient of heat transfer by natural convection.

Keywords: Nanofluid, Nano-particles, natural Convection, Thermal-hydraulic, Coefficient ofheat transfer

1. INTRODUCTION

Safety is a major concern in the design, operation and development of a nuclear reactor. Therefore, the analysis method used in all these activities must be thorough and reliable so as to predict a wide range of operating conditions of the reactor, both under normal operating conditions and in the event of an accident [1]. In addition neutrons aspect, thermal-hydraulic aspect is an important aspect for the safety of the design and operation of a nuclear reactor. Magnitude thermal-hydraulic such as pressure, coolant flow rate and temperature of the fuel needs to be known through predictive calculation [2].

Currently begun development of nuclear power plant or nuclear power plants around the world by type of Small Modular Reactor (SMR) is a mini nuclear reactor with power below 300 MW. Where an excess of Small Modular Reactor (SMR) is more flexible in matters of design, the cheaper the price development and also the price of treatment, as well as the advantages of SMR did pooling system thermal-hydraulic of 2-loop system cooling into the cooling loop in other words the use of integral system loop on SMR. SMR cooling fluid using fluid (H_2O) in the reactor coolant loop integral system. Studies related to the use of other types of nanofluid as coolant into the study and research are interesting and important to optimize heat transfer in the SMR, becoming thermal-hydraulic important aspect in the cooling system in a nuclear reactor [3]

One study that is currently a research priority in the cooling system is the use of nano-particles that are mixed with a fluid to improve the performance of decision-calorific. Theoretically nanofluid including nanoparticles has high thermal conductivity value than ordinary light fluid, so as to absorb and transfer heat better. Buongiorno and his team at the Massachusetts Institute of Technology (MIT) in the United States of America has conducted research related to nanofluid, it has been proven that the value of Critical Heat Flux (CHF) nanofluid greater than fluid water [4].

This research was developed for the use of nanofluid as coolant fluid in a nuclear reactor with the type SMR CAREM-25, made in Argentina. The research activities focused on assessing the deeper aspects thermal-hydraulic that occurred in the cooling system of a nuclear reactor CAREM-25 using nanofluid Al_2O_3-Water as coolant fluid. So that later acquired the flow characteristics of natural circulation cooling system used by the SMR, heat generation process from the beginning, the temperature distribution in the fuel element, in order to obtain an analytical model for the safety performance of the reactor type CAREM SMR-25.

2. RESEARCH METHOD

2.1. Modeling of Core Nuclear Reactor of Small Modular Reactor (SMR)

It needs to be done first before doing the simulation process is to create a model that happens to sub-channel arrangement of hexagons. In this case the model is made is in the form of volume models. Simplifying assumptions made model that is considered a model sub-channel arrangement of hexagons similar to the SMR reactor core (Small Modular Reactor) type hexagonal and uniform heat flux generated. In the manufacture of this model using CAD program as modeling. And models to be simulated in this research are as follows in Figure 1.

2.2. Meshing

Meshing is the process by which the overall geometry is divided into small elements this will act as a control surface or volume in the calculation process and then each of these elements will be input to the element next to it. This will happen over and over again until the fullest domain. In meshing elements that would have been tailored to the needs and geometric shapes. In this thesis meshing application used is CAD. In this study all types of configuration elements are simulated using hybrid or tetrahedron elements. Below are images meshing with volume meshing configuration and size interval by 5 (Figure 2).

2.3. Numerical modeling with CFD

CFD is one type of CAD program that uses finite volume method. CFD mesh provides complete flexibility, so as to resolve the case of fluid flow with unstructured mesh even with a relatively easy way. CFD mesh types supported are the type of triangular-quadrilateral 2D, 3D tetrahedral-hexahedral-pyramid-wedge, and mesh mixture (hybrid). CFD also provides facilities to refine or enlarge an existing mesh. In general, the steps in performing analysis by using CFD.

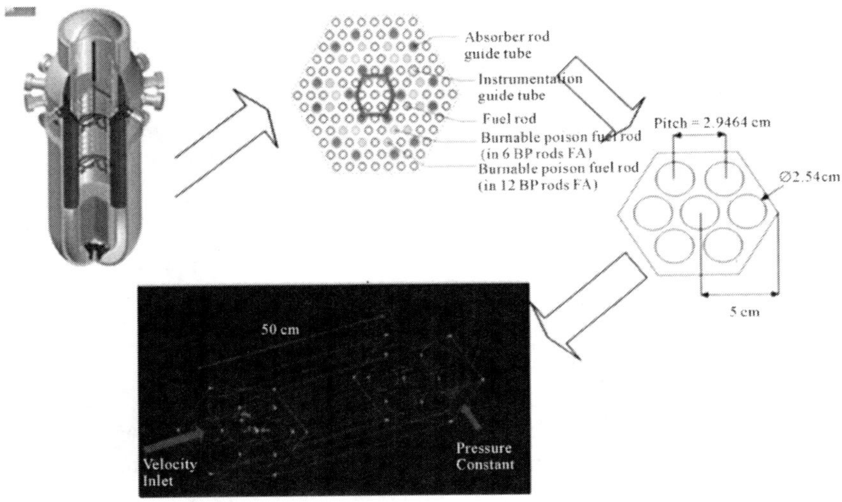

Figure 1. Modeling the SMR reactor sub channel of hexagon.

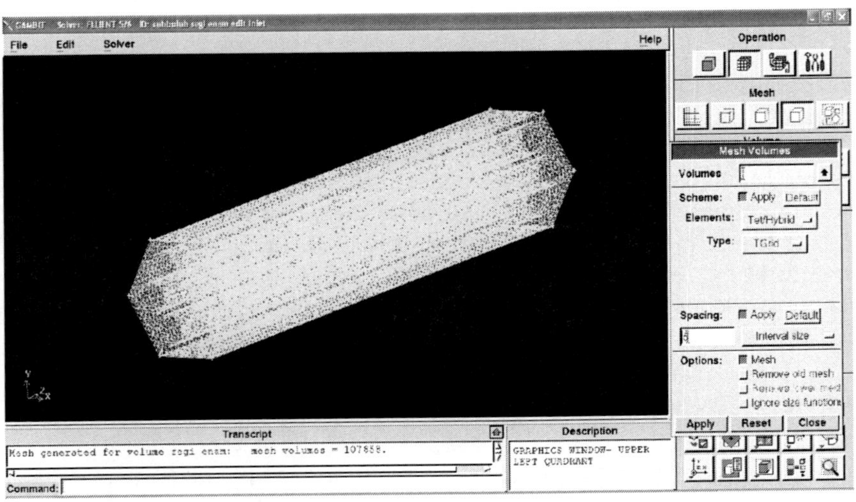

Figure 2. Results meshing in CAD.

2.4. Input Parameters on CFD

Numerical modeling in this study is done by using the CFD. Numerical simulations of the volume set have been created with the CAD program carried out for steady state. In the numerical modeling, it takes some assumptions, among others:

- Heat conduction to the walls inside and outside the sub model of channel hexagonal ignored.
- Radiation heat transfer from the walls inside and outside the sub model of channel hexagonal ignored.

The stages are carried out in numerical modeling is as follows:

1) Defining of model

  - Solver

In this experiment, based pressure because the flow is analyzed incompressible at low velocity.

  - • Energy

Energy equation must be enabled in order to model the heat transfer process that occurs in the system.

2) Determine the properties of the material

The type of fluid used is nanofluid. Nanofluid physical properties such as density, viscosity, thermal conductivity, and thermal types are modeled as a function of temperature (**Table 1**).

3) Determine the operating conditions

The influence of the gravitational acceleration is also included in the system with values 9.8 m/s. Gauge Pressure = 0 Pa.

4) Determine of boundary condition

In the numerical modeling, setting the boundary conditions on the volume set is done by inserting a quantitative value of the parameters related to the type of boundary, such as a large heat flux on the surface of the heating cylinder. Configuration of heat on the surface approximated by a constant heat flux. The values of the parameters as set the volume boundary conditions are shown in **Table 2**.

3. RESULTS AND DISCUSSIONS

3.1. Velocity Flow Characteristics on Nuclear Reactor Core of Small Modular Nuclear Reactor

Characteristic patterns of movement flow of nanofluid as coolant fluid by convection naturally occurring in a nuclear reactor core types Small Modular Reactor (SMR) with sub channel hexagon, can be seen in **Figures 3-5**.

Figure 3 shows the condition of nanofluid of Al2O3-Water treatment for the condition is given heat flux of 1000 W/m2 with a volume fraction of 1%, and for **Figure 4**. A nanofluid of Al2O3-Water with the volume fraction of 3%, and for **Figure 5** shows a comparison of cooling fluid usually used in nuclear reactor cooling water is H_2O. **Figure 3** and **Figure 5** is a flow velocity contour of nanofluid of Al_2O_3-Water and fluid H_2O water at the center position of the sub channel on the hexagon in a nuclear reactor core types Small Modular Reactor (SMR).

Table 1. Properties of nanofluid Al_2O_3-Water [5].

Properties	Al_2O_3-Water (1%)	Al_2O_3-Water (3%)
Density [kg/m^3]	1021.7	1073.8
C_p (Heat Capacity) [J/kgK]	4.149	4.081
Thermal Conductivity [W/mK]	0.620	0.656
Viscosity [kg/ms]	8.17×10^{-4}	8.56×10^{-4}

Table 2. Input parameters values boundary conditions.

Type of Boundary	Parameter	Value
Inlet	Velocity inlet	0.1 m/s
	Temperature	300 K
Outlet	Gauge Pressure	0 Pa
	Backflow temperature	300 K
Wall of Cylinder	Heat Flux	100; 1000 W/m^2

The picture shows the pattern on the flow velocity in the direction of the z-axis (from bottom to top) shows a decrease in the velocity of the movement began in the early flow of 0.1 m/s decline at the end of the sub channel amounted to 0.083 m/s. This proves that what happened is the natural convection, i.e. the heat transfer due to the influence of the buoyant force of the fluid without being influenced by the velocity. And, also shows that the flow velocity profile of each nanofluid and also an increase in water fluid flow velocity at the beginning is marked in yellow as the initial velocity is given, then decreased with the simulation results are marked in light blue.

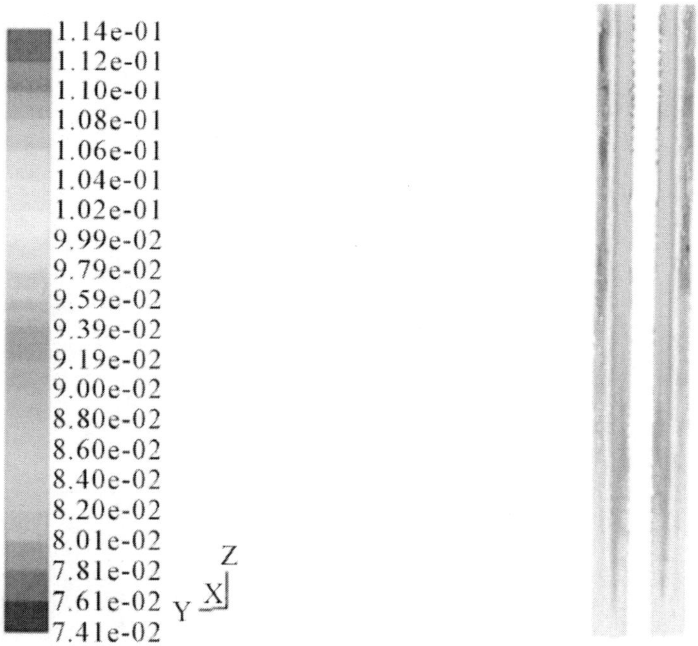

Figure 3. Contour of fluid flow of nanofluid Al_2O_3-Water (1%) with heat flux of 1000 W/m^2.

3.2. Cylinder Wall Temperature Distribution at the Sub channel on Nuclear Reactor Core of Small Modular Reactor

Furthermore, after knowing the flow patterns that occur on the patio nuclear reactor types Small Modular Reactor (SMR) with the arrangement of sub channel hexagon, can be analyzed temperature distribution on the cylinder walls of fuel when given the fluid coolant is nanofluid of Al_2O_3-Water and fluid water H_2O, can be seen in Figures 6-8 the following:

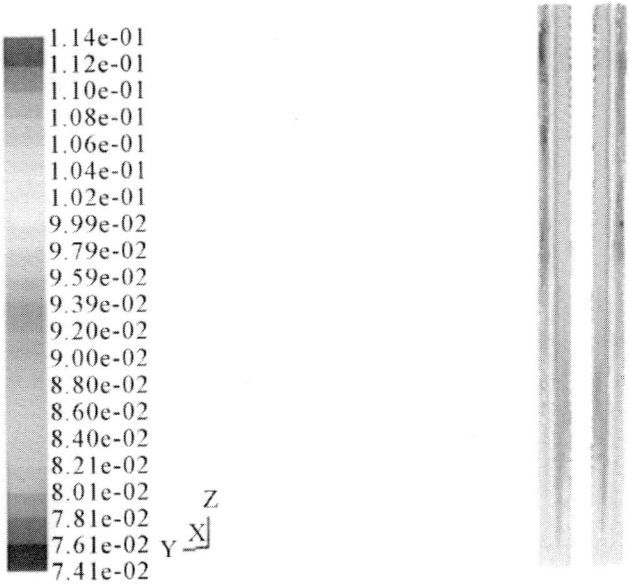

Figure 4. Contour of fluid flow of nanofluid Al_2O_3-Water (3%) with heat flux of 1000 W/m^2.

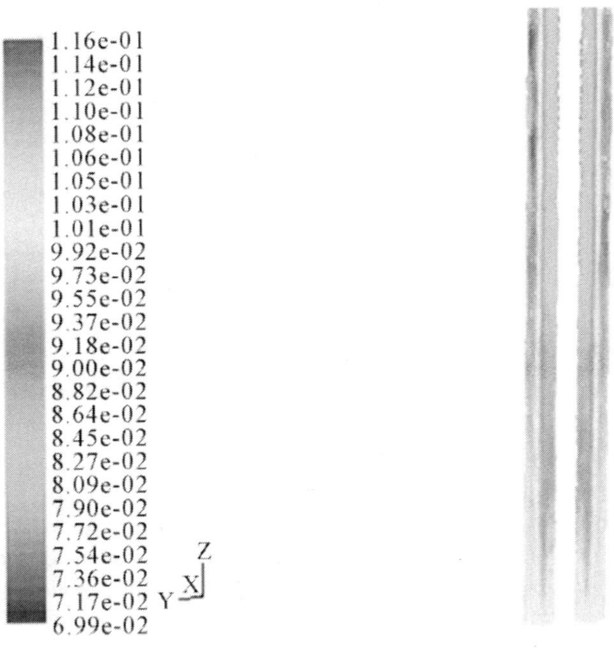

Figure 5. Contour of fluid flow of fluid flow of H_2O- with heat flux of 1000 W/m^2.

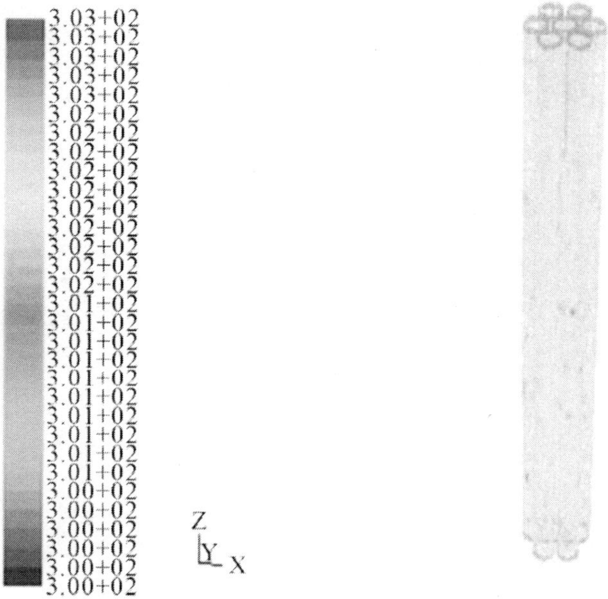

Figure 6. Contour of temperature distribution of nanofluid Al_2O_3-Water (1%) with heat flux of 1000 W/m^2.

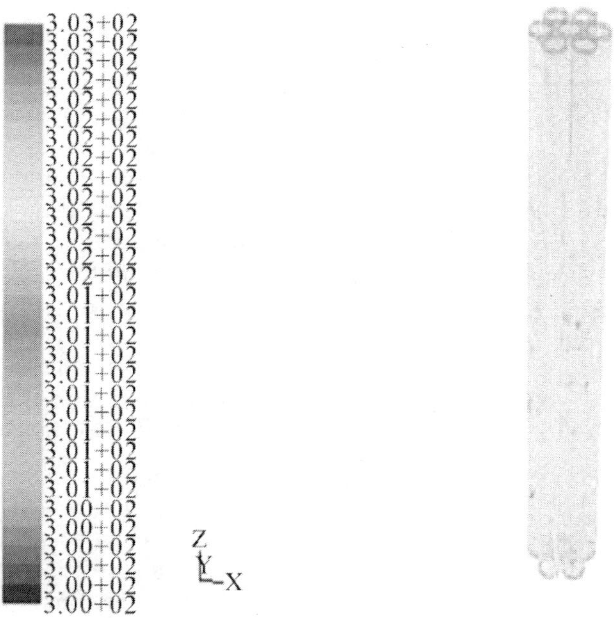

Figure 7. Contour of temperature distribution of nanofluid Al_2O_3-Water (3%) with heat flux of 1000 W/m^2.

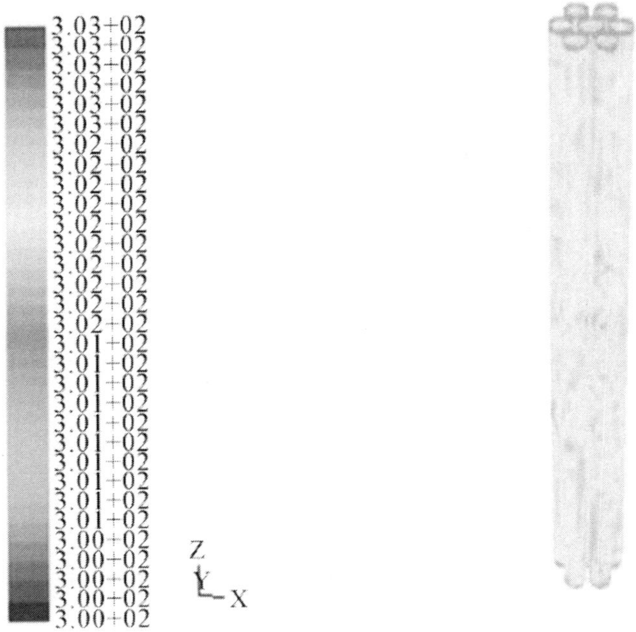

Figure 8. Contour of temperature distribution of H_2O with heat flux of 1000 W/m^2.

Figures 6-8, the contours of the temperature distribution of each of the cooling fluid, i.e. nanofluid of Al_2O_3- Water and fluid H_2O water under conditions of heat flux of 1000 W/m^2 to 7 fuel elements in the composition of the sub channel hexagon in a nuclear reactor core Small Modular Reactor (SMR). From Figure above shows that the distribution pattern of temperature in each cooling fluid have similar temperature distribution in the cylinder element of the fuel, which is the spread of the temperature on the cylinder early in the direction of the z-axis (at 0 m to 0.5 m) will increase significantly and approach the linear pattern. Therefore, when made a Curve of the temperature distribution of the cylinder wall to the position (m) for each of the cooling fluid with the condition of the heat flux of 100 W/m^2 and 1000 W/m^2, can be seen in Figure 9 and Figure 10.

Figure 9 and Figure 10 shows for each cooling fluid that nanofluid Al_2O_3-Water and fluid water H_2O with each condition of the heat flux in the cylinder fuel elements show patterns upward trend in line with the position of the cylinder, starting at position 0 m to the position 0.5 m (end cylindrical fuel elements). And, showed also that nanofluids have a wall temperature is smaller than the fluid water (H_2O). This proves that the given velocity of 0.1 m/s did not significantly affect the temperature of the wall of the cylinder fuel elements. Due also the factor of the thermal properties of nanofluid of Al_2O_3-Water higher than the fluid water. This proves that the results obtained by Pandey, et al., as shown in Figure 11.

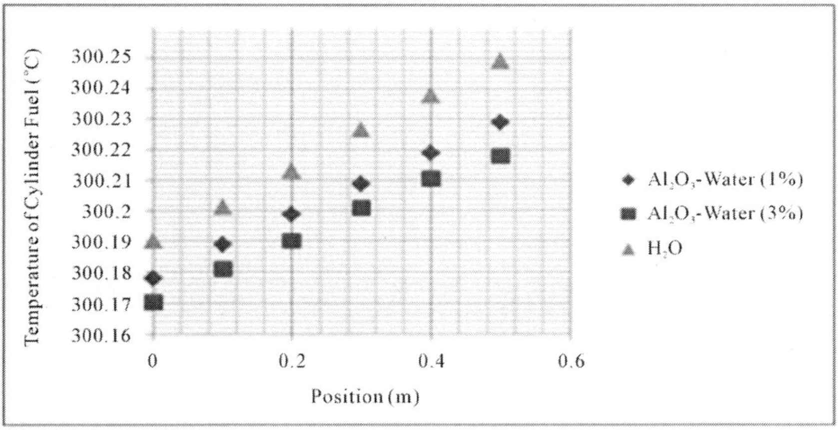

Figure 9. Relationship massive cylinder wall temperature distribution of the position (m) in the sub channel hexagon in nuclear reactor core SMR for nanofluid of Al_2O_3-Water and Water H_2O with a heat flux of 100 W/m^2.

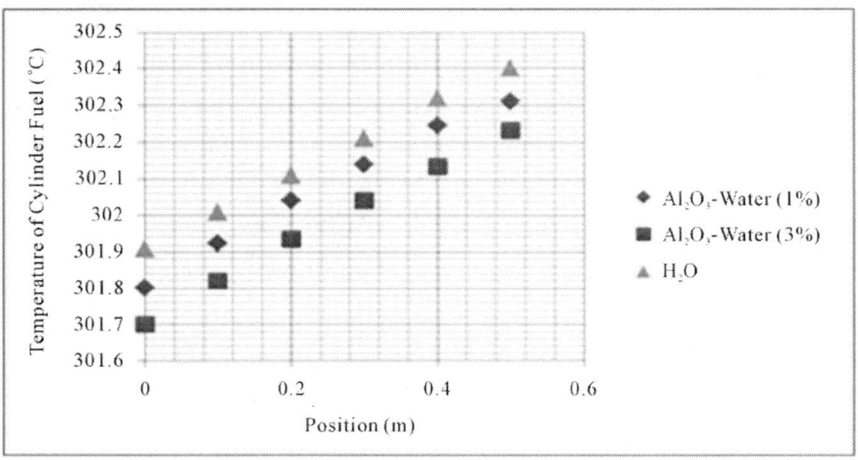

Figure 10. The relationship on the distribution of the temperature of the cylinder wall to the position (m) in the sub channel hexagon in nuclear reactor core SMR for nanofluid of Al_2O_3-Water and H_2O with a heat flux of 1000 W/m^2.

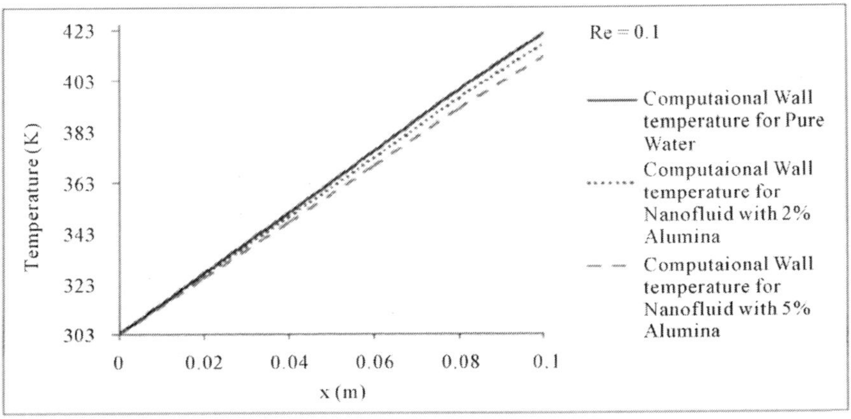

Figure 11. The relationship of the distribution of the wall temperature of the position (m) in the micro channel [6].

By comparing the Curves obtained between Figures 9-11 showed the same pattern, namely the fluid light water has a value of wall temperature greater than the temperature of the wall for nanofluid of Al_2O_3-Water, it is influenced by the movement of the flow velocity in the sub channel so it looks boosts the pattern of the distribution of temperature in the cylinder wall. Furthermore, it can be seen the results of numerical simulation on natural convection heat transfer coefficient by using CFD is as follows:

Figure 12 explains that for nanofluid of Al_2O_3-Water (1%) and (3%) had the same pattern of the static pressure, which decreased pressure based on the effect on the point position of cylinder fuel elements in the model.

Figure 13 shows that nanofluid of Al_2O_3-Water (1%) and (3%) have a static relationship between the temperature of the fuel element cylinder position on raising test model suffered linearly on each of the fuel elements.

Figure 14 shows that nanofluid of Al_2O_3-Water (1%) and (3%) had the relationship between calorific heat transfer coefficient against the position of the cylinder fuel elements in the test model has decreased in each of the fuel elements.

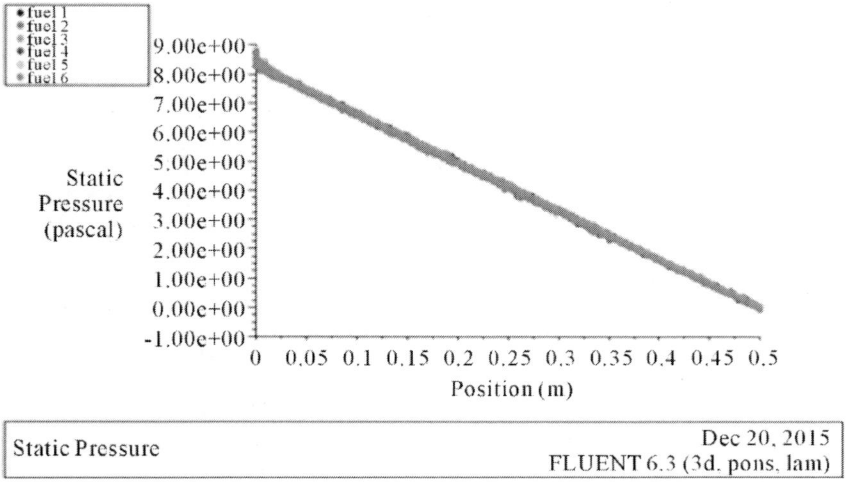

(a)

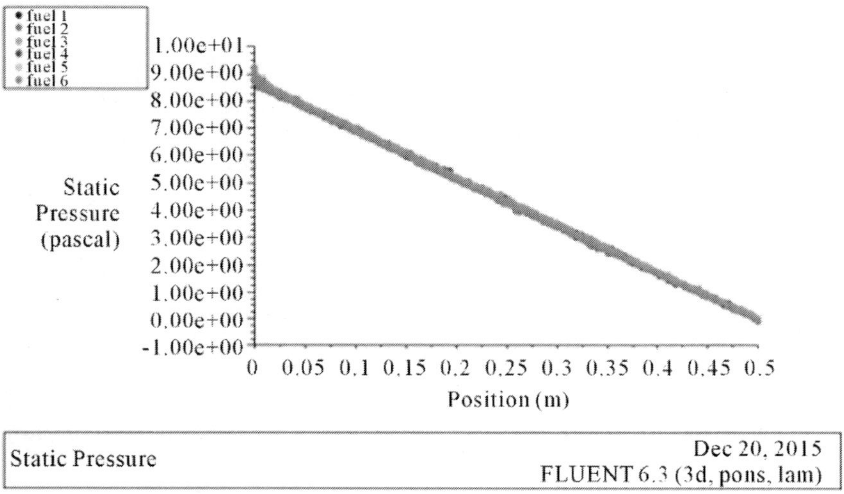

(b)

Figure 12. Curve the relationship between the static pressure of the fuel element positions. (a) Nanofluid Al_2O_3-Water (1%); (b) Nanofluid Al_2O_3-Water (3%).

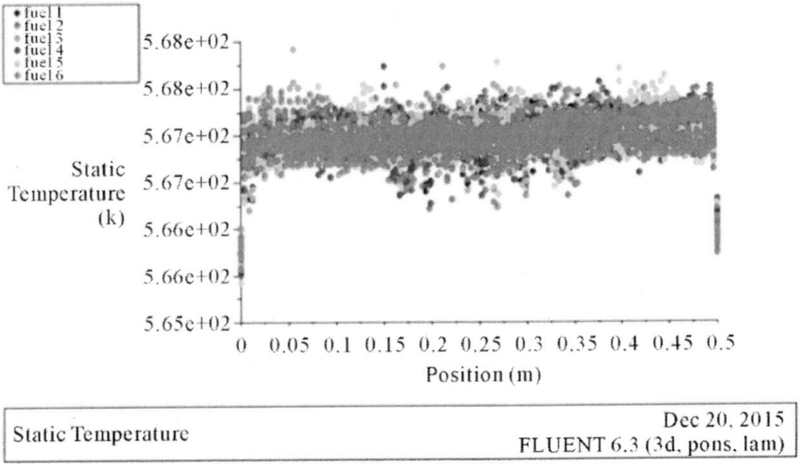

(a)

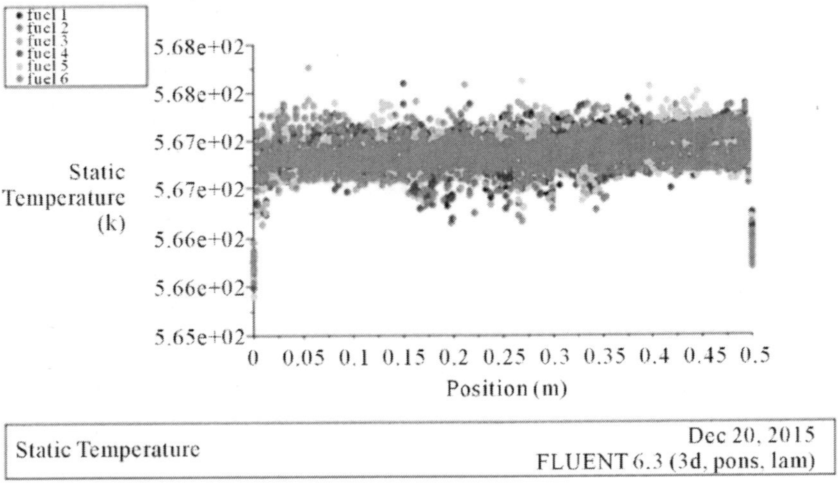

(b)

Figure 13. Curve the relationship between the static temperature of the fuel element positions. (a) Nanofluid Al_2O_3-Water (1%); (b) Nanofluid Al_2O_3-Water (3%).

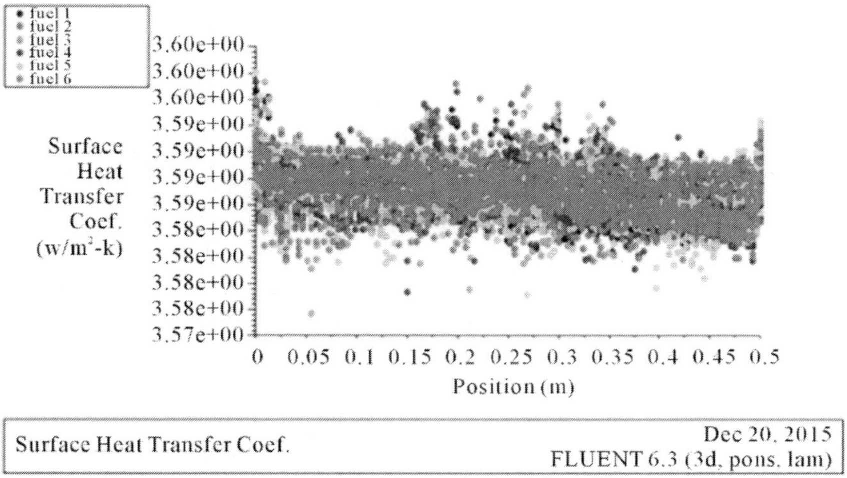

(a)

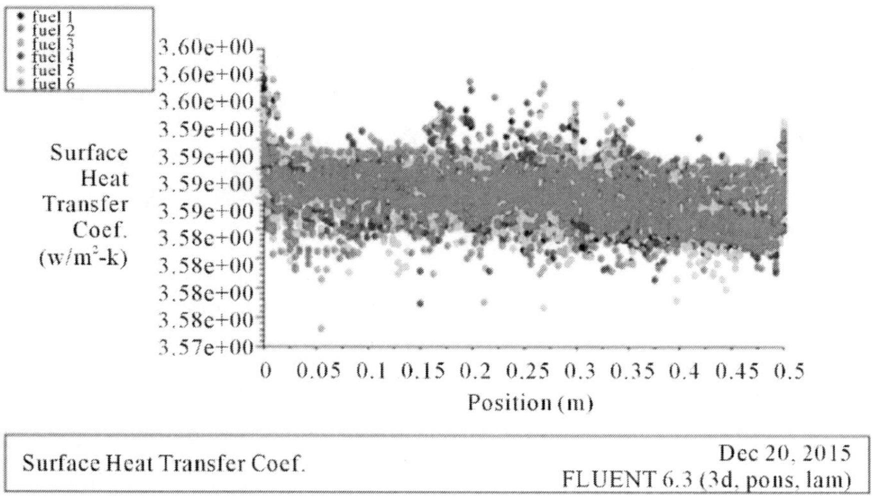

(b)

Figure 14. Curve the relationship between the coefficient of heat transfer to the position of fuel elements. (a) Nanofluid Al_2O_3-Water (1%); (b) Nanofluid Al_2O_3-Water (3%).

4. CONCLUSIONS

From these results provide the following conclusions:

1) The pattern of the spread of the flow rate in each cooling fluid (H_2O) and nanofluid of Al_2O_3-Water (1% and 3%) had a similar pattern, which is beginning to experience a great velocity, along with the altitude is naturally decreasing velocity tip flow cylinder. So what happens is proving convection natural convection.

2) Distribution of temperature on the wall on the analysis of heat transfer occurs in sub channel hexagon on Small Modular Reactor (SMR) shows a tendency to rise significantly and approached the rise linearly in the direction of the z-axis position, and proves also that water has a value of fluid temperature distribution wall larger than nanofluid of Al_2O_3-Water (1% and 3%).

3) The results of numerical simulation using saw that for the coefficient of heat transfer by natural convection is affected by the value of the volume fraction of the nanoparticles in nanofluid of Al_2O_3-Water, a value of 3% volume fraction has a heat transfer coefficient greater than the volume fraction of 1%.

ACKNOWLEDGEMENTS

The authors would like to thank PAKARTI (Center of Affiliate for Study and Research Technology) Faculty of Engineering, Universitas Muhammadiyah Jakarta, which has provided a research grant to develop this research in 2015.

REFERENCES

1. Puradwi and Juarsa, M. (2009) Literature Review of Nuclear Reactor Safety Innovation through Technology Use as a Nano Fluid as Cooling Fluid. Journal of Sigma Epsilon, 13, 49-53.
2. Ramadhan, A.I. and Diniardi, E. (2014) Analytical Study of Velocity Flow and Temperature Distribution of Fuel Elements Cylinder in Nuclear Reactor Core. National Symposium on Applied Technology (SNTT), Surabaya.
3. Gimenez., M.O. (2011) CAREM Technical Aspects, Project and Licensing Status. Interregional Workshop on Advanced Nuclear Reactor Technology, Vienna.
4. Buongiorno, et al. (2008) Alumina Nanoparticles Enhance the Flow Boiling Critical Heat Flux of Water at Low Pressure. Jurnal of Heat Transfer, 130.

5. Ramadhan, A.I., Lasman, A.N., Septilarso, A. (2015) Correlation Equations of Heat Transfer in Nanofluid Al2O3-Water as Cooling Fluid in a Rectangular Sub Channel Based CFD Code. International Journal of Science and Engineering, 8, 120-124.
6. Pandey, A.K. (2011) A Computational Fluid Dynamics Study of Fluid Flow and Heat Transfer in a Micro Channel. Thessis Programme Master, National Institute of Technology Rourkela, India.

CHAPTER 15

On the Nature of the Average Transverse Momenta of the Low-p_T Secondaries in Some High Energy Nuclear Collisions

Amar Chandra Das Ghosh[1], Goutam Sau[2], Pradeepta Guptaroy[3], Subrata Bhattacharyya[4]

[1]*Department of Microbiology, Surendranath College, Kolkata, India*

[2]*Beramara RamChandrapur High School, South 24 Parganas, India*

[3]*Department of Physics, Raghunathpur College, Raghunathpur, India*

[4]*Physics and Applied Mathematics Unit (PAMU), Indian Statistical Institute, Kolkata, India*

ABSTRACT

That the values of average transverse momenta $(\langle p_T \rangle)$ of the secondaries produced in high energy collisions rise very slowly with energy is modestly well-known and accepted. We would like to probe into this aspect of the problem for production of the main variety of the "soft" secondaries in two high energy symmetric nuclear collisions with the help of two non-QCD models. Our model-based results are found to be quite consistent with the anticipated behaviours and also with the observations.

Keywords: Relativistic Heavy Ion Collision; Inclusive Production with Identified Hadrons; Inclusive Cross-Section; Meson Production

1. INTRODUCTION

Amidst the observables measured by the high energy experiments, “average transverse momentum”, denoted normally by $\langle p_T \rangle$, is one of great prominence and importance, so much so, that it is, at times, juxtaposed at par with “average multiplicity” of the secondary particles detected in the high energy accelerators and colliders. Even in Cosmic ray phenomenon and at cosmic ray energies the observable has some special physical significance and bearing. This observable is generally defined as

$$\langle pT \rangle_c = \frac{\int_0^{p_{T\max}} p_T E \left.\frac{d^3\sigma}{dp^3}\right|_c d^2 p_T}{\int_0^{p_{T\max}} E \left.\frac{d^3\sigma}{dp^3}\right|_c d^2 p_T} \tag{1}$$

where p_T is the transverse momentum of the “c”-type of secondary; where c could any of the pions, kaons, baryon-antibaryons or any other.

Quite spectacularly, this observable, by definition, is tied up with the measurements/model-based formulas for inclusive cross-section for any specific variety of the secondary particles. The letter “c” in the expression (1) indicates only the detected particle among the secondaries which comprise of both neutral and charged pions, kaons, baryons etc. Thus, any study on the nature of the average transverse momenta cannot be delinked from the studies on the nature of or expression for inclusive cross-sections, or from the model-based fits to them.

2. OUTLINE OF THE THEORETICAL FRAMEWORK

We will present here the synopsis of the Hagedorn’s Power law model which would lead us to calculate the average transverse momentum values in a somewhat simple manner. This is being delineated here blow.

2.1 Hagedorn's Power Law Model

Our objective here is to study the inclusive P_T -spectra of the various secondaries of main varieties produced in *PP* collisions. The kinematics of an inclusive reaction $h_a h_b \rightarrow hX$ is described by Lorentz invariants. These are e.g. the center-of-mass energy squared $s = (P_a+P_b)^2$, the transverse momentum transverse (squared) $t = (P_h-P_a)^2$ and the missing mass M_x. It is common to introduce the dimensionless variables ($u = M^2_x - s - t$),

$$x_1 = -\frac{u}{s}, \quad x_2 = -\frac{t}{s} \tag{2}$$

where *s, t, u* are called Mandelstam variables. These variables are related to the rapidity y and radial scaling factor, x_R of the observed hadron by

$$y = \frac{1}{2}\log(x_1/x_2) \tag{3}$$

$$x_R = \frac{2|\overline{p_{cm}}|}{\sqrt{s}} = 1 - \frac{M_X^2}{s} \approx x_1 + x_2 \tag{4}$$

Since most of existing data are at $y = 0$ where $x_R = x_T = 2p_T/\sqrt{s}$, one often refers to the scaling of the invariant cross section as "x_T scaling". Ford $y \neq 0$, we find the variable x_R more useful than x_T, since x_R allows a smooth matching of inclusive and exclusive reactions in the limit $x_R \rightarrow 1$.

We will assume that at high p_T, the inclusive cross section takes a factorized form. And one such a factorized form was given by Back *et al.* [1].

$$\frac{dN}{dp_T} = \frac{p_T(n-2)(n-1)}{p_0^2}\left(1 + \frac{p_T}{p_0}\right)^{-n} \tag{5}$$

where n and p_0 are adjustable parameters. The values of the exponent n are just numbers. p_0 is a critical value of transverse momentum of the secondaries below which no secondary could be detected by the present day detectors set to measure data on hard interactions. So the factor p_T/p_0 is a dimensionless quantity.

With the simplest recasting of form the above expression (5) and with replacements like $p_T = x$, $p_{0} = q$, $C = a$ normalisation factor, and in the light of the definitions of the inclusive cross-sections, we get the following form as the final working formula in the applied form of the Power Law.

$$y = f(x) = \frac{C}{q^2}\left(1 + \frac{x}{q}\right)^{-n} \tag{6}$$

Our first task here would be to check up the efficacy of this working expression, for which at the very beginning, we provide some model-based fits to the invariant crosssections for the production of the major varieties of secondaries in Pb + Pb collisions at various respectively low energies (17.3 GeV, 20A GeV, 30A GeV, 40A GeV) and Au + Au reaction at 19.6 GeV. Besides, some important ratio-behaviors based on the same expression have also been studied in the work depicted by several figures. With the fundamental definition of the average transverse momentum given by Equation (1) above and with the acceptance of the d*N*/d*p* for invariant cross-section, Equation (1) can be transformed in the form

$$\langle p_T \rangle = \frac{\int_0^{p_{T\max}} p_T \frac{dN}{dp_T} dp_T}{\int_0^{p_{T\max}} \frac{dN}{dp_T} dp_T} \tag{7}$$

In this power law scaling form it is generally found that q and n are intrinsically related by the average transverse momentum, $\langle p_T \rangle$, with the undernoted relationship :

$$\langle p_T \rangle = \frac{2q}{n-3} \tag{8}$$

Actually ,this work could attain added significance, had we been able to compute the average transverse momentum values based on Equation (1) and compare them with the values obtained by the somewhat empirical formula given by Equation (8). But due to gross uncertainty in the values of q (which would in all cases be entirely arbitrary) we keep ourselves limited to expression (8) alone for the calculational purposes of the average transverse momentum ($\langle p_T \rangle$)values.

The work presented here is essentially based on some fundamental physical ideas: 1) the idea of large- $p_T$scaling (or x_T-scaling as was defined earlier), 2) the validity of the factorization hypothesis, 3) the constraint of exclusivity of reactions on inclusivity as was studied and pointed out by Brodsky *et al.* [2] who set the limit on the value of the exponent *n*, given by $n \leq 20$.

3. RESULTS

The results are presented here in graphical plots and the accompanying tables for the values of the used parameters. In **Figures 1(a)** and **(b)** the differential cross-sections for negative and positive pion, kaon and proton production cases in Pb + Pb collisions at SPS energies (at 20A GeV) are reproduced by the used theoretical framework. The figures in all the cases have been appropriately labeled and the parameters are shown in the table **(Table 1)**. The plots in **Figures 2(a)** and **(b)** are for positive and negative pion, positive and negative kaon in Pb + Pb collisions at 30A GeV. The plots presented in **Figure 2(c)** for proton-antiproton production in Pb + Pb interaction at SPS energies, specifically at 17.3 GeV. The parameter values used to obtain the nature of fits for **Figure 2** are shown in **Table 2**. The graphs in **Figures 3** and **4** present the fits to the invariant cross-sections for the secondaries, $\pi^{\pm}$, $K^{\pm}$ and proton-antiproton produced in the Au + Au collisions at 19.6 GeV and the corresponding parameters are depicted in **Table 3**. Against the data points, the plots shown in **Figure 5** by solid lines depict the fits based on the same model to the invariant cross-sections for various major varieties of secondaries produced in Pb + Pb collisions at 40A GeV, though data on production of $\pi^{\pm}$ were not available. The parameter values used are given in **Table 4**. In **Figure 6**, we demonstrate the fits to the invariant cross-sections for various major varieties of the secondaries produced in central (0% - 10%) Au + Au collisions at 19.6 GeV and the corresponding parameters are depicted in **Table 5**. The fits to the charge ratio-values of the pionic, kaonic and baryonic particles are drawn in **Figure 7**. These charge-ratio values offer a cross-check of the results arrived at for the invariant cross-sections. **Tables 6** to **10** indicate the calculation of average transverse momentum in Pb + Pb collisions at 20A GeV, 30A GeV, 17.3 GeV, 19.6 GeV, 40A GeV respectively and **Table 11** in Au + Au collisions at 19.6 GeV. At last, the plots shown in **Figures 8(a)** and **(b)** (**Tables 12** and **13**) show the nature of the average transverse momentum values, which form the core of this work and its central theme.

Table 1. Numerical values of the fit parameters of power law equation for Pb + Pb collisions (Reference Figure 1).

Energy	Productions	c	q	n	χ^2/ndf
20A GeV	π^-	43.935 ± 0.015	1.197 ± 0.037	10.010 ± 0.234	0.612/7
20A GeV	π^+	38.950 ± 0.233	1.178 ± 0.087	9.410 ± 0.531	2.628/6
20A GeV	K^-	1.690 ± 0.013	2.404 ± 0.014	14.003 ± 0.054	0.806/4
20A GeV	K^+	5.515 ± 0.136	2.001 ± 0.022	11.888 ± 0.180	4.894/4
20A GeV	P	4.390 ± 0.313	1.541 ± 0.065	4.645 ± 0.282	3.092/5

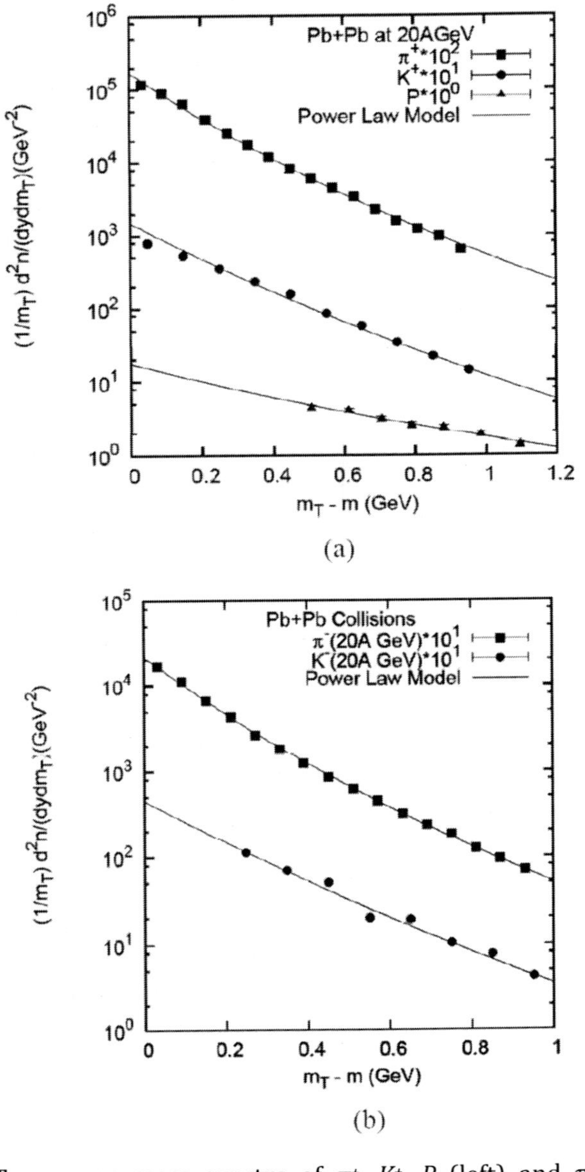

Figure 1. Transverse mass spectra of π^+, K^+, P (left) and π^-, K^- (right) produced in central Pb + Pb collision at 20A GeV. The lines are fits of equation power law model. The statistical errors are smaller than the symbol size, for which no errors are shown in the figure. Parameter values are taken from Table 1. The experimental data are taken from References [3] and [4].

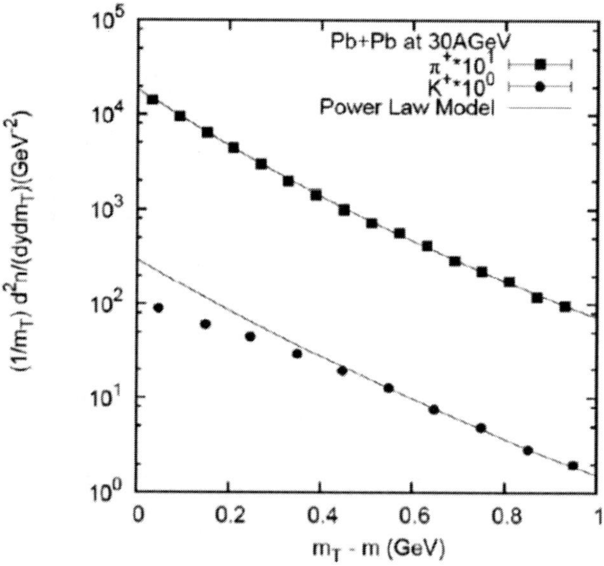
Pb+Pb at 30AGeV
π⁺*10¹
K⁺*10⁰
Power Law Model
(1/mT) d2n/(dydmT)(GeV-2)
mT - m (GeV)

(a)

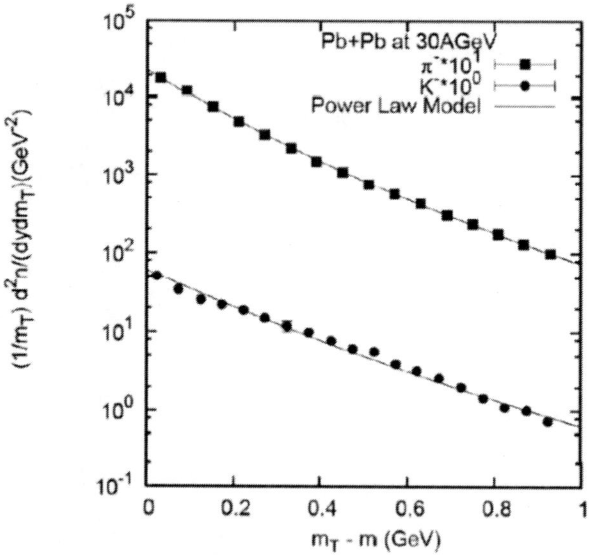
Pb+Pb at 30AGeV
π⁻*10¹
K⁻*10⁰
Power Law Model
(1/mT) d2n/(dydmT)(GeV-2)
mT - m (GeV)

(b)

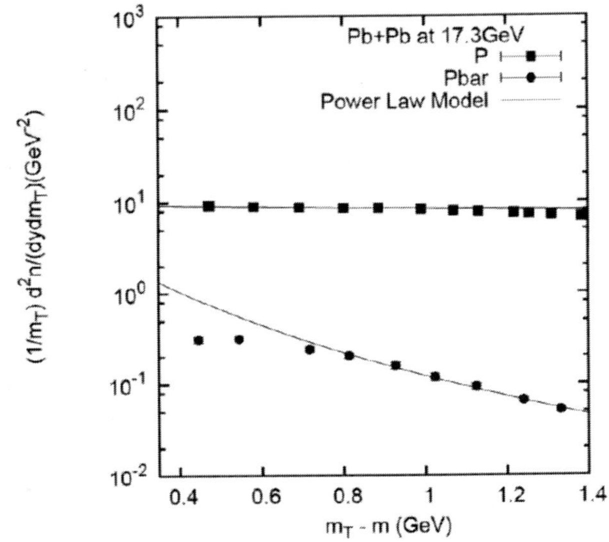

(c)

Figure 2. Transverse mass spectra of π+, K+ (left) and π−, K− (middle) and P, $\bar{P}$ (right) produced in central Pb + Pb collision at 30A GeV and 17.3 GeV. The lines are fits of equation of power law model. The statistical errors are smaller than the symbol size, for which no errors are shown in the figure. Parameter values are taken from Table 2. The experimental data are taken from References [3] and [5].

Table 2. Numerical values of the fit parameters of power law equation for Pb + Pb collisions (Reference Figure 2).

Energy	Productions	c	q	n	χ^2/ndf
30A GeV	π^-	52.202 ± 0.307	1.158 ± 0.073	9.775 ± 0.438	1.158/5
30A GeV	π^+	47.041 ± 0.637	1.344 ± 0.211	10.026 ± 1.166	5.790/8
30A GeV	K^-	2.551 ± 0.040	2.001 ± 0.021	11.240 ± 0.115	3.519/3
30A GeV	K^+	8.962 ± 0.341	1.999 ± 0.017	12.986 ± 0.293	3.697/4
17.3 GeV	P	41.249 ± 3.03	2.000 ± 0.002	0.401 ± 0.029	1.014/5
17.3 GeV	$\bar{P}$	0.450 ± 0.042	0.439 ± 0.049	3.999 ± 0.037	1.338/5

A. C. DAS GHOSH *ET AL.*

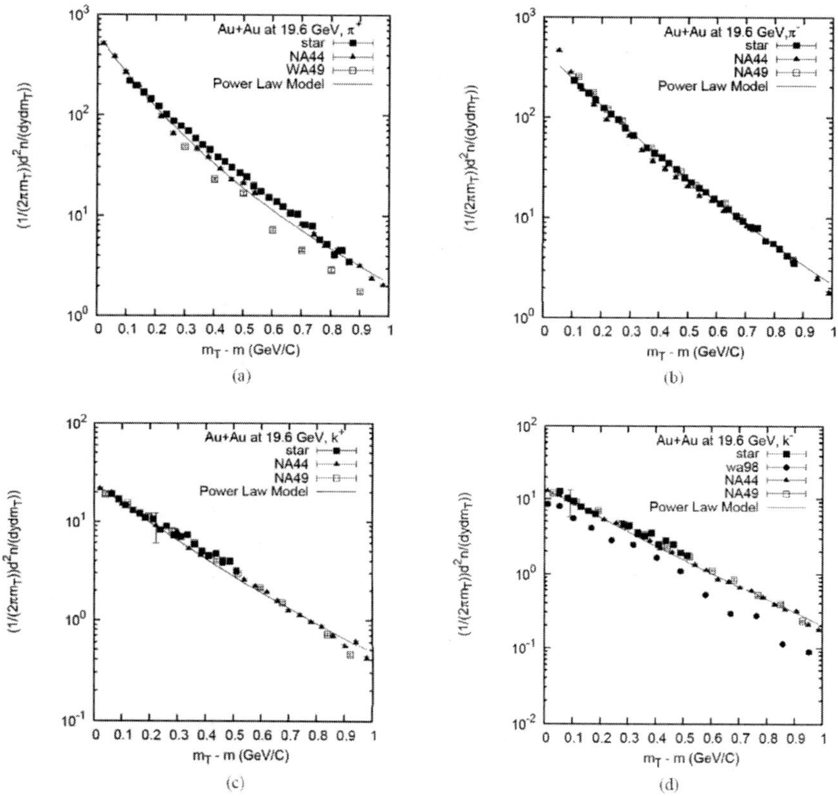

Figure 3. The transverse mass spectra of π^+ (upper left), π^- (upper right) and K^+ (lower left), K^- (lower right) from STAR experiment at 19.6 GeV in Au + Au collisions and the results of SPS experiments NA44, NA49, WA98 at 17.3 GeV in Pb + Pb collisions. The line is fit of power law model with all the STAR and SPS experiment. Parameter values are taken from Table 3. The experimental data are taken from Reference [6] and all errors are only of statistical nature.

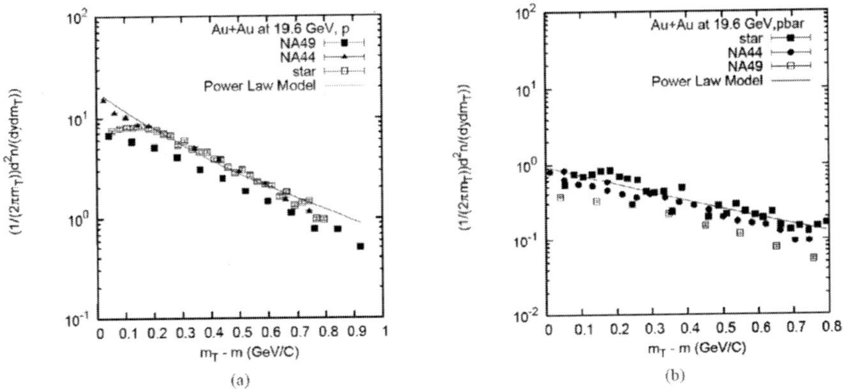

Figure 4. The transverse mass spectra of P (left) and $\overline{P}$ (right) from STAR experiment at 19.6 GeV in Au + Au collisions and the results of SPS experiments NA44, NA49, WA98 at 17.3 GeV in Pb + Pb collisions. The line is fit of power law model with all the STAR and SPS experiment. Parameter values are taken from Table 3. The experimental data are taken from Reference [6]. The statistical errors are smaller than the symbol size, for which no errors are shown in the figure.

Table 3. Numerical values of the fit parameters for pion, kaon, proton and antiproton using power law model for Au + Au collisions at 19.6 GeV (Reference Figures 3 and 4).

Production	c	q	n	χ^2/ndf
π^+	12.588 ± 0.183	0.559 ± 0.061	5.547 ± 0.358	6.518/7
π^-	13.443 ± 0.089	2.001 ± 0.011	13.192 ± 0.065	27.759/29
K^+	1.455 ± 0.017	2.002 ± 0.256	9.803 ± 0.109	13.526/11
K^-	0.742 ± 0.007	6.640 ± 0.038	29.998 ± 0.050	27.027/18
P	1.995 ± 0.027	0.887 ± 0.103	4.257 ± 0.233	18.933/11
$\overline{P}$	0.188 ± 0.005	2.440 ± 0.057	7.001 ± 0.044	2.741/06

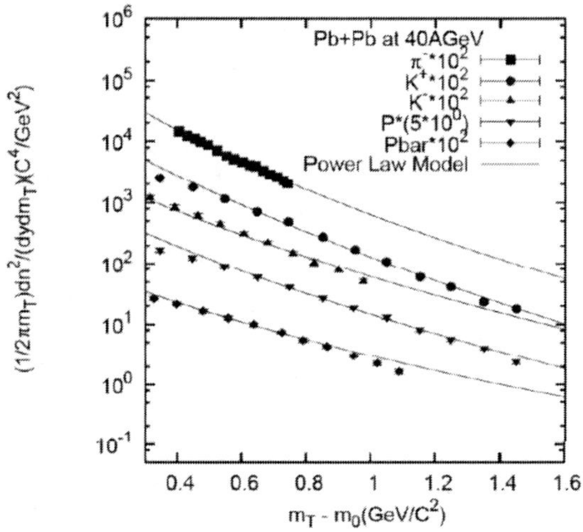

Figure 5. Transverse mass spectra of K+, K−, P, and π− produced in central Pb + Pb collision at 40A GeV. The lines are fits of equation of power law model. The statistical errors are smaller than the symbol size, for which no errors are shown in the figure. Parameter values are taken from Table 4. The experimental data are taken from [4].

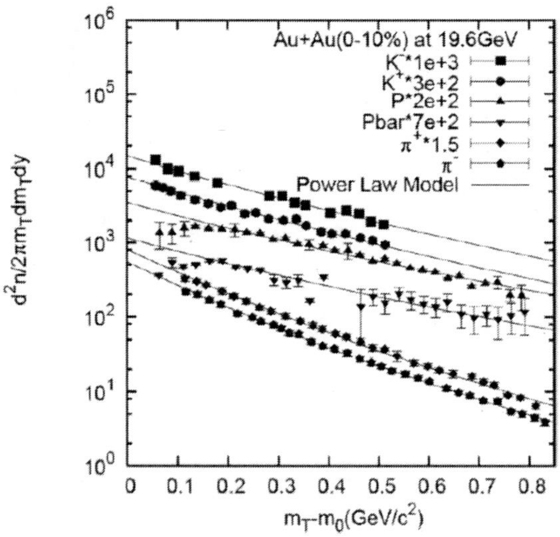

Figure 6. Transverse mass spectra of identified hadrons measured at midrapidity $(|y| < 0.1)$. The results at $\sqrt{s_{NN}}$ = 19.6 GeV for the production of π^+, π^-, P, $\bar{P}$, K^+ and K^- for 0% - 10% centrality in Au + Au collisions. The solid curves provide the power law model based results. Parameter values are taken from Table 5. The experimental data are taken from Reference [7].

Table 4. Numerical values of the fit parameters for negative pion, positive and negative kaon, proton and antiproton using power law model for Pb + Pb collisions at 40A GeV (Reference Figure 5).

Production	c	q	n	χ^2/ndf
π^-	54.328 ± 1.296	1.198 ± 0.022	10.003 ± 0.042	0.103/15
K^+	9.176 ± 0.166	1.999 ± 0.007	13.856 ± 0.132	0.617/07
K^-	2.429 ± 0.067	1.734 ± 0.048	10.031 ± 0.210	1.114/11
P	12.827 ± 0.589	1.629 ± 0.043	10.002 ± 0.070	0.433/08
$\bar{P}$	0.078 ± 0.002	1.382 ± 0.054	7.022 ± 0.178	0.800/08

Table 5. Numerical values of the fit parameters for pion, kaon, proton and antiproton using power law model for Au + Au collisions at 19.6 GeV for 0% - 10% centrality (Reference Figure 6).

Production	c	q	n	χ^2/ndf
π^+	14.002 ± 0.204	1.367 ± 0.020	10.015 ± 0.102	0.823/28
π^-	13.382 ± 0.133	1.321 ± 0.012	10.009 ± 0.063	0.385/31
K^+	1.686 ± 0.028	1.833 ± 0.070	8.648 ± 0.234	0.841/19
K^-	0.999 ± 0.039	1.99 ± 0.032	9.123 ± 0.264	0.700/12
P	1.672 ± 0.016	1.567 ± 0.025	6.587 ± 0.066	10.212/16
$\bar{P}$	0.159 ± 0.002	1.568 ± 0.018	6.587 ± 0.052	8.980/21

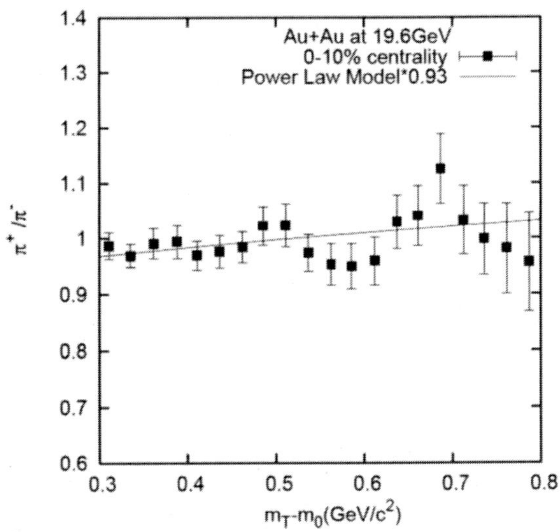

(a)

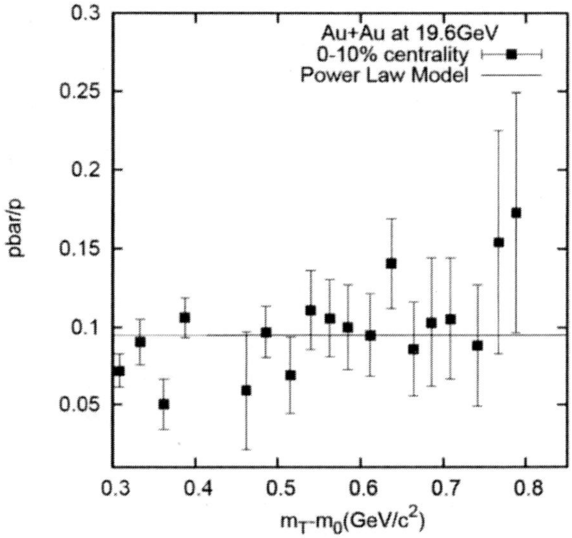

(b)

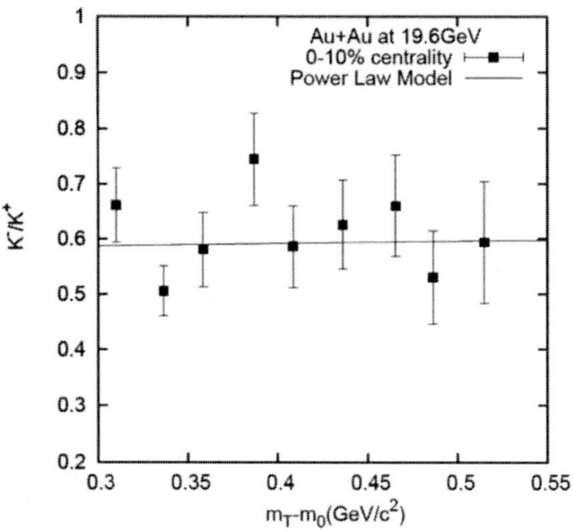

(c)

Figure 7. π^-/π^+, $\bar{P}/P$, and K^-/K^+ ratios vs. $m_T - m_0$ for 0% - 10% centrality in Au + Au collisions at $\sqrt{s_{NN}}$ = 19.6 GeV $(-0.1 < y < 0.1)$. The solid curves provide the power law model based results. Data are taken from Reference [7]. All errors are only of statistical nature.

Table 6. Calculation of average transverse momentum for Pb + Pb collisions at E_{lab} = 20A GeV (Data are taken from Table 1).

Production	q	n	$\langle p_T \rangle = 2q/(n-3)$
π^-	1.197	10.010	0.342
π^+	1.178	9.410	0.368
K^-	2.404	14.003	0.437
K^+	2.001	11.888	0.450
P	1.541	4.645	1.874

Table 7. Calculation of average transverse momentum for Pb + Pb collisions at E_{lab} = 30A GeV (Data are taken from Table 2).

Production	q	n	$\langle p_T \rangle = 2q/(n-3)$
π^-	1.158	9.232	0.372
π^+	1.344	10.026	0.383
K^-	2.001	11.240	0.486
K^+	1.999	12.986	0.364

Table 8. Calculation of average transverse momentum for Pb + Pb collisions at $\sqrt{s_{NN}}$ = 17.3 GeV (Data are taken from Table 2).

Production	q	n	$\langle p_T \rangle = 2q/(n-3)$
P	2.000	0.401	1.539
$\overline{P}$	0.439	3.999	0.879

Table 9. Calculation of average transverse momentum for Au + Au collisions at $\sqrt{s_{NN}}$ = 19.6 GeV (Data are taken from Table 3).

Production	q	n	$\langle p_T \rangle = 2q/(n-3)$
π^+	0.559	5.547	0.439
π^-	2.001	13.192	0.393
K^+	2.002	9.803	0.589
K^-	6.640	29.998	0.492
P	0.887	4.257	1.411
$\overline{P}$	2.440	7.001	1.220

Table 10. Calculation of average transverse momentum for Pb + Pb collisions at E_{lab} = 40A GeV (Data are taken from Table 4).

Production	q	n	$\langle p_T \rangle = 2q/(n-3)$
π^-	1.198	10.003	0.342
K^+	1.999	13.856	0.368
K^-	1.734	10.031	0.493
P	1.629	10.002	0.465
$\overline{P}$	1.382	7.022	0.687

Table 11. Calculation of average transverse momentum for Au + Au collisions at $\sqrt{s_{NN}}$ = 19.6 GeV for 0% - 10% centrality (Data are taken from Table 5).

Production	q	n	$\langle p_T \rangle = 2q/(n-3)$
π^+	1.367	10.015	0.390
π^-	1.321	10.009	0.377
K^+	1.833	8.648	0.649
K^-	1.999	9.123	0.653
P	1.567	6.587	0.874
$\overline{P}$	1.568	6.587	0.874

Table 12. Data for drawing graph $\sqrt{s}$ vs. average transverse momentum ($\langle p_T \rangle$) (Data are taken from Tables 6, 7 and 9) for π^+ (Reference Figure 8(a)).

$\sqrt{s}$	$\langle p_T \rangle$
4.472	0.368
7.746	0.383
19.6	0.439

Table 13. Data for drawing graph $\sqrt{s}$ vs. average transverse momentum ($\langle p_T \rangle$) (Data are taken from Tables 6, 7 and 9) for π^- (Reference Figure 8(b)).

$\sqrt{s}$	$\langle p_T \rangle$
4.472	0.342
7.746	0.342
8.944	0.342
19.6	0.393

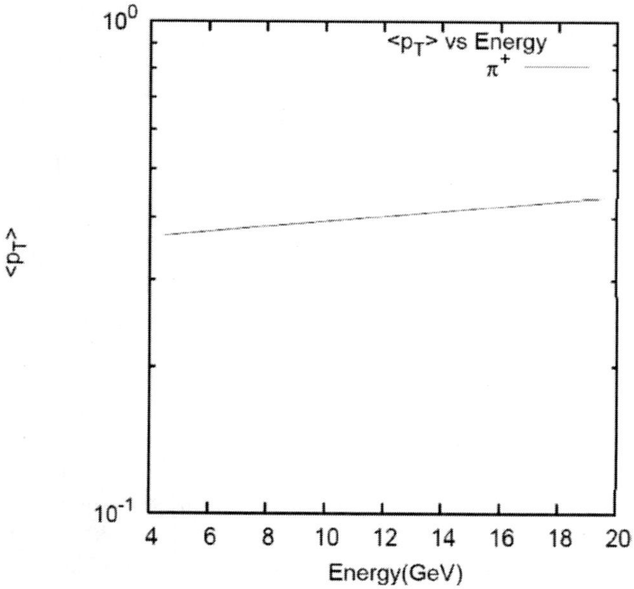

(a)

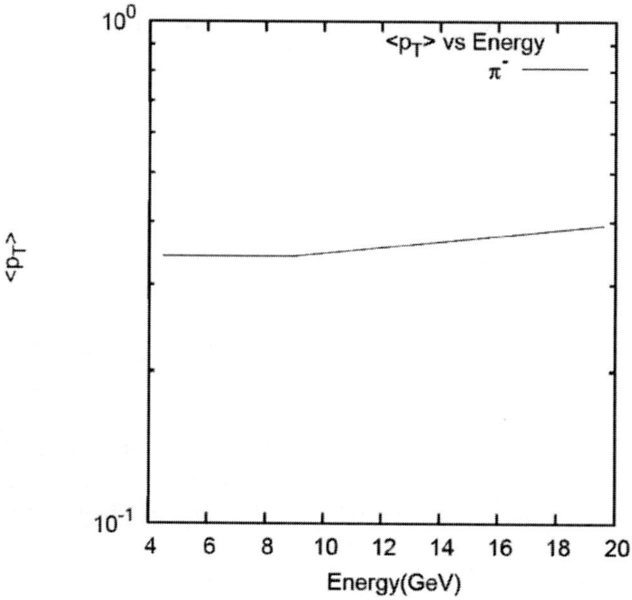

(b)

Figure 8. Average transverse momenta vs. c.m. energy plots. Parameter values are taken from Tables 12 and 13.

4. DISCUSSION AND CONCLUSIONS

By all indications, the results manifested in the measured data on the specific observables chosen here are broadly consistent with the power law model put into work here. This is modestly true of even the nature of charge-ratios which provide virtually a cross-check of the model utilised here. Of course, some comments on our modelbased plots, especially the plots on the charge-ratios are in order here. The lack of the predictivity of the used model is caused only by the circumstances, *i.e.* the lack of measured data at the successive and needed intervals; the problem can be remedied by supplying the necessary and reliable data from the arranged laboratory experiments at high-to-very high energies. However, the problem of constraining the parameters still remains. The other observations are: as is expected for two symmetric collisions of neighbouring values of mass numbers at very close energies do not reveal any significant differences with respect to the observables chosen by the experimental groups. This work demonstrates somewhat convincingly that the power law models can easily take care of data; so the notion of compartmentalisation between the possible applicability of the power law models and of the exponential models is only superficial and also questionable. Thus, the power law models which establish them as more general ones obtain a clear edge over the exponential models. Reliable data on various other related observables are necessary for definitive final conclusions. By all indications, the experimentally observed nature of p_T -scaling is found to remain valid in the studied low- p_T range of this paper.

Now follow a few comments on the nature of the average transverse momenta in nucleus-nucleus collisions: Firstly, the average p_T values for the secondaries, especially for the pion secondaries, show very slow rising nature with center of mass energy. Secondly, the ranges of the averages p_T – values do not differ any prominently between simple PP reactions and heabvy nucleus-nucheus collisions at the similar ranges of energy. Thirdly, differences between the Au+Au reactions at 19.6 GeV and Pb + Pb interaction at 17.3 GeV, 20A GeV, 30A GeV, 40A GeV are quite insignificant. This is in fact, the main content of the "universality" property of high energy interactions, Fourthly, the nature of avegage transverse momenta of the secondaries does not show any major difference between "soft" (small-p_T) and "hard" (large- p_T) reactions. Lastly and finally, that the values of $\langle p_T \rangle$ are insensitive to the collisions-specifies like Au + Au or Pb + Pb at the neighboring energies is only natural. Such minor differences between "A"-values do hardly introduce any significant changes in the magnitudes of the average transverse momenta.

REFERENCES

1. B. B. Back, "Consequences of Energy Conservation in Relativistic Heavy-Ion Collisions," Physical Review C, Vol. 72, No. 6, 2005, Article ID: 064906. doi:101103/PhysRevC.72.064906
2. S. J. Brodsky, H. J. Pirner and J. Raufeisen, "Scaling Properties of High p_T Inclusive Hadron Production," Physics Letters B, Vol. 637, No. 1-2, 2006, pp. 58-63. doi:10.1016/j.physletb.2006.03.079
3. C. Alt, et al., "Pion and Kaon Production in Central Pb + Pb Collisions at A-20 and A-30 GeV: Evidence for the Onset of Deconfinement," Physical Review C, Vol. 77, No. 2, 2008, Article ID: 024903. doi:10.1103/PhysRevC.77.024903
4. C. Alt, et al., "Bose-Einstein Correlations of π-π-Pairs in Central Pb + Pb Collisions at A-20, A-30, A-40, A-80 and A-158 GeV," Physical Review C, Vol. 77, No. 6, 2008, Article ID: 064908. doi:10.1103/PhysRevC.77.064908
5. C. Alt, et al., "High Transverse Momentum Hadron Spectra at $\sqrt{s_{NN}}$ = 17.3 GeV in Pb + Pb and p + p Collisions," Physical Review C, Vol. 77, No. 3, 2008, Article ID: 034906. doi:10.1103/PhysRevC.77.034906
6. D. Cebra, "Charged Hadron Results from Au + Au at 19.6 GeV," Invited Talk Presented at Quark Matter, 2008.
7. R. Picha, "Charged Hadron Distribution in 19.6 GeV Au + Au Collision," Ph.D. Thesis, University of California, Davis, 2005.

CHAPTER 16

Shortening transmutation time by using the molten salt reactor

Klaus Hendrik Biss, Bruno Thomauske

ABSTRACT

This paper investigates the duration of the Partitioning and Transmutation process for different efficiencies in the total mass reduction of high-level waste by describing the mass flow during the process. Besides the transmutation and partitioning efficiency per fuel cycle, the impact of the deployed thermal power and the refreshing time, which is the sum of cooling, reprocessing, and fabrication time, was analyzed. It was discovered that besides the transmutation efficiency the refreshing time has a significant impact on the duration. The molten salt reactor concept is proposed in order to realize this potential for shortening the time consuming P&T process.

Keywords: Partitioning and Transmutation (P&T), Molten salt reactor, Duration, P&T efficiency, P&T implementation

1. INTRODUCTION

The utilization of nuclear power by fission leads to a build-up of radioactive nuclides. While most of the fission products are stable or short lived, the long-lived ones are considered as responsible for the effective dose by a direct disposal in deep geological structures (Renn, 2014; Von Lensa et al., 2008). Together with transuranic elements (TRU), they represent the high level waste (HLW). While plutonium as major part of TRU can be used nowadays as Mixed oxide (MOX) fuel (Carbajo *et al.*, 2001 ; Ramirez *et al.*, 2012), the minor actinides (MA) such as americium, curium, and neptunium are not considered as fuel for present nuclear power plants.

The further use of TRU and long-lived fission products after Partitioning (P) has been investigated for different fuel cycle strategies (Von Lensa *et al.*, 2008; OECD, 2002 ; Hyland and Gihm, 2011). Due to its neutronic properties, the usage of MA in critical reactors is limited (Zhang et al., 2013). Therefore, only small MA quantities could be used in homogeneous transmutation cores. This would, however, prevent a build-up of most of the transuranic elements (Ochoa *et al.*, 2013; Song *et al.*, 2008 ; van Rooijen and Kloosterman, 2009).

A sub-critical reactor, realized by an accelerator-driven system (ADS), allows better core loading flexibility, especially to accommodate very high minor actinide inventories (Malambu *et al.*, 2004 ; Artioli *et al.*,2008). Therefore, it is a useful option to incinerate MA efficiently, as suggested in the double-strata fuel cycle (Oigawa et al., 2011).

In conclusion, Partitioning and Transmutation (P&T) offers an opportunity to reduce the quantity of TRUs by converting it into shorter-lived fission products. This could reduce volume, decay heat, and radiotoxicity of the HLW in a long-term disposal. Maximize P&T efficiency requires a high burn-up to minimize the number of reprocessing cycles, in which losses occur. Assuming a high burn-up of 200 MWd/kg_{HM} and a mass loss of 0.1% in each reprocessing cycle, a theoretical P&T efficiency of 99.5% in mass reduction is possible (Magill et al., 2003).

The major source for losses during hydrometallurgical processing, like the PUREX process, occurs by shredding and dissolving the spent fuel rods (Head–End) (Renn, 2014, p.107). Since the molten salt reactor (MSR) is based on liquid fuel, this process step is not required leading to a short reprocessing time. Therefore, MSR, which is part of the generation IV reactor designs (Gen IV International Forum, 2013), should be of particular interest for P&T, as shown and discussed in this paper.

The potential of a high P&T efficiency has been the subject of numerous investigations (Delpech et al., 1999). Although the number of facilities for the P&T process has been mentioned in some studies, no particular attention has been paid to the duration (OECD, 2009). The reason for this might be that the benefit of P&T for final disposal has been of primary interest (Nishihara *et al.*, 2008 ; González-Romero, 2011).

The relation between the duration and the efficiency of the P&T process is analyzed in the present paper. In addition to burn-up and reprocessing losses, the refreshing time for the fuel and the deployed thermal power of the facility are considered.

In order to calculate the P&T efficiency, the mass flow within the P&T process is described by five sectors which are defined in the method section. Then, values for the parameters of the calculation are determined. Next, the influence of each parameter is studied by varying one parameter while the others are kept constant. By calculating all possible combinations for the variations, the upper and lower duration for a certain P&T efficiency is obtained. The reasons for the large variation in duration are discussed and, as a result, the molten salt reactor is introduced as a plausible solution to realize the potential of a short duration for any P&T efficiency.

2. METHOD

In order to calculate the duration necessary to achieve a specific P&T efficiency, the fraction of the mass which has been incinerated has to be documented. For this reason, the mass flow of the P&T process was tracked by five material sectors. According to each sector an equation describes the mass quantity over time.

In contrast to studies for a specific core or fuel design where the irradiated fuel has been recorded with isotopic resolution (Hyland and Gihm, 2011 ; Malambu *et al.*, 2004), the mass in each sector was considered as sum of all elements. This means that all isotopes were incinerated equally. This simplification is a consequence of investigating not a specific core design, reactor type, or fuel composition. As a result neither the decay of radioactive isotopes nor the influence of the changing fuel composition was investigated.

Instead, all requirements for proper operation were assumed to be fulfilled and since this paper focuses on the duration, the P&T process was influenced by specific parameters only. These parameters are the transmutation and partitioning efficiency, the deployed thermal power, and the refreshing time. Further constant parameters used in the equations are the cycle length and the availability of the facilities.

Using above-mentioned parameters, five mass sectors were characterized. These sectors are the storage of manufactured nuclear fuel for use in the reactor m_S, the inventory within the reactor m_I, the fuel in the refreshing process m_R, the overall losses m_L, and the incinerated material m_T (Fig. 1). The refreshing time was defined as the time for cooling and reprocessing of the spent fuel and fabrication of new fuel assemblies.

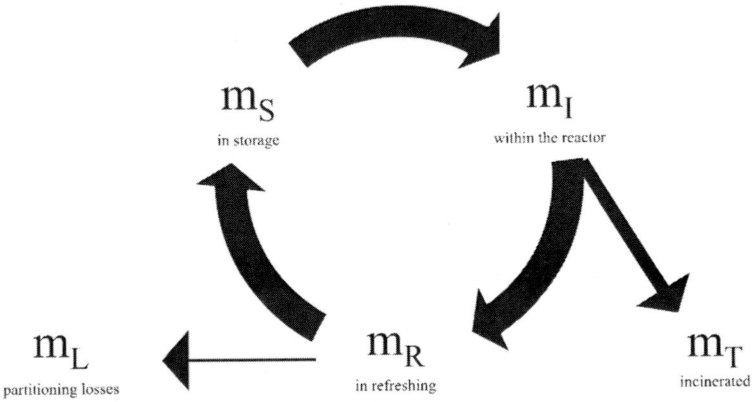

Figure. 1. Scheme of the material flow represented by material sectors.

The section for the manufactured nuclear fuel in storage was the starting point for material flow development. At time step zero it was assumed that all TRU elements are available as manufactured nuclear fuel. The development of the stored nuclear fuel $m_s(t)$ is described by equation(1)

$$m_s(t) = m_s(t-1) - m_I(t-1)/CL + m_I(t-1-t_R) * \epsilon_P * (1-\epsilon_T)/CL + (m_I(t-1) - m_I(t)) \quad (1)$$

The additional mass sections used are m_I, which represents the mass of TRU elements in the reactors, and m_R, which is the mass fed back from the refreshing process. The first term represents the mass at the previous time step. The second term describes the extraction of fuel due to cycle length (CL) in the core. CL was chosen to be four years according to previous transmutation investigations (Biss, 2014). The third term represents the recirculation from the refreshing process from mass entered into refreshing at time t-1-t_R. This mass is reduced by the transmutation efficiency $\in_T$ and partitioning efficiency $\square_P$. The fourth and last term considers changes in the usage of the deployed power plants and therefore the change of the inventory mass of the reactor cores. For example, a reduced usage of the deployed power leads to an increase in the storage mass since the inventory mass is reduced.

This handling is based on the assumption that burn-up and cycle length remain unchanged. This means that changing fuel composition and its impact on the reactor system were compensated. This is a challenge as in fertile free fuel like it is the case for TRU fuel reactivity loss is high, which reduces the cycle length. On the other side, including fertile material would decrease the burn-up rate, but stabilize the cycle length. One strategy might be the usage of a fertile free fuel in the first stage. Reactivity loss could be compensated by an increase in the proton beam in ADS. In the second stage, a switch to another fuel composition with thorium as fertile material could ensure longer irradiation time, but decreasing transmutation efficiency for TRU elements (Biss, 2014). Technical challenges like higher neutron flux, higher neutron source in case of an ADS, the switch to another reactor core, or the reduction of the transmutation zone by replacing fuel assemblies with reflector material were not discussed, since the present paper focuses on the duration of the P&T process.

One important fact for the application of the formula is that mass sectors are changed at the beginning of the time period. For example, at time step t equals zero the reactors are loaded, but there is no mass which has been incinerated or is in the refreshing process. Therefore, m_R and m_T are zero.

The inventory mass m_I is derived by

$$m_I(t) = (P_i * AV * CL/\epsilon_T) * (uta * P_m); uta \in [0, P_i/P_m] \subset \mathbb{N} \quad (2)$$

The minimal thermal power P_m also represents the smallest possible change of the deployed thermal power P_i. AV stands for availability of the facility and was set to 0.85 (OECD, 2009, p. 48). During the burn-up time in the core, the fraction ϵ_T is incinerated by nuclear fission. Altogether, the factor within the first brackets represents the fissionable mass by using the conversion of 1 MW_{th}d equals 1 g of incinerated fissile material. The second factor describes the utilization of the deployed power capacity by multiplying P_m by its integer multiple, designated uta . Since a time step describes the situation at the beginning of the time period, m_I represents the amount of fissile material in a fully loaded reactor. It should be noted again that at that point the transmutation efficiency ϵ_T corresponds to the fuel burn-up. As a consequence, this description of m_I is only valid for fertile free systems, because additional heavy metal for a breeding process is unnecessary.

The mass in the refreshing process is described by

$$m_R(t) = m_R(t-1) + m_I(t-1) * (1-\epsilon_T) * \epsilon_P/CL - m_I(t-1-t_R) * (1-\epsilon_T) * \epsilon_P/CL \quad (3)$$

The first term describes the mass that was in the refreshing process in the previous time step. Additionally, the second term adds the amount of material which went into the refreshing process at the end of the previous time step. The inventory mass from the previous time step is taken as starting point for calculation. The fraction of $1/CL$ is replaced from the reactor in every time step. Furthermore, only fissile material is fed into the reprocessing process. Since it was assumed that all fuel assemblies have the same burn-up characteristics, the factor $(1-\epsilon_T)$ is considered to be constant for all refreshing processes. The fraction which is lost during the partitioning is immediately excluded by multiplying the term by the partitioning efficiency ϵ_P. Finally, the third term represents the recirculation into m_s after the refreshing time t_R. It describes the same situation as the second term, but at time $t-1-t_R$.

The mass which is transmuted into fission products is described by

$$m_T(t) = m_T(t-1) + m_I(t-1) * \epsilon_T/CL \tag{4}$$

In addition to the already transmuted mass $m_T(t-1)$, the mass transmuted in the previous time step is added with the second term. This is realized by multiplying the inventory mass m_I by ϵ_T/CL. The P&T efficiency ϵ_{PT} for transuranic elements is calculated by dividing m_T by the initial fissile mass.

The last mass sector represents the losses due to the refreshing process, which is the process chain of cooling and reprocessing spent fuel and fabrication of new fuel assemblies.

$$m_L(t) = m_L(t-1) + m_I(t-1)/CL * (1-\epsilon_T) * (1-\epsilon_P) \tag{5}$$

It was assumed that losses only occur during reprocessing. This process can be characterized by three steps: shredding and dissolving, separation of the elements, and fabrication of new fuel assemblies. In this paper only the overall losses were considered. There was no differentiation between different partitioning processes such us hydrometallurgical or pyro-chemical. The first term for the description of losses during P&T represents existing losses. The second term is similar to the second term of Eq. (3) and describes the mass which has been unloaded from the reactor for reprocessing. Instead of the partitioning efficiency ϵ_P, the factor $(1-\epsilon_P)$ is multiplied by this amount. Since there is no decreasing term, m_L increases monotonously and limits the P&T efficiency.

Besides the proper description of the material flow, controllability of the deployed thermal power was necessary for the simulation. As introduced in Eq. (2), *uta* allows an integer multiple of the minimal possible power P_m to be used. In each time step power can be increased or reduced. For simulation, the increase or decrease was calculated by the available mass. Reduction of used deployed power was caused by an insufficient amount of fuel needed for current power.

3. CALCULATION

In order to run proper simulations to estimate the duration of the partitioning and transmutation process, a set of values for the parameters described in the method section has to be chosen. The parameters transmutation efficiency $\square_T$ and partitioning efficiency $\square_P$ were selected on the basis of literature values. In addition to these parameters, the set of parameters for deployed thermal power and refreshing time were chosen. After input data are described in the present section, the simulation run is briefly presented leading directly to the results section.

The theoretical maximal efficiency of the partitioning and transmutation process has been calculated as 99.5% (Magill et al., 2003). This calculation has been based on a transmutation efficiency $\square_T$ of 0.20 and a partitioning efficiency $\square_P$ of 0.999. Both values are best case assumptions if there is a high share of minor actinides. Especially for the transmutation efficiency, the value of 0.20 has been assumed to be a reasonable limitation, which is caused by neutronic behavior like neutron flux, neutron effluences, embrittlement, or decay heat. Therefore, lower efficiencies of 0.10 and 0.15 were included in the set for the burn-up and have been achieved in special reactor designs (Artioli et al., 2008).

Even if the partitioning efficiency of 0.999 has already been established for the PUREX process (Salvatores and Palmiotti, 2011), this value is rather high for fuel containing a high share of minor actinides. Especially the transmutation of Am-241 leads to a build-up of Cm-242, which causes a high decay heat and alpha emission due to its half-life of 183 days. Furthermore, emission of neutrons, which are harder to shield, hinders reprocessing. However, these are only physical challenges. The chemical ones are those which determine the efficiency. In the case of MA separation, the difficulty lies in the chemical similarity to the lanthanides (Salvatores and Palmiotti, 2011). An overview of R&D for the pyro-chemical processes is given in Nawada and Fukuda (2005). In order to show the impact of the efficiency, a high efficiency of 0.9995 and a lower one of 0.995 were included in the set for the partitioning efficiency. Although in liquid fuel concept like in the MSR shredding, additional dissolving, and fabrication of new fuel assemblies would be redundant, there was no differentiation in the efficiency between the partitioning processes. This choice was justified by lower experience in pyro-chemical processes.

Another parameter is the thermal power deployed for the transmutation process. Besides possible economic considerations, this power was limited by the inventory mass m_I. For simulation, a deployed thermal power was assumed, which leads to a first loaded inventory of 1/3 and 1/6 of the total transuranic mass. The deployed power varies due to the fact that m_I is dependent on the transmutation efficiency (see Eq. (2)). Nevertheless, the choice of expressing the deployed power in terms of the fraction of the total inventory rather than the installed power in thermal units was due to the fact that this present paper investigates the duration of the P&T process independent of the absolute mass, which is going to be transmuted.

The refreshing time t_R is the last parameter for the calculation. It is the time for cooling and reprocessing of spent fuel and fabrication of new fuel assemblies. The cooling and reprocessing time depends on fuel type and burn-up, because they determine the level of alpha, gamma, and neutron emissions of spent fuel. A minimum cooling time of 5 years before reprocessing is currently required, followed by a fuel ageing time of 2 years before repeated irradiation in a reactor (Grouiller *et al.*, 2003 ; Camarcat *et al.*, 2011). For a fuel type with higher content of minor actinides, as in the case of spent MOX fuel, a refreshing time of 10 years has been assumed (Ishida and Sekimoto, 2010). However, a refreshing time of 3 years has been considered feasible with a cooling time of 2.5 years and a reprocessing and manufacturing time of six months (Fiorina et al., 2013). For this reason the values for the refreshing time were chosen as 3, 7, and 10 years.

Additionally to this set, a calculation was performed for a refreshing time of one year and no refreshing time at all. While no refreshing time is considered as the optimal value, the refreshing time of one year is based on the assumption to use the molten salt reactor (LeBlanc, 2010). By using liquid fuel, reprocessing is a continuous process. Helium bubbling allows a gaseous extraction of all gaseous fission products and noble metals (Delpech et al., 2009). Although the removal of most fission products and actinides is taken place off-line, the continuous reprocessing allow the reprocessing of the entire core volume within 450 days (Heuer et al., 2014). In the US ORNL–MSBR concept the reprocessing time of less than 10 days were considered to achieve better breeding performances (Robertson, 1971).

Table 1. Set of parameters used to investigate the duration of the P&T process.

Parameter	Set of values
Transmutation efficiency	10%, 15%, 20%
Partitioning efficiency	99.50%, 99.90% 99.95%
Deployed thermal power	1/6 and 1/3 of initial transuranic mass
Refreshing time	0, 1, 3, 7, 10 years

Table 1 presents an overview of all parameter sets. The duration for seven efficiencies is presented in figures in which one parameter is flexible and the other three are kept constant. This is done in order to discuss the impact of the flexible parameter.

4. RESULTS

The transmutation of transuranic elements has been considered in studies about a sustainable nuclear fuel cycle (closed fuel cycle) (González-Romero, 2011; Camarcat *et al.*, 2011 ; Fiorina *et al.*, 2013). In these studies, the duration has not been of specific interest since a long-term use of nuclear power has been assumed. In the case of nuclear power being phased out, the duration of the transmutation of the final HLW has been investigated in principle rather than in detail. Instead, the benefit of the P&T process for a final repository has been discussed.

With the parameters described in the methods section, the present paper focuses on the duration of the P&T process, which was investigated independent of a specific reactor or fuel concept. The duration was calculated for 90 different combinations for the partitioning and transmutation process. The impact of each parameter is discussed separately in the following.

The influence of partitioning efficiency is shown in Fig. 2. Up to a transmutation of 80%, differences between partitioning efficiencies are insignificant. At 90% the duration of the two higher efficiencies is still similar while the deviation of the lowest one rises by 5 and 16 years for P&T efficiencies of 90 and 95%, respectively. P&T efficiencies higher than 95% are only possible for higher partitioning efficiencies. This is consistent with the calculation of the maximal P&T efficiency

$$\epsilon_{PT} = \frac{\epsilon_P \epsilon_T}{1 - (1 - \epsilon_T)\epsilon_P} \tag{6}$$

derived by Magill et al. (2003). The P&T efficiency ϵ_{PT} depends on the transmutation efficiency ϵ_T and the partitioning efficiency ϵ_P. For a P&T efficiency close to 1, the difference at the duration due to the partitioning efficiency becomes substantial. There, difference in the available mass between the efficiencies is found to be significant.

Furthermore, it can be seen from the Eq. (6) that also the transmutation efficiency influences the achievable P&T efficiency at a given partitioning efficiency. However, for reasons of clarity, the other parameters are kept constant while discussing one specific parameter. As a consequence, the transmutation efficiency variation is varied next.

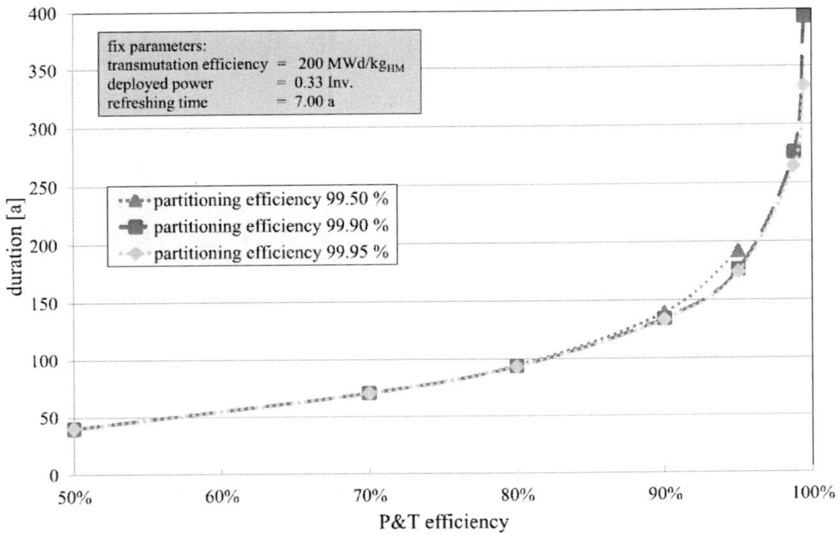

Figure. 2. Duration for achieving a certain P&T efficiency by varying the partitioning efficiency ϵ_P and maintaining values for transmutation efficiency, deployed power and refreshing time.

The transmutation efficiency is one of the most important parameters for the reactor design. It is, however, difficult to achieve high burn-up levels because the irradiation causes an increasing He pressure and decay heat in the solid fuel as well as swelling or deteriorating of it. A burn-up of 200 MWd/kg_{HM} was considered as the highest value in multirecycling. This level is necessary to achieve a P&T efficiency of 99.5% as shown in Fig. 3.

Apart from the highest P&T efficiencies, the duration is almost anti-proportional to the burn-up levels. This is expected since the burn-up represents the amount incinerated per fuel cycle. As a result, duration is halved when burn-up level is doubled. Due to the refreshing time and the partitioning efficiency, however, duration for lower burn-up levels increases faster and the anti-proportional relation between different burn-up levels weakens. This makes a higher burn-up level even more desirable. Consequently, a high burn-up is not only important for realizing high P&T efficiency, but for a short P&T process as well.

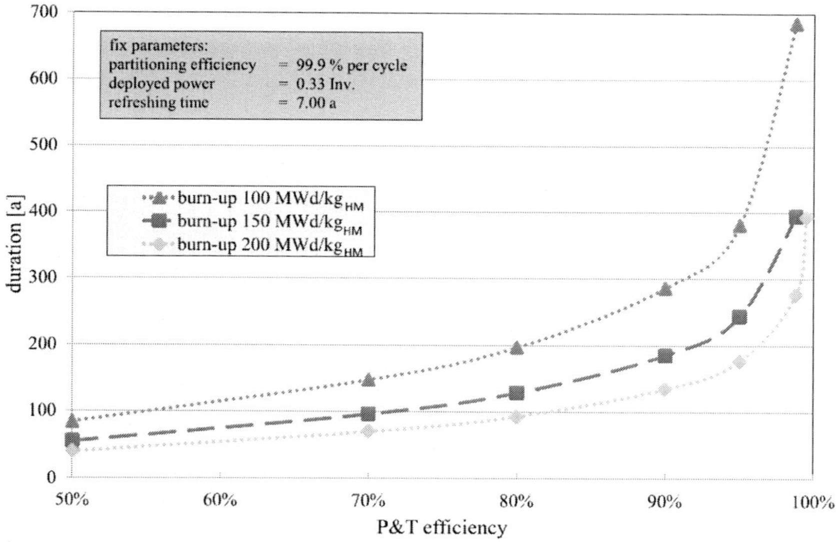

Figure. 3. Duration for achieving a certain P&T efficiency by varying the transmutation efficiency η_T and maintaining values for partitioning efficiency, deployed power and refreshing time.

Another parameter that influences the duration is the thermal power deployed at the beginning of the P&T process. As mentioned above, the power is expressed by the fraction of the loaded inventory and the total available inventory at the beginning of the P&T process. The values of 0.17 and 0.33 are quite high and were chosen to minimize duration as much as possible by this parameter. As shown in Fig. 4, the impact on duration does not seem to be appreciable. Only for lower P&T efficiencies does the influence of doubling the deployed power seem to be noteworthy. In addition, the time difference between the two cases seems to be time independent.

The reason for the low and constant impact of the deployed power is its utilization. In the case of higher deployed power, the inventory is used up after two fuel cycles. After that, refreshing time of 7 years prevents full utilization of deployed capacity. As a consequence, refreshing time is set to zero in order to maximize the period where the deployed power can be fully used (Fig. 5). As a result of this change, the reduction in the duration for a P&T efficiency of 50% is almost proportional to the ratio of deployed powers. For a P&T efficiency of 70%, the gap between the two curves increases to 42 years and remains constant for higher efficiencies. This is due the fact that the higher deployed power is only an advantage at the beginning of the process. After a few fuel cycles, utilization decreases because there is not enough fuel for full utilization. However, as long as utilization is larger than for the lower deployed power the duration of the P&T process is reduced in comparison. In case of no refreshing time, this reduction is 42 years.

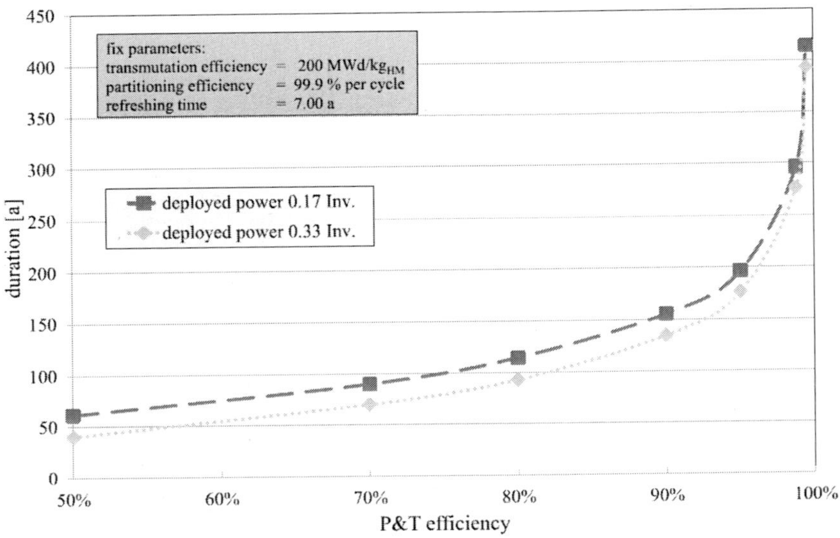

Figure. 4. Duration for achieving a certain P&T efficiency by varying the deployed power and maintaining values for transmutation and partitioning efficiency and refreshing time.

The last parameter investigated is the refreshing time. It covers cooling and reprocessing of spent fuel and fabrication of new fuel assemblies. The direct influence of this parameter on the duration of the P&T process is shown in Fig. 6. Although the impact is not as great as for the transmutation efficiency, the reduction is appreciable, especially for higher P&T efficiencies.

Neglecting the impact of the refreshing time, the latter will probably increase to 10 years due to the higher burn-up and MA content. In this way, the P&T process is extended by about almost a century for a P&T efficiency of 95% in comparison to the refreshing time of 3 years. The value of 10 years for refreshing describes the business as usual scenario while the value of 3 years stands for the technically feasible refreshing time for solid fuel, according to the discussion in the calculation section.

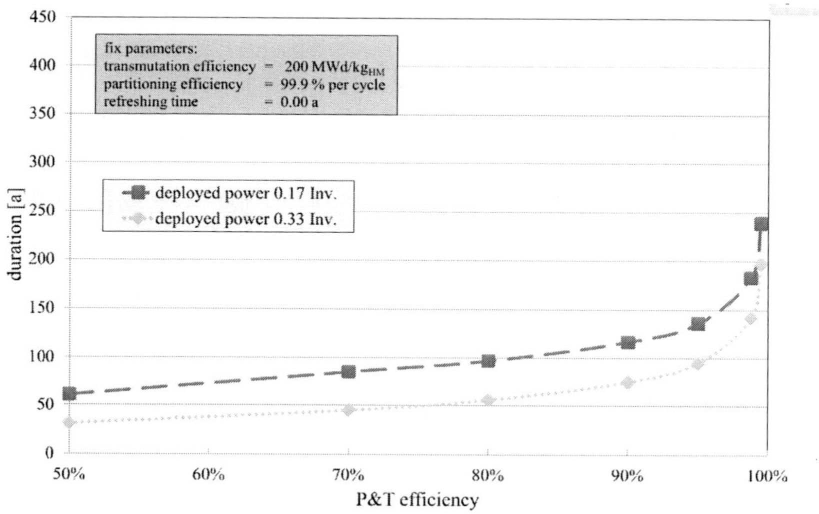

Figure. 5. Duration for achieving a certain P&T efficiency by varying the deployed power and maintaining values for transmutation efficiency and deployed power. The refreshing time is not considered.

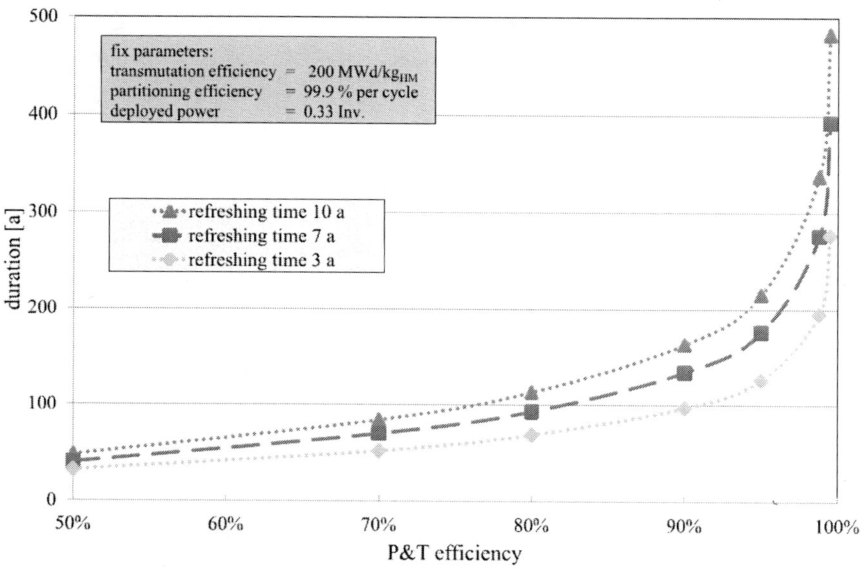

Figure. 6. Duration for achieving a certain P&T efficiency by varying the refreshing time for solid fuel and maintaining values for transmutation and partitioning efficiency and deployed power.

In addition, the use of liquid fuel with on-line reprocessing could make a cooling and new fabrication of fuel assemblies unnecessary. The reprocessing time in a molten salt reactor has been estimated to be 1 year (Sheu et al., 2013), which is more a theoretical value because the reprocessing takes place continuously. With such a short refreshing time, a duration of less than 230 years for achieving a P&T efficiency of 99.5% is possible as can be seen in Fig. 7. Therefore, the refreshing time has to be considered as an important parameter in order to realize high P&T efficiencies in a shorter time. Especially for these efficiencies, the refreshing time becomes an essential factor for the duration.

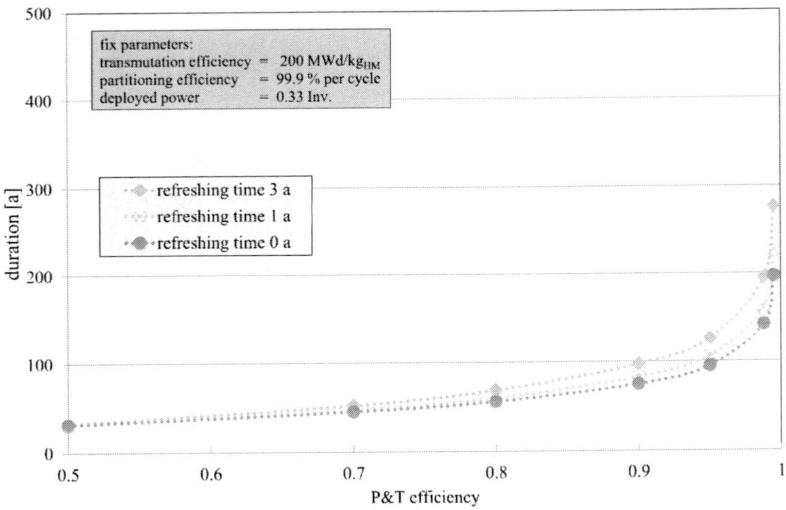

Figure. 7. Duration for achieving a certain P&T efficiency by varying the refreshing time for solid (3 years) and liquid fuel (0, 1 year) and maintaining values for transmutation and partitioning efficiency and deployed power.

5. DISCUSSION

In the following approximations made for the calculations are discussed. Moreover, the molten salt reactor is described in more detail because this reactor concept could be the key for a short P&T process.

One approximation was constant transmutation efficiency during the P&T process. Especially for the ambitious rate of 20% per cycle, several challenges were assumed to be solved. For starters, only fertile free fuel consisting of TRU was used. By that, a high content of plutonium would be necessary to reach criticality. The reactivity loss would be high, since no breeding of new fissile material was assumed. The transmutation of americium into plutonium might stabilize the k_{eff} but only if americium content is high enough (Artioli et al., 2008). Based on the composition of TRU, this is not the case in a phase-out scenario (Biss, 2014). Reactivity loss would make it necessary to feed new fissile material which shortens the cycle length. Conversely, the cycle length was kept constant too. Despite the cycle length, the changing plutonium vector would decrease the reactivity gained by refueling reprocessed fuel into a core. As a consequence of a constant burn-up and cycle length approximation, the calculation for a given transmutation efficiency underestimates the duration.

A solution might be the usage of ADS at beginning of P&T as a single-facility type solution because a high burn-up for TRU can be achieved per fuel cycle (OECD, 2009). By increasing proton beam, reactivity loss can be compensated to a certain extend allowing additional flexibility for ensuring the assumed conditions for burn-up and cycle length. Alternatively, the MSR could be used already in this phase because it has also high fuel flexibility regarding to fuel composition (Ignatiev et al., 2014) and additionally provide a low refreshing time. At a certain point the unfavorable remaining plutonium vector would make it necessary to switch to fuel with breeding material. In this second phase, the molten salt reactor could be used to incinerate the remaining TRU and breeding new fissile material by thorium. The usage of thorium would prevent the build up of new TRU. But also in this solution, it needs to be kept in mind that with decreasing amount of TRU in the core more and more fission taking place by the breeded material causing a longer irradiation time to achieve aspired burn-up for TRU. Avoiding of running a breeder reactor at the end limits this phase as well. Furthermore, left over of breeded U-233 has to be used abroad or the target of reducing the long term effects of TRU is corrupted.

The reason why molten salt reactor is favoured as key for shorting the P&T process lies not only in liquid fuel. Also the safety characteristics are conclusive. The continuous processing and feed in of new fissile material avoid the necessity of initial reactivity reserves (Delpech et al., 2009) and as consequence no burnable poison has to be added. Additionally, as shown for the MOSART design, the strong negative reactivity coefficient of the fuel salt temperature leads to an inherently stable reactor design for all basic transient conditions (LeBlanc, 2010 ; Zhang *et al.*, 2009). Furthermore, the partitioning losses remain in the fuel. Since in reprocessing the fission products and lanthanides are extracted, losses of such elements remain in the fuel. This reduces the transmutation performance but not overall transmutation efficiency.

Despite the advantages of molten salt reactor, there are still open challenges. For instance, the pyro-chemical process needs still intensive research. Furthermore, "closed safety approaches for liquid fuel reactors in conjunction with the fuel processing has not been developed up to now " (Merk et al., 2014).

Another improvement which could be achieved in the molten salt reactor due to the unnecessary fuel change is the enhancement of the availability. This, however, would only be the case if stability of the structural material is not diminished by the influence of high temperature, high neutron irradiation, and exposition to corrosive media (Merk et al., 2014), which would lead to more frequent breaks caused by maintenance.

6. CONCLUSION

Partitioning and Transmutation (P&T) offers an opportunity to reduce the quantity of transuranic elements by converting it into fission products. Although decay heat and radiotoxicity would increase in the short term, on the long term P&T would reduce volume, decay heat, and radiotoxicity of long lived HLW in a long-term repository. While these aspects have been investigated extensively, duration of the P&T concept has not been considered as a significant part of the discussion. This is most likely due to the fact that P&T is seen as a part of a long-term strategy for the use of nuclear power. In the case of nuclear power being phased out, however, the duration should be considered in P&T implementation.

Therefore in the present paper, besides the transmutation and partitioning efficiency which determine the P&T efficiency, the parameters refreshing time and deployed power were additionally included in order to estimate a minimal duration. As a result of varying the parameters, the duration can be reduced significantly. In line with previous investigations, transmutation efficiency is the crucial parameter, not only for P&T efficiency but for duration as well.

In addition, it is shown that the refreshing time (sum of cooling, reprocessing, and fabrication time) is the second significant parameter for shortening the duration. It not only reduces the duration directly but also implicitly by increasing the use of a higher deployed capacity at the beginning of the P&T process.

In view of the importance of the refreshing time, the duration is illustrated in Fig. 8 with the best case values for the other parameters. There is quite significant potential for reducing the duration for solid fuel, which follows from a reduction of refreshing time from 7 to 3 years. However, only for liquid fuels a refreshing time under 1 year seems to be achievable, corresponding to a shorter duration. In case of a long-term use of nuclear power, the use of the molten salt reactor concept should strongly be considered.

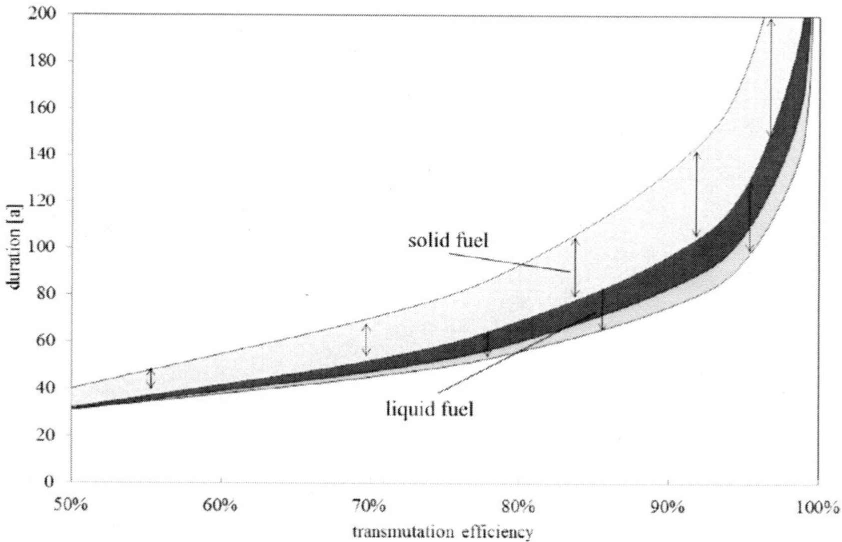

Figure. 8. Duration for achieving a certain P&T efficiency for different refreshing times and the maintained parameters burn-up (200 MWd/kg_{HM}), partitioning efficiency (99.9%), and deployed thermal power (0.33 Inv.). The refreshing time (sum of cooling, reprocessing and fabrication time) is in the range of 3–7 years for solid fuel. For liquid fuel, a refreshing time of 1 year (blue area) and none at all (green area) was considered. For interpretation of the references to color in this figure legend, the reader is referred to the web version of this article.

Since the benefit of P&T has been characterized by different indicators for the final repository such as heat load, volume and radiotoxicity reduction, it has to be kept in mind that in this paper the P&T efficiency only describes the mass reduction. The reduction for the other indicators might be similar but not the same, especially for different periods of storage.

Furthermore, the used parameters were kept constant for the entire calculation, which is especially at the end phase of P&T an approximation. As a consequence of this approximation the calculated duration most likely underestimates the real time for the P&T process.

As a result of the findings in this paper, future work should concentrate on analyzing a proper implementation of the P&T process by considering in particular transmutation efficiency as well as the refreshing time. Additionally, the relation between P&T efficiency, duration, and costs should be investigated to develop a effective strategy for P&T. This includes number, controllability, and installed power of reactors to be built.

REFERENCES

1. Artioli, C., Chen, X.-N., Gabrielli, F., Glinatsis, G., Liu, P., Maschek, W., Petrovich, C., Rineiski, A. Sarotto, M., Schikorr, M., 2008. Minor actinide transmutation in ADS: the EFIT core design. International Conference on the Physics of Reactors Nuclear Power: A Sustainable Resource.
2. Biss, K., 2014. Transmutation von Transuranen in einem gasgekühlten beschleunigergetriebenen System, Dr., RWTH Aachen, Jülich, Diss. (2014). URL http://juser.fz-juelich.de/record/156033.
3. Camarcat, N., Garzenne, C., Mer, J.L., Leroyer, H., Desroches, E., Delbecq, J.-M., 2011. Industrial research for transmutation scenarios. Comptes Rendus Mécanique 339 (4), 209–218. http://dx.doi.org/10.1016/j.crme.2011.01.006.
4. Carbajo, J.J., Yoder, G.L., Popov, S.G., Ivanov, V.K., 2001. A review of the thermophysical properties of MOX and UO2 fuels. J. Nucl. Mater. 299 (3), 181–198. http://dx.doi.org/10.1016/S0022-3115(01)00692-4.
5. Delpech, M., Tommasi, J., Zaeta, A., Salvatores, M., Mouney, H., Vanbenepe, G., 1999. The Am and Cm transmutation, physics and feasibility. In: Proc. Int. Conf. Future Nuclear Systems, GLOBAL'99.
6. Delpech, S., Merle-Lucotte, E., Heuer, D., Allibert, M., Ghetta, V., Le-Brun, C., Doligez, X., Picard, G., 2009. Reactor physic and reprocessing scheme for innovative molten salt reactor system. J. Fluorine Chem. 130 (1), 11–17. http://dx.doi.org/ 10.1016/ j.jfluchem.2008.07.009.
7. Fiorina, C., Stauff, N.E., Franceschini, F., Wenner, M.T., Stanculescu, A., Kim, T.K., Cammi, A., Ricotti, M.E., Hill, R.N., Taiwo, T.A., Salvatores, M., 2013. Comparative analysis of thorium and uranium fuel for transuranic recycle in a sodium cooled Fast Reactor. Ann. Nucl. Energy 62, 26–39. http://dx.doi.org/10.1016/j.anucene.2013.06.001.
8. Gen IV International Forum, Annual Report, 2013. URL https://www.gen-4.org/gif/jcms/c_64097/2013-gif-annual-report.
9. González-Romero, E., 2011. Impact of partitioning and transmutation on the high level waste management. Nucl. Eng. Design 241 (9), 3436–3444. http://dx.doi.org/ 10.1016/j.nucengdes.2011.03.030, Seventh European Commission conference on Euratom research and training in reactor systems (Fission Safety 2009).
10. Grouiller, J., Pillon, S., Saint Jean, C., Varaine, F., Leyval, L., Vambenepe, G., Carlier, B., 2003. Minor actinides transmutation scenario studies with PWRs, FRs and moderated targets. J. Nucl. Mater. 320 (1-2), 163–169. http://dx.doi.org/10.1016/S0022-3115(03)00184-3, 2nd Seminar on European Research on Materials for Transmutation, Karlsruhe, Germany, Sep 26–27, 2002.
11. Heuer, D., Merle-Lucotte, E., Allibert, M., Brovchenko, M., Ghetta, V., Rubiolo, P., 2014. Towards the thorium fuel cycle with molten salt fast reactors. Ann. Nucl. Energy 64 (0), 421–429. http://dx.doi.org/10.1016/j.anucene.2013.08.002.
12. Hyland, B., Gihm, B., 2011. Scenarios for the transmutation of actinides in CANDU reactors. Nucl. Eng. Design 241 (12), 4794–4802. http://dx.doi.org/10.1016/ j.nucengdes.2011.03.039, the 18th International Conference on Nuclear Engineering (ICONE-18).
13. Ignatiev, V., Feynberg, O., Gnidoi, I., Merzlyakov, A., Surenkov, A., Uglov, V., Zagnitko, A., Subbotin, V., Sannikov, I., Toropov, A., Afonichkin, V., Bovet, A., Khokhlov, V., Shishkin, V., Kormilitsyn, M., Lizin, A., Osipenko, A., 2014. Molten salt actinide recycler and

transforming system without and with Th-U support: fuel cycle flexibility and key material properties. Ann. Nucl. Energy 64, 408–420. http://dx.doi.org/10.1016/j.anucene.2013.09.004.
14. Ishida, S., Sekimoto, H., 2010. Finding the best fuel assemblies shuffling scheme of ADS for MA transmutation using Dynamic Programming. Nucl. Eng. Design 240 (10), 3645–3653. http://dx.doi.org/10.1016/j.nucengdes.2010.07.008, 4th International Topical Meeting on High Temperature Reactor Technology (HTR 2008), with Regular Papers.
15. LeBlanc, D., 2010. Molten salt reactors: a new beginning for an old idea. Nucl. Eng. Design 240 (6), 1644–1656. http://dx.doi.org/10.1016/j.nucengdes.2009.12.033.
16. Magill, J., Berthou, V., Haas, D., Galy, J., Schenkel, R., Wiese, H., Heusener, G., Tommasi, J., Youinou, G., 2003. Impact limits of partitioning and transmutation scenarios on the radiotoxicity of actinides in radioactive waste. Nucl. Energy J. Brit. Nucl. Energy Soc. 42 (5), 263–277.
17. Malambu, E., Haeck, W., Sobolev, V., 2004. MA and LLFP Transmutation Performance Assessment in the MYRRHA Small-scale ADS. In: Eighth Information Exchange Meeting, Las Vegas, Nevada.
18. Merk, B., Rohde, U., Glivici-Cotruta, V., Litskevich, D., Scholl, S., 2014. On the use of a molten salt fast reactor to apply an idealized transmutation scenario for the nuclear phase out. PLoS One 9 (4), e92776. http://dx.doi.org/10.1371/journal.pone.0092776.
19. Nawada, H.P., Fukuda, K., 2005. Role of pyro-chemical processes in advanced fuel cycles. J. Phys. Chem. Solids 66 (2-4), 647–651. http://dx.doi.org/10.1016/j.jpcs.2004.07.022, Proceedings of the 11th International Conference on High Temperature Materials Chemistry (HTMC-XI).
20. Nishihara, K., Nakayama, S., Morita, Y., Oigawa, H., Iwasaki, T., 2008. Impact of partitioning and transmutation on LWR high-level waste disposal. J Nucl. Sci. Technol. 45 (1), 84–97. http://dx.doi.org/10.3327/jnst.45.84.
21. Ochoa, R., Vázquez, M., Álvarez Velarde, F., Martín-Fuertes, F., García-Herranz, N., Cuervo, D., 2013. A comparative study of monte carlo-coupled depletion codes applied to a sodium fast reactor design loaded with minor actinides. Ann. Nucl. Energy 57 (0), 32–40. http://dx.doi.org/10.1016/j.anucene.2013.01.039.
22. OECD, Accelerator-driven Systems (ADS) and Fast Reactors (FR) in Advanced Nuclear Fuel Cycles, vol. NEA3109, NEA-OECD, 2002. URL http://www.oecd-nea.org/ndd/reports/2002/nea3109.html.
23. OECD, Nuclear Fuel Cycle Transition Scenario Studies, 2009. Nuclear Science NEA No. 6194 p. 39–54, ISBN 978-92-64-99068-5. http://dx.doi.org/10.1787/9789264107083-en.
24. Oigawa, H., Tsujimoto, K., Nishihara, K., Sugawara, T., Kurata, Y., Takei, H., Saito, S., Sasa, T., Obayashi, H., 2011. Role of ADS in the back-end of the fuel cycle strategies and associated design activities: The case of Japan. J. Nucl. Mater. 415 (3), 229–236. http://dx.doi.org/10.1016/j.jnucmat.2011.04.032, proceedings of the International DEMETRA Workshop on Development and Assessment of Structural Materials and Heavy Liquid Metal Technologies for Transmutation Systems.
25. Ramirez, J.R., Enriquez, P., Castillo, R., Alonso, G., 2012. MOX fuel use in a BWR with extended power up-rate. Ann. Nucl. Energy 50 (0), 63–70. http://dx.doi.org/10.1016/j.anucene.2012.07.008.
26. Renn O., 2014. Partitionierung und Transmutation Forschung – Entwicklung – Gesellschaftliche Implikationen. URL http://www.acatech.de/de/publikationen/publikationssuche/detail/artikel/partitionierung-und-transmutation-forschungentwicklung-gesellschaftliche-implikationen.html.
27. Robertson, R.C., 1971. Conceptual design study of a single fluid molten-salt breeder reator, OAK Ridge National Laboratory.
28. Salvatores, M., Palmiotti, G., 2011. Radioactive waste partitioning and transmutation within advanced fuel cycles: Achievements and challenges. Progr. Particle Nucl. Phys. 66 (1), 144–166. http://dx.doi.org/10.1016/j.ppnp.2010.10.001.
29. Sheu, R., Chang, C., Chao, C., Liu, Y.-W., 2013. Depletion analysis on long-term operation of the conceptual Molten Salt Actinide Recycler & Transmuter (MOSART) by using a special

sequence based on SCALE6/TRITON. Ann. Nucl. Energy 53 (0), 1–8. http://dx.doi.org/10.1016/j.anucene.2012.10.017.
30. Song, H., Kim, S.-J., Jang, J., Kim, Y.-I., 2008. Core performance and safety implications of TRU burning: Medium to large fast reactor core concepts. In: 10th Information Exchange Meeting, Mito, Japan.
31. W. van Rooijen, J.L. Kloosterman, Gas-Cooled Fast Reactor: A Historical Overview and Future Outlook, Science and Technology of Nuclear Installations ID 965757 (2009) 11. http://dx.doi.org/10.1155/2009/965757.
32. Von Lensa, w., Nabbi, R., Rossbach, M., 2008. RED-IMPACT: Impact of Partitioning, Transmutation and Waste Reduction Technologies on the Final Nuclear Waste Disposal, Vol. 15 of Schriften des Forschungszentrums Jülich: Energie & Umwelt/Energy & Environment, Forschungszentrum, Zentralbibliothek, Jülich, record converted from JUWEL: 18.07.2013. URL http://juser.fz-juelich.de/record/1315.
33. Zhang, D., Qiu, S., Su, G., 2009. Development of a safety analysis code for molten salt reactors. Nucl. Eng. design 239 (12), 2778–2785. http://dx.doi.org/ 10.1016/j.nucengdes.2009.08.020.
34. Zhang, Y., Wallenius, J., Jolkkonen, M., 2013. Transmutation of americium in a large sized sodium-cooled fast reactor loaded with nitride fuel. Ann. Nucl. Energy 53(0), 26–34. http://dx.doi.org/10.1016/j.anucene.2012.08.009.

CHAPTER 17

Safety of Future NPPs Must Not Be in Conflict with Economics

Vladimir Petrochenko[1], Georgy Toshinsky[1,2*], Oleg Komlev[1]

[1]*JSC "AKME-Engineering", Moscow, Russia*

[2]*JSC "SSC RF-IPPE", Obninsk, Russia*

ABSTRACT

The conflict between safety and economics requirements is peculiar to the present nuclear power (NP). The main point of the conflict is that for traditional type reactors the increase of requirements to safety of nuclear power plants (NPP) worsens their economical characteristics. This is caused by large potential energy accumulated in reactor coolant. In the presented paper the opportunity and expediency of changeover to reactors with heavy liquid-metal coolants (HLMC) in future NP is grounded. First of all, this refers to lead-bismuth coolant (LBC) mastered in the process of operating nuclear submarines (NS) reactors. The reactor facilities (RFs) of that type cannot cause destruction of defense barriers and make possible deterministic elimination of severe accidents with catastrophic radioactivity release. So it will make possible to eliminate the highlighted conflict and reasons for existence of population's radiophobia. Lead-bismuth fast reactor SVBR-100 with electric power of 100 MWe is the reactor facility of that type. The effect of accumulated in coolant potential energy on safety and economics is considered. Main specific features of SVBR-100 technology providing a high level of inherent self-protection and passive safety are presented.

Keywords: SVBR-100, Reactor, Lead-Bismuth Coolant, Nuclear Power Plant, Inherent Self-Protection, Passive Safety

1. INTRODUCTION

The problem of NPP safety assurance that slightly quietened after happening of Chernobyl disaster has become more acute after the accident occurred at NPP Fukushima 1. In certain countries that problem was put at the political level.

At the same time, further increase in safety requirements, which important quantity criterion is a value of probability of happening of the severe accident requiring population evacuation, can cause loss in competitiveness of NP based on use of traditional type reactors. Safety of such type reactors is mainly proved by probabilistic analysis methods (PSA).Just those reasons result in the necessity for future changeover to the reactors with a much higher level of inherent self-protection. In such reactors the severe accidents, which require population evacuation, will be deterministically eliminated.

First the necessity to develop those reactors was highlighted in 1985 in paper [1], in which such type reactors were named as inherently safe reactors. Detailed justification in the necessity to develop those reactors and general principles of their designing was summarized in 1990 in paper [2] (after happening of Chernobyl disaster). In that paper it was pointed out that for the population the opportunity of catastrophic consequences caused by the nuclear accident was much more important than extremely low possibility of accident occurrence. That is considered as a radiophobia effect. However, in accordance with the reliable statistical data [3], man-caused risks arisen due to operation of industrial enterprises and their fuel-energy infrastructure are by many orders higher than corresponding risks from the NP.

It can be presumed it is indeed the population radiophobia, which increased after the Fukushima-1 accidenthappened, was the reason of the fact that the NP was not mentioned as the carbon free option in the Global Agreement on Climate, which was signed by 196 parties on 12.12.2015 at UN Climate Change Conference held in Paris and purposed to replace the Kyoto Protocol. Nevertheless, in fact, taking into account inevitable future increases in electricity consumption due to population growth and scarcity of fresh water, there is no way to reduce significantly carbon emissions, but to develop the NP.

From the standpoint of nuclear community and educated people that sensing of NP is not rational. Nevertheless, radiophobia must be considered as an objective one and high safety of the NPP must be assured for the population, whose judgement is final, by "transparent" arguments and, if possible, without application of probabilistic analysis methods. It is much easier to convince the population in the NPP safety if it is provided by nature laws (e.g. lack of pressure in the reactor means that explosion cannot occur and so on), which people can understand on the basis of their own experience. It is more easily understood by the most of people than the results of probabilistic safety analyses.

Touching upon Russia reaction to the accident happened at NPP Fukushima 1, Director General of Rosatom State Atomic Energy Corporation Sergey Kirienko in the address to the Special Plenary Session of the International Congress on Advances in Power Plants, ICAPP-2011 (May 2-5, 2015, Nice, France) clearly argued for three stages of response actions [4].

The first near-future stage includes performance of stress-tests for all operating NPPs in Russia and their equipping with transportable sources of emergency electric and water supply, as well as proper training of the operating personnel. Now that task has been completed successfully.

The second medium-term stage consists in correction the NPP projects being under development and construction for the purpose of further enhancement of their safety level in conditions similar to those at NPP Fukushima 1.

The third long-term stage consists in changeover to construction of NPPs with natural safety reactors, in which severe accidents with catastrophic consequences have been deterministically eliminated by nature laws.

The term "natural safety" of the NPP that was offered in paper [5] with reference to lead cooled fast reactors (FR) is not quite appropriate as it dulls vigilance of both NPP developers and operating personnel. For developers and operating personnel the basic principle must be presumption of potential hazard in each nuclear reactor because of huge amount of radioactivity accumulated in the reactor.

The reactors, in which severe accidents requiring population evacuation have been deterministically eliminated, i.e. there are no reasons for their realization, should be rather named reactors with a high level of inherent self-protection. As it is shown below, the fast reactors with HLMC belong to such type reactors.

The technological basis for designing of those reactors is experience of development of two nuclear power technologies, namely: technology of fast reactors with sodium coolant (SFR) for civilian nuclear power, which could provide obtaining of a short doubling time of plutonium, and reactor technology developed in Russia for NS and based on application of HLMC-lead-bismuth eutectic alloy.

In the course of designing and operating the LBC cooled RFs at NSs a package of principal scientific and engineering problems on mastering of LBC technologies was solved [6] , namely: assurance of corrosion resistance of structural materials; control of LBC quality and control of mass-transfer processes in the reactor circuit; assurance of personnel's radiation safety while carrying out works with the equipment contaminated with polonium-210 radionuclide; multiple LBC "freezing-unfreezing" in the RF.

These two already mastered technologies coupled in a single innovative technology could ensure elimination of the progressive conflict between safety requirements and economics requirements that is typical for NPPs with traditional type reactors.

The highlighted conflict is expressed in such a way that for enhancement of safety of such reactors it is required to increase a number of safety systems and defense-in-depth barriers that is conditioned by natural properties of used coolants. That results in decrease of investment attractiveness of NPP projects due to growth of the value of specific capital costs, and for their restraint it is required to increase the reactor unit capacity that is a cause of raise in the total costs of NPP construction and construction terms.

In the paper the following issues are considered: the opportunity and expedience of changeover in the future NP to technologies of FRs with HLMC, effect of potential energy accumulated in the RF coolant on NPP safety and economics and typical features of technology of modular Lead-Bismuth Fast Reactors (SVBR-100).The paper reveals due to what technical solutions realized in the SVBR-100 design the high level of inherent self-protection and passive safety is assured that makes possible elimination of the conflict between safety and economics requirements peculiar to traditional type RFs.

2. THE OPPORTUNITY AND EXPEDIENCE OF CHANGEOVER TO NEW NUCLEAR TECHNOLOGIES IN THE FUTURE NP

Use of HLMC in FRs provides the reactor with important properties of inherent self- protection regarding to the certain severe accidents requiring the population evacuation. This is determined by those nature properties of HLMC as high boiling point and chemical inertness upon interaction with water and air, and thus being coupled with a proper design of the reactor, it makes possible deterministically eliminate such accidents.

At this point, the backgrounds for simplification of the reactor facility (RF) design arise due to elimination of the number of safety systems required in the RFs with other coolants. So it allows for NPPs realized on the basis of HLMC cooled FRs to be not only safer, but more economical as compared with NPPs based on traditional thermal reactors.

It is proved by the results of development of conceptual design of the large power NPP with SVBR-100 type reactors [7]. In Table 1 the technical and economical characteristics of different type electric power plants are compared [8].

Therefore, in conditions of existing low costs on natural uranium and services on its enrichment the new incentive for development of NPP s with fast reactors can appear: instead of the "plutonium factory" building up plutonium at high paces, that is the construction of NPPs on the basis of LBC cooled FRs, which are self-protected against the severest accidents, with lower values of the specific capital costs and cost of produced electricity. It can be expected that at the first stage of implementation of FRs with LBC in the NP the opened nuclear fuel cycle (NFC) with postponed reprocessing of the spent nuclear fuel (SNF) will be economically expedient regardless of the significantly higher (approximately by a factor of 2.5) specific (per 1 kW-h) consumption of natural uranium. As an example, that opportunity is demonstrated by the results given in Table 1. It could be expected that those reactors would be competitive with NPPs on the basis of thermal reactors and fossil power plants (FPP), because NP development in market conditions is not possible without successful competitiveness.

Of course, in the far future upon significant increase of the cost of natural uranium and costs of SNF storage, changeover to the closed NFC will become economically expedient. At this point, FRs with HLMC will operate in a mode of fuel self-providing with a core breeding ratio that slightly exceeds 1. Only after experience of operation of those reactors at the demonstrational prototypes is gained, it would be possible to talk about their wide use in the NP. Using the expression of American Admiral Hyman G. Rickover known as the "Father of the Nuclear Navy", now these are "paper" reactors, though LBC was mastered in conditions of operating NS reactors.

The time for economically justified wide implementation of any type FRs and the structure of future NP i.e. proportion of thermal and fast reactors will be determined by economic indices of the latter (firstly by a value of specific capital costs).

Table 1. Comparative parameters of different power plants.

Parameter name and measurement unit	NPP based on SVBR-75/100	NPP based on VVER-1500	NPP based on VVER-1000	NPP based on BN-1800	FPP based on PGU-325
1. Installed power of the unit, MWe	1625	1550	1068	1780	325
2. Number of units at the plant	2	2	2	2	10
3. Share of electric power used to operate the plant, %	4.5	5.7	6.43	4.6	4.5
4. Net power plant unit efficiency, %	34.6	34.4	33.3	43.6	44.4
5. Specific capital cost, $/kW	610*) 550**)	625	819.3	860	600
6. Design-based cost of produced electricity, cent/kW·h	1.3	1.35	2.02	1.6	1.75

*) The additional margin cost of ~17% (over the normative one) has been introduced that is 60 % of the cost of the RF equipment.

**) With due account of realizing the opportunities to changeover to the over-heated steam or to increase the temperature of the fuel elements claddings up to 650°C.

Provided that FRs are cheaper in comparison with VVER (PWR) type reactors (most likely it is expected for FRs with HLMC), the NP structure will gradually become a single-component one that is with FRs only and their implementation will be determined by the level of technological readiness. If the breeding ratio of those FRs is of about 1, then in case the own plutonium or plutonium from TR SNF (or warehouse reserves) is in deficiency, the required pace of NP development before changeover to the closed NFC must be provided by the starting loads of such FRs, which can be manufactured on the basis of enriched uranium.

In case FRs will be more expensive than VVER (PWR) type reactors with due account of the NFC cost, thus in conditions of deficiency of cheap natural uranium it will be expedient to have a two-component structure of the NP. Then to lessen the share of more expensive FRs a high value of breeding ratio is required (in other words, a minimal doubling time of plutonium that can be only provided by SFR type reactors).

The priority level of requirements of economic competitiveness for innovative nuclear energy systems (INES) is emphasized in the Report of Phase 1A of the International Project on Innovative Nuclear Reactors and Fuel Cycles (INPRO) [9]. In the proposed basic principles, it is highlighted that with account of all costs and credits, the cost of energy from innovative nuclear energy systems must be competitive with that from alternative energy sources; otherwise the alternative technologies will squeeze nuclear technologies out of the market. The comparative assessment of investment attractiveness of the innovative nuclear energy systems must take into account the total costs on the life-cycle including the costs on technology demonstration.

The higher NP competitiveness is the more opportunities for NP development will be both due to own resources (an opportunity to increase the investment share in the tariff) and borrowed financing resources (reduction of the capital repayment term).

From that point of view, the "early start" of FRs with a closed NFC under low costs of natural uranium can result in worsening of economic characteristics, reduction of NP investment opportunities and, as a consequence, decrease in paces of NP development. At the same time, for reasons of power safety the governments in certain countries can support development of the NP and FRs with a closed NFC in particular even with a high cost of electricity i.e. similar to development of renewable energy sources.

As safety characteristics and economic indices of the NPP strongly depend on potential energy accumulated in the reactor [10] , it should consider that effect in details.

3. EFFECT OF POTENTIAL ENERGY ACCUMULATED IN COOLANT ON SAFETY AND ECONOMICS

Over the historically short time of NP development, operation of NPPs with different type reactors was followed by a number of low probability accidents of the various severity extent. These accidents caused strong radioactivity release into the environment and/or considerable economical losses. These were such accidents as Three Mile Island Unit 2 accident in 1979 (the USA) that occurred at the pressurized water reactor, Chernobyl disaster at Unit 4 that occurred on 26 April 1986 (the former USSR), fire that happened in 1995 at SFR "Monju" (Japan) as a result of non-radioactive sodium leak in the intermediate circuit pipeline, disaster that happened in 2011 at NPP Fukushima 1 (Japan) at the boiling water reactors.

The initial events for the listed accidents are very unlike. These are such as personnel's errors, technical failures, defects in the reactor design and extremal external impacts. However, there is a common cause of severe consequences of all mentioned accidents. They are the result of release of various types of potential energy accumulated in different materials, first of all, in the RF coolant [10]:

- Compression energy of water coolant;
- Chemical energy of interaction of water steam with zirconium;
- Chemical energy of interaction of hydrogen produced by steam-zirconium reaction with air oxygen;
- energy of chemical interaction of sodium with air oxygen.

Due to those accidents, in many countries the population trust in the NP safety was lost. In order to return the population confidence to the NP, the necessity to make corrections in the strategy of NP development, bearing in mind demands for large-scale NP development in the future.

In the system approach, safety and hazard are considered as interconnected concepts. Therefore, in order to understand better how to achieve a high safety level for the RF, the nature of hazard peculiar to nuclear reactors should be analyzed.

The hazard from the RF is determined by three factors:

1) radiation potential accumulated, i.e., total radioactivity (more exactly, radiotoxicity) contained in the RF;
2) residual heat release of fission products radioactive decay;

3) amount of radiotoxic products released into the environment upon different initial events.

The first and second hazard factors do not depend strongly on the RF type, because total radioactivity contained in the RF and determined mainly by the amount of fission products is associated primarily with thermal power of the reactor and fuel lifetime, i.e. with energy production. These factors are "inherently peculiar" to each RF and cannot be reduced.

The third factor depends strongly on the RF type and is determined by reactivity margin, reactivity feedbacks, design features, and potential energy accumulated in the RF materials (nuclear energy, internal thermal energy, coolant compression energy, chemical energy), which in an event of damage of defense barriers can cause huge exhaust of radiotoxicity into the environment. Therefore, the hazard associated with the RF (for the identical power levels and fuel lifetime) will be determined by the third factor that can be reduced at the stage of project designing.

That factor requires detailed consideration. The nuclear fission energy, which can be released under conditions of reactivity accidents (such as unauthorized insertion of positive reactivity higher than effective fraction of delayed neutrons, β_{eff}), can and must be minimized as early as at the stage of reactor designing by limitation of the reactivity margin, use of negative reactivity feedbacks, and various engineering solutions in the control and protection system (CPS), which should exclude an opportunity of prompt neutrons runaway in the reactor.

At the same time the potential (non-nuclear) energy accumulated in coolant used in the RF primary circuit is an inherent property of the coolant material and cannot be changed by engineering solutions.

Therefore, upgrade of safety of NPPs based on RFs with high amount of potential energy accumulated in coolant requires increase of the number of safety systems and defense-in-depth barriers, which decrease the probability of severe accidents (but do not eliminate deterministically the reasons for their occurrence) and weight of their consequences. Upon assessment of that probability, failures in the basic equipment, safety systems, defense barriers, and personnel's errors are considered as random events.

However, the results of safety substantiation by methods of PSA legalized in regulation documents as applicable to severe accidents, their probability being very low (~ 10^{-6} per reactor-year and less) do not possess the credibility value as necessary. This is conditioned by high diversity and complication of the processes developing in conditions of the severe accident, lack of the certain initial data required for calculation, and many uncertainties in the available data.

Moreover, use of probabilistic analysis methods makes no sense upon consideration of such initial events, in which all active safety systems (or passive ones, which have been initiated by actuation of some mechanic devices), which are in a standby mode, and defense barriers can be disabled. For example, the over-normal external environmental effects or man-caused impacts, ill-intended people's actions can result in disastrous radioactivity release. A well-known fact is the explosions performed by terrorists at NPP "Lemoniz" (Spain) in 1978 and 1979 [11]. The importance to consider such scenarios is verified by the fact that they have also been addressed by the IAEA [12] because the NPP with potential energy accumulated in the RF coolant in great amounts could be used by terrorists as an instrument of political blackmail.

After the accident happened at NPP Fukushima 1, several European countries put forward the following approach to safety requirements to IV Generation reactors: "Never More" i.e. a conception of near-zero radiologic risk for the population at any external initiating events [13]. However, in case of realization in practice, that approach, which eliminates the necessity to take protection measures and evacuate the population beyond the NPP site even in events of hypothetical ultimate accidents with total NPP damage and further total radioactivity release into the environment (the worst), will mean the NP end. Being unrealizable for reactors of all types, that requirement must not be accepted by countries and organizations responsible for NP development including the IAEA. This is confirmed by the fact that in the last version of IAEA safety requirements issued in 2012 after the Fukushima accident happened, in the list of external initiating events such requirement is absent [14]. Moreover, the approach to the risk of radiation influence on the population known as ALARA (as low as reasonable achievable) was replaced by principle ALARP (as low as reasonable practicable), which is reflecting more accurately the real abilities.

For that reason, for the future NP many countries began to consider the following approach to substantiation of NPP safety as the most expedient one. That is deterministic elimination of the necessity for population evacuation in events of any **technically possible** severe accidents in combination with a probabilistic approach for assessment of the frequency of accident occurrence including such severe accidents as core damage unless it is accompanied by release of potential energy that can cause destruction of protective barriers and subsequent release of radioactivity.

If we accept the same values for the risk of radioactivity release from different type RFs to be at a socially acceptable level, the number of safety systems and defense-in- depth barriers, which strongly determine the NPP technical and economical characteristics, can be reduced provided of lower accumulation of potential (non-nuclear) energy stored in the RF, mainly in coolant. At this point, it is important that the high safety level at low value of potential energy stored in coolant can be achieved with absence of corresponding safety systems (and improvement of technical and economical parameters of the NPP) due to elimination of the causes for those severe accidents, i.e., deterministically.

The values of potential energy stored in different coolants, which were calculated using the reference data, are summarized in Table 2 [10].

When computing assessments were performed, the following components of potential energy were taken into account:

1) for water coolant some amount of stored thermal energy is converted into kinetic energy of steam expansion (assessment in Table 2 is performed for the adiabatic process) and a certain part is used for water evaporation that can cause loss of core cooling. Moreover, in an event of the severe accident while steam is chemically interacting with zirconium, thermal energy and hydrogen are released in large quantities. hydrogen, in its turn, is a high-rating source of hazard;
2) for sodium coolant the release of stored chemical potential energy can be realized both while contacting with air, and while contacting with water (in that case the thermal energy and hydrogen are released);
3) for heavy liquid-metal coolants (lead-bismuth alloy, lead) and molten salts the stored thermal potential energy cannot be converted into kinetic energy, there is no significant release of energy in an event of coolant's chemical interaction with air, water or structural materials, there is no loss of core cooling in an event of tightness failure in the gas system.

Nevertheless, taking into account the material presented above, we cannot approve a priori that the RFs with coolants with large amount of stored potential energy are less safe than RFs with HLMC. The required level of safety is also achieved due to the NPP equipped with a necessary number and efficiency of defense-in-depth barriers and safety systems. However, these differently affect the NPP technical and economical parameters.

As a competitive price of electricity is an only criterion for access to the electricity market, because its quality is always standard and does not depend on the NPP safety level, obtaining of the highest (therefore, more competitive) economical characteristics must be the principal task of the NPP design. And safety must be at a socially acceptable level that is specified by regulation documents and is the same for all NPP types. Increase of safety characteristics over that level is justified if it is only obtained due to inherent self-protection properties without detriment to economic characteristics.

Table 2. The values of specific (per a volume unit) stored potential energy for different coolants.

Coolant	Water	Sodium	Lead, Lead-bismuth
Parameter	P = 16 MPa, T = 300°C	P = 0.1 MPa, T = 500°C	P = 0.1 MPa, T = 500°C
Maximum potential energy, GJ/m^3, including:	~21.9	~10	~1.09
Thermal energy	~0.90	~0.6	~1.09
Potential compression energy	~0.15	None	None
Potential chemical energy of interaction	With zirconium ~11.4	With water ~5.1 With air ~9.3	None
Potential chemical energy of interaction of hydrogen released with air	~9.6	~4.3	None
Potential compression energy and chemical energy	**~21**	**~9.4**	**None**

Along with that, proofs of safety for RFs with low value of potential energy stored in coolant will be much more accessible for population understanding than for RFs, which high safety level is justified by PSA methods. That will result in a higher level of social acceptability of NPPs with such RFs. For that reason, NPP projects with a higher (and more "transparent") level of inherent self-protection will have a better chance to win the tender.

Therefore, the NPPs based on RFs with HLMC possess prerequisites for finding the solution to the basic problems of NP, such as safety and economics. The other problems, namely: efficient use of power potential of natural uranium upon operating in the closed NFC, reduction of the risk of unauthorized proliferation of nuclear fissile materials (NFM), management of long-lived radioactive waste can be solved not only upon use of RFs with HLMC, but upon use of industrially mastered SFRs, which are technologically ready for wide implementation.

4. BRIEF DESCRIPTION OF THE RF SVBR-100 DESIGN [15]

The flow diagram of RF SVBR-100 is shown in Figure 1. Hydraulic links over the LBC path are formed in the RMB without pipelines and valves (weak places). Four loops of the second circuit with vertical separators and multiple force circulation (MFC) pumps are shown. In the RMB the primary circuit gas system is designed to maintain the inert atmosphere (argon) above LBC free levels in the primary circuit. There is a passive heat removal system (PHRS).In an event of steam generator (SG) tube rupture the gas system condenser will be a steam condensate, and two bubblers are provided to receive a steam-gas mixture from the RMB gas volume over LBC level for protection of RMB vessel from steam pressure increase above the permissible limit.

The basic components of RF SVBR-100 are as follows: the RMB with a removable shielding plug and core; a set of control and protective system (CPS) drives; 2 main circulation pumps (MCP) of the primary circuit with gas-tight electric motors; 12 SG modules generating dry saturated steam; LBC technology equipment; internal equipment; four-loop circuit with MFC of the secondary circuit working medium; four-loop system of the PHRS; primary circuit gas system; steam-gas mixture receiving system. General layout of the RMB of SVBR-100 is shown in Figure 2. General layout of the RF is shown in Figure 3.

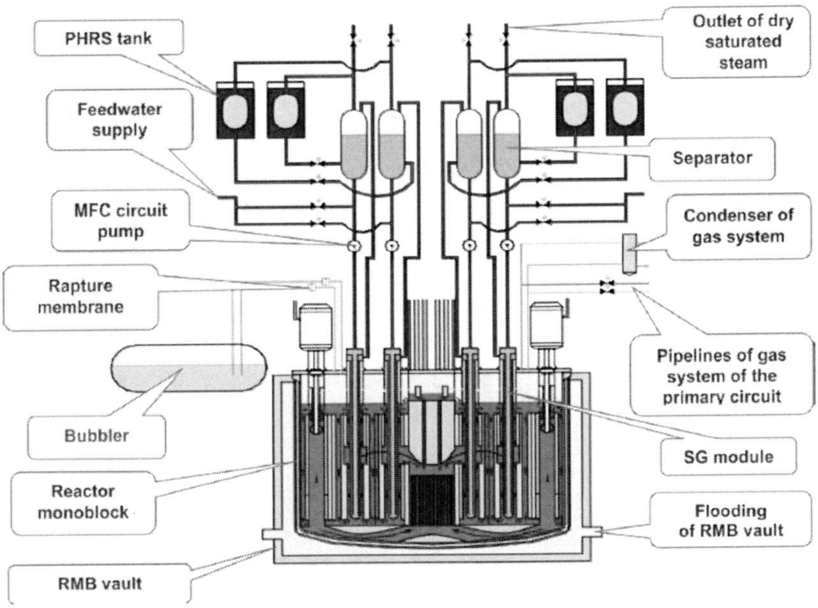

Figure 1. The flow diagram of RF SVBR-100.

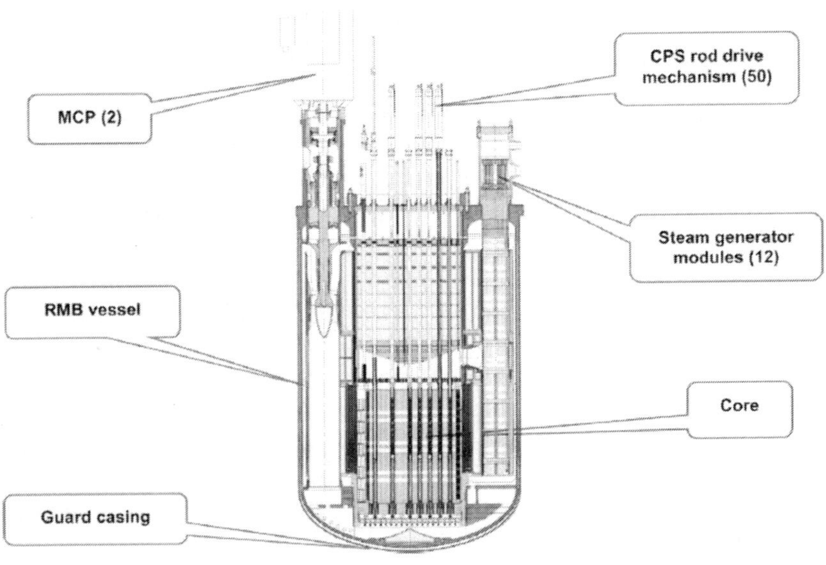

Figure 2. Layout of the SVBR-100 RMB.

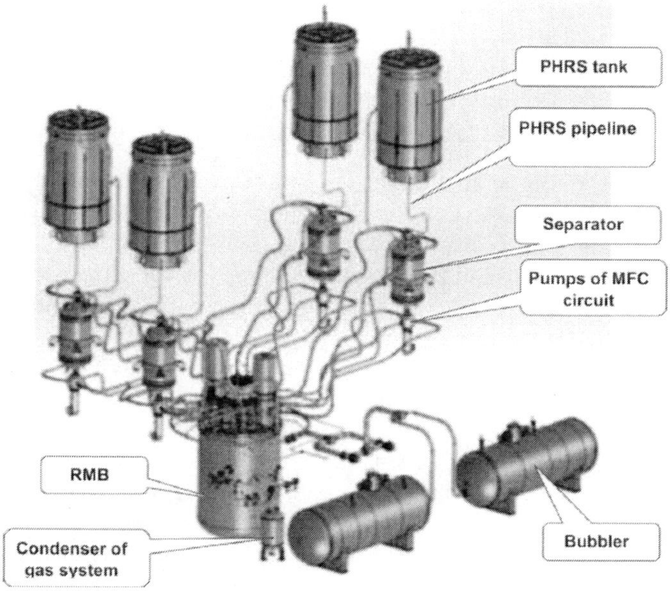

Figure 3. General layout of RF SVBR-100.

5. MAIN SPECIFIC FEATURES OF THE SVBR-100 TECHNOLOGY PROVIDING A HIGH LEVEL OF INHERENT SELF-PROTECTION AND PASSIVE SAFETY

- The high safety level resulting from natural properties of LBC (lack of excessive pressure and chemical inertness in events of contacting with water and air that is possible in accidental conditions). Upon monoblock design of the RF with the RMB having a protective casing, this eliminates an opportunity of LBC loss (loss of coolant accident) with core melting, reactor explosions and fires (lack of hydrogen release), which can be caused by internal reasons.
- Chemical compatibility of fuel (UO_2) with LBC eliminates the accidental situation with damage of fuel element cladding that can be developed in an event of large radioactivity release into coolant.
- Fuel pellets retain the main part of accumulated fission products. Fuel density approximately equals to coolant density (in case of melt, fuel does not sink). Therefore, there is no need in a core catcher.

- LBC retaining iodine, cesium and other fission products (with an exception of gaseous ones), which can ingress in it in an event of tightness failure in the fuel element cladding.
- Polonium is defining the radiation situation in an event of tightness failure in the RF gas system and requires providing of corresponding radiation safety measures. The concentration of polonium-210 formed in LBC under irradiation by bismuth neutrons is very low (10^{-6} %) and it forms thermodynamically resistant intermetallic compound with lead. Those factors reduce evaporation of polonium from LBC by a factor of 10^9 and that provides a comparatively favorable radiation situation in an event of postulated tightness failure in the primary circuit or gas system pipelines operating without excess pressure. In the postulated events of LBC leakage radioactivity release significant decrease of due to "freezing" of coolant.Those measures were developed and realized while operating of LBC cooled RFs at the NSs. Developed radiation safety measures were very effective as nobody of the personnel (both military and civilian ones), who took part in the elimination of accidents consequences (about 20 t of radioactive LBC leaked in the reactor compartment of facility 27/VT that was the first in the world LBC based reactor), got the polonium in-take dose that exceeded the permitted one [16].
- In all heat-removal circuits the level of coolant natural circulation sufficient enough for removal of residual heat release is provided. Heat removal via the SG is provided by four independent channels of the PHRS due to evaporation of water in the PHRS tanks with steam discharge into the atmosphere, grace period is 72 hours. In an event of postulated failure of four channels, it is provided that the reactor monoblock vault is flooded by water. Management of that accident that is considered as the accident being beyond the design basis is provided by feed of PHRS tanks or RMB vault from emergency sources of water(~2.5 m^3/hour)and electricity supply (for example, fire engines and so on).
- The reactor possesses a negative void reactivity effect and negative temperature reactivity coefficient. In addition to emergency protection (EP) rods actuating by electric signals, the reactor is equipped with directly acting additional emergency protection (AEP) rods, which actuate according to increase in LBC temperature (fusible locks).The temperature margin of SVBR-100 is over 50 percent in power for the limiting core parameters. In the accidental conditions (reactor loop test) the fuel element with cladding made of ferrite-martensitic type steel that is corrosion-resis- tant in LBC can withstand emergency

overheating up to 900°C without damage during 5 minutes. These all eliminate the unprotected transient overpower type accidents.

- In an event of simultaneous shutdown of both MCP of the and failure of the main EP system, self-protection of the RF is provided passively due to actuation of the AEP rods, inertial rundown of the MCP and natural circulation of coolants in heat-removal circuits. That excludes the unprotected loss of flow accidents.
- To localize the accident with leak in SG tubes, the steam condensers are provided in the primary circuit gas system. In an event of their failure it is provided that steam-gas mixture is passively discharged to the bubbler via rupture membranes. The scheme of LBC circulation in the RMB provides effective gravitational separation of steam bubbles on the LBC free level under the RMB lid.
- At the stage of storage of SNF elimination of radioactivity release is provided as follows: after removal from the reactor the fuel sub-assembly (FSA) is imbedded in a steel case filled with liquid lead, which then is put into the storage cell where removal of residual energy-release is realized passively due to natural circulation of atmospheric air. At this point, there are four safety barriers on the path of radioactivity release into the environment, namely: fuel pellet, fuel element cladding, hardened lead and leak proof case.
- Actually, in the process of operation no liquid radioactive wastes (LRW) are produced as refueling is performed without removal of coolant from the primary circuit and its further decontamination, which is a cause of formation of LWR in large quantities.
- In an event of the shutdown reactor and low level of residual energy-release, self-protection against unauthorized LBC “freezing” in the RF is provided by zero change of the LBC volume upon transition from a liquid state into solid one. Upon “freezing-unfreezing” of LBC, maintaining of the equipment operability is verified not only experimentally at large-scale prototypes but in conditions of operating of the NS RFs.
- To assess the safety potential of reactor SVBR-100, the preliminary calculation analysis of the consequences caused by a postulated severe accident was performed [17] under combination of such events as:

    1) destruction of the protective shell of the reactor building,
    2) damage of the reinforced concrete overlapping of the reactor box,
    3) rupture of gas system pipelines in the reactor monoblock installed in the concrete vault below the ground level with direct contact of

the free surface of lead-bismuth coolant under the monoblock lid and atmospheric air,

4) total blackout of the NPP.

That combination of initial factors is only possible in extreme events, such as military actions, terror attacks, nature disasters, which occur very rarely, and so on. The results of the performed calculation analysis have revealed that even in an event of extremely unfavorable atmospheric conditions, no population evacuation beyond a three-kilo- meter zone is required. For reactors with water or sodium coolants such combination of initial events can result in catastrophic consequences.

The performed analysis has revealed that RF SVBR-100 is not an amplifier of external impacts and, therefore, the scale of damages will be only determined by the energy of external impacts. Those type RFs assure their high resistance not only in events of single failures of the equipment and personnel errors but in events of premeditated ill-intentioned actions when all special safety systems operating in a standby mode can be intentionally disabled. At those reactors such catastrophic accidents as Chernobyl or Fukushima disasters as well as fires similar to that occurred at reactor “Monju” are impossible in principle or can be easily localized with a purpose to prevent population’s exposure to irradiation beyond the NPP site (loss of heat sink type accidents). This is extremely viable for realization of NPP construction in developing countries where the level of terroristic threat is high.

6. CONCLUSIONS

The following conclusions can be made on the basis of the material presented above:

- Fast reactors with HLMC, in which there is no potential energy stored in the RF coolant that under the certain initial events can cause destruction of defense barriers makes possible deterministic elimination of severe accidents with catastrophic radioactivity release (that is caused by release of coolant potential energy) requiring population evacuation. That assures the higher level of social acceptability of the NP and gives additional competitive advantages.

- Construction of FRs as well as NP development on the whole can be only justified provided they produce electricity that is cheaper than electricity produced by the alternative power sources. With due account of larger duration of mastering and launching of new nuclear technologies, it can be expected that till the end of the century the VVER

(PWR) type reactors operating in an open NFC with postponed reprocessing of the SNF will play the key role in the NP structure.

- First of all, large-scale development of NP is impeded by the population radiophobia and low level of competitiveness. These problems can be solved by use of FRs with HLMC that are the point of the present paper. The other impeding problems are as follows: very low efficiency of use of natural uranium energy potential in thermal reactors, postponed problem of handling the increasing amount of SNF, assurance of the NFM nonprolife ration mode. These problems can be solved not only FRs with HLMCs but also with mastered sodium FRs, which operate in the closed NFC.
- Use of HLMC cooled FRs in the NP can ensure elimination of the progressive conflict between safety requirements and economics requirements that is typical for traditional type reactors as improvement of safety is not required increasing of the number of safety systems and defense-in-depth barriers, but it is achieved due to the higher level of inherent self-protection, i.e. without detriment to economic characteristics.SVBR-100 using LBC mastered in conditions of operating NSs is the fast reactor of that type.
- It is planned that the technology of SVBR type reactors will be realized in construction of a first-of-a-kind experimental-industrial power-unit (EIPU). The Project is being realized within the framework of state-private partnership with joint venture JSC "AKME-engineering" established on a parity basis by the State Atomic Energy Corporation "Rosatom" and Limited Liability Company "Irkutskenergo". At present JSC "AKME-engineering" has received a "Rostechnadzor" license for placing of the EIPU in Dimitrovgrad (Ulyanovsk region).
- In the nearest future the tasks are as follows: development of the EIPUSVBR-100 design to the stage of receiving of Construction License; development of proposals for serial production of the NPP, in which the opportunities for improvement of technical and economical characteristics providing competitiveness of small and medium power plants with alternative power sources must be realized to the full extent. That can be achieved due to elimination of excessive conservatism of technical solutions adopted in the EIPU design and use of progressive technical solutions (namely: changeover to once-through SGs generating overheated steam and so on), which require development of additional R&D.

ACKNOWLEDGEMENTS

The authors would like to thank SSC RF-IPPE employees S.V. Budarina, K.G. Melnikov, I.V. Tormyshev for the assistance in preparation of the present paper.

REFERENCES

1. Spievak, I. and Weinberg, A.M. (1985) Inherently Safe Reactors. Annual Review of Energy, 10, 431-462. http://dx.doi.org/10.1146/annurev.eg.10.110185.002243
2. Forsberg, C.W. and Weinberg, A.M. (1990) Advanced Reactors, Passive Safety, and Acceptance of Nuclear Energy. Annual Review of Energy, 15, 133-152. http://dx.doi.org/10.1146/annurev.eg.15.110190.001025
3. Arutyunyan, R.V. (2013) Have We Got Safety Culture? Atomnaya Strategiya XXI, 83, 9-11 (in Russian) http://www.proatom.ru/ (2011) Nucleonics Week, Vol. 52, No. 19, 12 May 2011, 6-7.
4. Orlov, V.V., Avrorin, E.N., Adamov, E.O., et al. (1992) Nontraditional Concepts of NPPs with Natural Safety. Atomnaya Energiya, 72, 317-328 (in Russian).
5. Toshinsky, G.I., Komlev, O.G., Stepanov, V.S., et al. (2008) Innovative Nuclear Technology Based on Modular Multi-purposed Lead-Bismuth Cooled Fast Reactors. Progress in Nuclear Energy, 50, 170-178. http://dx.doi.org/10.1016/j.pnucene.2007.10.025
6. Toshinsky, G.I., Komlev, O.G., Stepanov, V.S., et al. (2006) Nuclear Power Development in Market Conditions with Use of Multi-Purpose Modular Fast Reactors SVBR-75/100. Nuclear Engineering and Design, 236, 1490-1502.
7. Zrodnikov, A.V., Chitaykin, V.I., Toshinsky, G.I., et al. (2001) NPP Based on Reactor Modules SVBR-75/100. Atomnaya Energiya, 91, 415-425 (in Russian).
8. Report on Phase 1A of the International Project on Innovative Nuclear Reactors and Fuel Cycles (INPRO). IAEA-TECDOC-1362, 2004.
9. Toshinsky, G.I., Komlev, O.G., Tormyshev, I.V. and Petrochenko, V.V. (2013) Effect of Potential Energy Stored in Reactor Facility Coolant on NPP Safety and Economic Parameters. World Journal of Nuclear Science and Technology, 3, 59-64. http://www.scirp.org/journal/wjnst http://dx.doi.org/10.4236/wjnst.2013.32010
10. Nuclear Terrorism. http://www.gochs.info/p669.htm
11. IAEA (February 2006) Advanced Nuclear Plant Design Options to Cope with External Events. IAEA-TECDOC-1487. http://www-pub.iaea.org/MTCD/publications/ PDF/te_1487_web.pdf
12. Ammirabile, L. European Project SARGEN_IV: Safety Approach and Assessment of GEN IV Reactors. http://www.iaea.org/NuclearPower/Downloadable/Meetings/2013/2013-02-26-02-27-TM-SFR/day-1/4-L.Ammirabile-EC.pdf
13. IAEA, Vienna (2012) SAFETY OF NUCLAR POWER PLANTS: DESIGN, Specific Safety Requirements. IAEA Safety Standards Series No. SSR-2/1.
14. Dzangobegov, V.V., Stepanov, V.S., Dedul, A.V., et al. (2014) Reactor Plant SVBR-100 for Modular-Type Small-and-Medium Power Nuclear Power Plants. Proceeding of 4th Conference "Heavy Liquid-Metal Coolants in Nuclear Technologies (HLMC-2013)", Paper № 10, Vol. 1, 23-26 September 2014, Obninsk, 77-86 (in Russian).
15. Pankratov, D.V., Yefimov, Ye.I., Toshinskyand, G.I. and Ryabaya, L.D. (2005) Analysis of Polonium Hazard in Nuclear Power Installations with Lead-Bismuth Coolant. Proc. of International Congress on Advances in Nuclear Power Plants, CD-ROM, Paper No. 5497, Seoul, Korea.

16. Bolhovitinov, V.N., Pankratov, D.V., Toshinsky, G.I., et al. (2003) Assessment of Radiation Consequences Caused by Large Tightness Failure in the Primary Circuit Gas System of RF SVBR-75/100 with Simultaneous Coolant Heating up to 600 °C. Paper No. 2203 (CD-ROM), 2nd Conference "Heavy Liquid-Metal Coolants in Nuclear Technologies (HLMC-2003)" Held within the Frameworks of Russian Scientific & Technical Forum "Fast Neutron Nuclear Reactors" Devoted to the 100th Anniversary of A.I. Leypunsky, Obninsk, Russia (in Russian).

Index